Integrated Environmental Management

A Transdisciplinary Approach

Applied Ecology and Environmental Management

A SERIES

Series Editor

Sven E. Jørgensen

Copenhagen University, Denmark

Handbook of Environmental Engineering, *Frank R. Spellman*

Integrated Environmental Management: A Transdisciplinary Approach, *Sven E. Jørgensen, João Carlos Marques, and Søren Nors Nielsen*

Ecological Processes Handbook, *Luca Palmeri, Alberto Barausse, and Sven E. Jørgensen*

Handbook of Inland Aquatic Ecosystem Management, *Sven E. Jørgensen, Jose Galizia Tundisi, and Takako Matsumura Tundisi*

Eco-Cities: A Planning Guide, *Zhifeng Yang*

Sustainable Energy Landscapes: Designing, Planning, and Development, *Sven Stremke and Andy Van Den Dobbelsteen*

Introduction to Systems Ecology, *Sven E. Jørgensen*

Handbook of Ecological Models Used in Ecosystem and Environmental Management, *Sven E. Jørgensen*

Surface Modeling: High Accuracy and High Speed Methods, *Tian-Xiang Yue*

Handbook of Ecological Indicators for Assessment of Ecosystem Health, Second Edition, *Sven E. Jørgensen, Fu-Liu Xu, and Robert Costanza*

ADDITIONAL VOLUMES IN PREPARATION

Integrated Environmental Management

A Transdisciplinary Approach

Sven Erik Jørgensen
João Carlos Marques
Søren Nors Nielsen

CRC Press
Taylor & Francis Group
Boca Raton London New York

CRC Press is an imprint of the
Taylor & Francis Group, an **informa** business

CRC Press
Taylor & Francis Group
6000 Broken Sound Parkway NW, Suite 300
Boca Raton, FL 33487-2742

First issued in paperback 2019

ISBN-13: 978-1-4987-0510-3 (hbk)
ISBN-13: 978-0-367-87168-0 (pbk)

Visit the Taylor & Francis Web site at
http://www.taylorandfrancis.com

and the CRC Press Web site at
http://www.crcpress.com

Contents

Part III Sources and Causes

Preface

The first global green wave started in the early to mid-1960s as a result of Rachel Carson's book *The Silent Spring* (1962). Previously, pollution was considered entirely a technological problem that would have to be solved by the use of environmental technology, but Rachel Carson's book started a discussion about the deterioration of nature due to pollution and that nature should be the focus of our concern regarding the consequences of pollution. It increased the interest for a wider knowledge of how the natural systems—ecosystems—are working and how they are affected by pollution and various types of impacts. It implied that ecology came into focus. Before most were aware of the scientific field of ecology, but after the first green wave, everyone learned the meaning of the new word *ecology* and used it in discussions about pollution abatement. To make a long story short, during the last three decades of the twentieth century, several new ecological subdisciplines aimed toward using ecology to solve and identify environmental problems were developed: ecological modeling, ecological engineering or ecotechnology, ecological economics, assessment of ecosystem health by ecological indicators, and assessment of ecosystem services used by society and systems ecology. These subdisciplines built a bridge from ecology to environmental management (see Chapter 1). Ecological engineering or ecotechnology, which focused on how nature could help us to solve the environmental problems and how we could help nature resist the consequences of pollution, became an important tool box in our effort to solve the environmental problems, while the other subdisciplines became indispensable tools to identify and understand the environmental problems and develop detailed diagnoses.

Furthermore, other new tool boxes to find moderate costs and workable solutions to the environmental problems emerged. Forty-five years ago, environmental technology was the only applicable tool box that could offer a solution to the environmental problems, but in addition to ecotechnology, environmental legislation and cleaner technology/production developed as powerful tool boxes to be used to solve the environmental problems. We have as a consequence of this development many available tool boxes today to identify, understand, and solve environmental problems and all these tool boxes should be used, and often in new and unexpected combinations to find the best or close to the best solution.

Integrated ecological and environmental management means that the environmental problems are viewed from a holistic angle considering the ecosystem as an entity and considering the entire spectrum of solutions and all possible combinations of the proposed solutions. The experience gained from environmental management during the last 40 years has clearly shown that it is important not to consider solutions of single problems but to consider *all* the possible problems associated with a considered ecosystem simultaneously and evaluate *all* the solution possibilities proposed by the relevant disciplines at the same time, or expressed differently: to observe the forest and not the single trees. The experience has clearly underlined that there is no alternative to *integrated, multidisciplinary, intradisciplinary, ecological–environmental* management at least not on a long-term basis. Fortunately, as will be presented in this book, the development has, as mentioned above, offered tool boxes to perform integrated ecological and environmental management. The aim of this book is to present all the available tool boxes and explain how they can be integrated and combined to find an optimum solution to the ecological–environmental problems.

Sven Erik Jørgensen, João Carlos Marques,
and Søren Nors Nielsen
Copenhagen, Denmark and Coimbra, Portugal

Authors

Sven Erik Jørgensen is a professor emeritus in environmental chemistry at the University of Copenhagen (Denmark). He has a master of science degree in chemical engineering from the Danish Technical University (1958), a doctorate degree in environmental engineering from Karlsruhe University, and a doctor of science degree in ecological modeling from Copenhagen University. He is a honorable doctor at Coimbra University, Portugal and at Dar es Salaam University, Tanzania. He is a recipient of the Einstein Professorship of the Chinese Academy of Sciences. He founded the journal *Ecological Modelling* in 1975 and the International Society of Ecological Modelling (ISEM) in 1978. He has received several awards, including the Ruder Boskovic Medal, the Prigogine Prize, the Pascal Medal, the Santa Chiara Prize for Multidisciplinary Teaching, and the very prestigious Stockholm Water Prize. He was the editor of the *Encyclopedia of Environmental Management*, which received a publishing achievement of "Outstanding Academic Title" for 2013.

Professor Jørgensen has published 370 papers, of which 256 were published in peer-reviewed international journals. He has authored or edited 79 books, of which several have been translated into various languages (Chinese, Russian, Spanish, and Portuguese). He has authored a successful textbook in ecological modeling titled *Fundamentals of Ecological Modelling* (Elsevier, Amsterdam), which was published as a fourth edition along with Brian Fath in 2011. It has been translated into Chinese and Russian (Third Edition). He recently authored another well-received textbook in systems ecology titled *Introduction to Systems Ecology* (CRC Press, Boca Raton). It was published as an English edition in 2012 and as a Chinese edition in 2013. He was an editor in chief of the *Encyclopedia of Ecology* (Elsevier, Amsterdam), published in 2008, and of the *Encyclopedia of Environmental Management* (CRC Press, Boca Raton), published in December 2012. He has taught courses in ecological modeling and systems ecology in 33 different countries. He is an editorial board member of 20 international journals in the fields of ecology and environmental management. He is the president of ISEM and has been elected as a member of the European Academy of Sciences, for which he is the chairman of the Section for Environmental Sciences.

João Carlos Marques (Lisbon, Portugal), received an undergraduate degree in biology from the University of Lisbon in 1980, and earned a PhD in ecology from the University of Coimbra (Portugal) in 1989. He was president of the Institute of Marine Research (IMAR) Portugal from 1997 to 2006, and was vice rector for Scientific Research at the University of Coimbra from 2003 to 2007. Dr. Marques is presently vice director of the Marine and Environmental Sciences Centre (MARE), and is the scientific coordinator of the MARE pole at the University of Coimbra.

A marine and estuarine ecologist, he has worked on biological and ecological processes in marine and estuarine ecosystems, systems ecology, and ecological modeling. Since 1990, he has coordinated 15 large research projects, both funded by the Portuguese Agencies and European Union Programmes, and has also participated as a member of the research team in 20 others. Thus far, he has authored or coauthored more than 225 scientific papers in international refereed journals, 7 books, and 12 book chapters (international editions), and supervised 47 MSc and 23 PhD students in Portugal and abroad.

He is an associate editor of the *Journal of Ocean and Coastal Management* and *Ecological Indicators*, from Elsevier.

Søren Nors Nielsen (Vejle, Denmark), graduated with a degree in biology from the University of Copenhagen in 1984, earned his PhD in the structural dynamics of Danish shallow lakes from Risø National Laboratory and the National Environmental Research Institute in 1992, and received a Dr. agregado in ecology from the University of Coimbra in 2009. Since 1985, he has taught more than 60 courses on systems analysis, environmental modeling and management, ecosystem theory and history, cleaner production, industrial ecology, and ecological economics and indicators to students at various universities in Denmark and several other countries. An aquatic ecologist, since 1989 he has been working with the application of optimization principles to ecosystem development mainly from a thermodynamic view, and more lately has become interested in expanding the approach to society in order to provide an improved definition of sustainability rooted in the functional principles of nature. He has more than 17 years of work experience with projects mainly on aquatic systems with an emphasis on modeling and management of these types of ecosystems in several countries. He was an associate professor in thermodynamics of biological systems at the

Danish University of Pharmaceutical Sciences (2001–2010), where in addition to systems analysis and modeling he has been teaching physical chemistry, ecotoxicology, and philosophy of sciences. Dr. Nielsen has authored and coauthored 55 papers in international refereed journals, 5 books, 9 book chapters, as well as project reports.

1

Introduction: Integrated and Holistic Environmental Management in Seven Steps

1.1 The Seven Management Steps

Integrated ecological and environmental management means that the environmental problems are viewed from a holistic angle considering the ecosystem as an entity and considering the entire spectrum of solutions including all possible combinations of proposed solutions. The experience gained from environmental management over the last 40 years has clearly shown that it is important not to consider solutions to single problems but to consider *all* the problems associated with a considered ecosystem simultaneously and evaluate *all* the possible solutions proposed by the relevant disciplines at the same time, or expressed differently: to observe the forest and not the single trees. The experience has clearly underlined that there is no alternative to an *integrated* management, at least not on a long-term basis. Fortunately, as presented in this chapter, new ecological subdisciplines have emerged and they offer tool boxes to perform an integrated ecological and environmental management.

Currently, integrated ecological and environmental management consists of a procedure involving seven steps (Jørgensen and Nielsen 2012):

1. Define the problem
2. Determine the ecosystems involved
3. Find and quantify all the sources to the problem
4. Set up a diagnosis to understand the relationship between the problem and the sources
5. Determine all the tools we need to implement to solve the problem
6. Implement the selected solutions
7. Follow the recovery process

When an environmental problem has been detected, it is necessary to determine and quantify the problem and all the sources to the problem. It requires the use of analytical methods or a monitoring program. To solve the problem, a clear and unambiguous diagnosis has to be developed: What are the problems, that the ecosystems are actually facing and what are the relationships between the sources and their quantities and the determined problems? Or expressed differently: To what extent do we solve the problems by reducing or eliminating the different sources to the problems? A holistic integrated approach is needed in most cases because the problems and the corresponding ecological changes in the ecosystems are most often very complex, particularly of course when several environmental problems are interacting. When the first green wave started in the mid-1960s, the tools to answer these questions, which we today consider as very obvious questions in an environmental management context, were not yet developed. We could carry out the first three points on the above given list, but had to stop at point 4 and could at that time only recommend to eliminate the source completely or be closed completely by the methods, that were available at that time—which means environmental technology at a lower level than that existing today.

Due to the development of several new ecological subdisciplines, today it is possible to accomplish points 4–6. The tool boxes that we can apply today to carry out points 4–6 are presented in the next sections. They are the result of the emergence of six (or two times three) new ecological subdisciplines, for a better diagnosis: ecological modeling, ecological indicators, ecological services; and for more tools to solve the problems: ecological engineering (also denoted as ecotechnology), cleaner production, and environmental legislation.

1.2 The Tool Boxes Available Today to Develop an Ecological–Environmental Diagnosis

A massive use of ecological models as an environmental management tool was initiated in the early 1970s. The idea was to answer the question: What is the relationship between a reduction of the impacts on ecosystems and the observable, ecological improvements (see Figure 1.1)? The answer could be used to select the pollution reduction that society would require and could afford economically. Ecological models were already developed in the 1920s by Steeter-Phelps and Lotka Volterra (see for instance, Jørgensen and Fath [2011]), but in the 1970s, a much more consequent use of ecological models was started and many more models of different ecosystems and different pollution problems were developed. In the early 1970s, we only had environmental technology to use for impact reduction, while today we have more tool boxes as will be presented in the next section.

Today, we practically have at least a few models available for all combinations of ecosystems and environmental problems. The journal *Ecological*

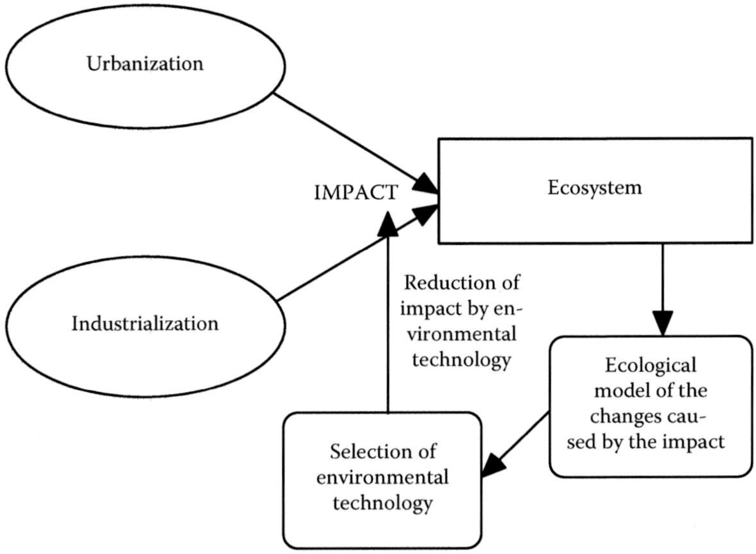

FIGURE 1.1
In the early 1970s, ecological models started to be used as illustrated above. The models express what changes the impacts (in model language called forcing functions) can cause in the ecosystems. The models contain a mathematical relationship between forcing functions and the state of the ecosystems described by the use of state variables. The model is used to decide on the required reduction of the forcing functions by selection of suitable environmental management solutions.

Modelling was launched in 1975 with an annual publication of 320 pages and about 20 articles. Today, the journal publishes 20 times as many articles. This means that ecological modeling has been adopted as a very powerful tool in ecological–environmental management to cover in particular point 4 in the integrated ecological and environmental management procedure proposed above. Chapter 8 presents more details about this tool box and its application in environmental management.

Ecological models are powerful management tools but they are not easily developed. They require in most cases good data, which are resource- and time-consuming to provide. If an economically important and environmentally crucial project is the focus, it is often very beneficial to develop a good ecological–environmental model, but if the project is minor it may be sufficient to consider other solutions. About 25 years ago, it was therefore proposed to use another tool box, that required fewer resources to provide a diagnosis, namely ecological indicators (see for instance, Costanza et al., 1992). Ecological indicators can be classified as shown in Table 1.1, according to the spectrum from a more detailed, or reductionistic view to a system or holistic view (see Jørgensen, 2002). The reductionistic indicators can for instance be a chemical compound that causes pollution or a specific species,

TABLE 1.1

Classification of Ecological Indicators

Level	Example
Reductionistic (single) indicators	PCB, species present/absent
Semiholistic indicators	Odum's attributes
Holistic indicators	Biodiversity/ecological network
"Super-holistic"	Thermodynamic indicators as eco-exergy (Work energy) and energy

that by their presence or absence can be translated to ecosystem quality. A holistic indicator could for instance be a thermodynamic variable or the biodiversity, determined for the focal ecosystem. The indicators can either be measured or they can be determined by the use of a model. In the latter case, the time consumption is of course not reduced by the use of indicators instead of models, but the models get a more clear focus on one or more specific state variable, namely, the selected indicator, which best describes the problems. In addition, indicators are usually associated with very clear and specific health problems of the ecosystems, which of course is beneficial in environmental management. Chapter 9 is devoted to the application of ecological indicators.

Over the past 15–20 years, the services offered by ecosystems to society have been discussed and calculating the economic value of these services has been attempted (Costanza et al., 1997). A diagnosis that would focus on the services actually reduced or eliminated due to environmental problems could easily be developed. Another possibility of using ecological services to assess the environmental problems and their consequences could be to determine the economic values of the overall ecological services offered by the ecosystems and then compare with what is normal for the type of ecosystems considered. Jørgensen (2010) has determined the values of all the services offered by various ecosystems by the use of the ecological holistic indicator eco-exergy (it is the work energy *including* the work energy of the information embodied in the organisms of the ecosystems). It is therefore expressing the total work energy offered by ecosystems per year. It is a good measure of the total amount of ecological services offered by the ecosystems, as all services require a certain amount of energy that can aid work. The values published in Jørgensen (2010) are shown in Table 1.2, and can be used for the above indicated comparison. The eco-exergy (work energy) is found as the sum of the β-values times the biomass for all the organisms. The β-values express the information the different ecosystem components carry (see Jørgensen et al., 2005). The average β-values for the various ecosystems are based on the living components that are present and dominant in typical ecosystems representing the various types of ecosystems. Note that detritus (dead organic matter) has the β-value = 1—which means that it does not contain information. Further details about these calculations (see Jørgensen, 2006) are given in Chapters 3 and 4.

Assessment of the values of the ecosystem services may also be coupled to sustainability, because it is crucial to maintain the many ecosystem services on which society is dependent. Environmental management scenarios, preferably developed by ecological models, are tested by means of the ecological sustainability trigon (EST); see Marques et al. (2009). The use of work energy (eco-exergy) as an indicator to find the value of the ecosystem services in this context is beneficial, because the development of sustainability can be determined as maintenance of the total work capacity that is at our disposal (Jørgensen, 2006). If we multiply the work energy capacity (the last column in Table 1.2) by the economic value of work energy which is about 4 euro cents or 5 U.S. cents per kWh, we can obtain the annual economic value of ecosystem services. Further details are presented in Chapter 10, which focuses on ecosystem services. The value found by this method is much higher than the value found by Costanza et al., because it expresses the total value of the services offered by the ecosystems—both the ones that we are using and the ones that we are not; while Costanza et al. (1997) only found the value of the services that we are actually using. For lakes, rivers, and coastal ecosystems, we use the ecosystem services much more than for forests and agricultural ecosystems, where we often only use the production directly by harvest of, for instance, cereals, vegetables, and wood.

Assessment of ecosystem services frequently requires the use of ecological indicators. The indicators are followed by the use of models, which can assess the reduced or lost ecological services of ecosystems. The three diagnostic tool boxes are in other words closely related and obviously the

TABLE 1.2

Work Energy Capacity Used to Express the Ecosystem Services for Various Types of Ecosystems

Ecosystem	Biomass Production MJ/m2 y	Information Factor β-Value	Work Capacity GJ/ha y
Desert	0.9	230	2,070
Open sea	3.5	68	2,380
Coastal zones	7.0	69	4,830
Coral reefs, estuaries	80	120	96,000
Lakes, rivers	11	85	9,350
Coniferous forests	15.4	350	53,900
Deciduous forests	26.4	380	100,000
Temperate rainforests	39.6	380	150,000
Tropical rainforests	80	370	300.000
Tundra	2.6	280	7,280
Croplands	20.0	210	42,000
Grasslands	7.2	250	18,000
Wetlands	18	250	45,000

Note: It is Calculated as Biomass* the Information Factor.

use of all three tool boxes will give the most complete diagnosis. However, the resources available for environmental management are always limited, which means that it is hardly possible to apply all three tool boxes in all cases, but it is necessary in many cases to a make a choice. If an ecological model is developed, in order to be able to give more reliable prognoses, it is of course natural to apply the developed model, and in addition it may be beneficial to select one or a few indicators to focus more specifically on a well-defined problem. If a model is not available but a monitoring program has to be developed, it would be natural to direct the observations to encompass the state variables that can be applied to assess the indicators that are closely related to the defined health problems. If society is dependent on specific ecological services of the ecosystem, it would be natural to assess to what extent these services are maintained, reduced, or lost, maybe supplemented with health indicators that are particularly important for the maintenance of these services. The choice of tool boxes is therefore a question about the available resources and the specific case and problem.

1.3 The Tool Boxes Available Today to Solve the Environmental Problems

The tool box of environmental technology was the only methodological discipline available to solve the environmental problems 45–50 years ago, when the first green waves started in the mid-1960s. This tool box was only able to solve the problems of point sources sometimes, but not always at a very high cost. Today, fortunately, we have additional tool boxes that can solve the problems of diffuse pollution or find alternative solutions at lower costs when the environmental technology would be too expensive to apply. As for the diagnostic tool boxes, these tool boxes are developed on the basis of new ecological subdisciplines.

To solve environmental problems, currently we have four tool boxes

1. Environmental technology
2. Ecological engineering, also denoted as ecotechnology
3. Cleaner production, and under this heading we would also in this context include industrial ecology
4. Environmental legislation

Environmental technology was available when the emergence of the first green waves started about 45–50 years ago. For an overview of the available methods of environmental technology, see Chapter 11. Since then, several new environmental–technological methods have been developed and all the methods have been streamlined and are generally less expensive to apply today. There is and has been however an urgent need for alternative methods to solve the entire spectrum of environmental problems at an acceptable cost. Environmental management today is more complicated than it was 50 years ago because of the many more tool boxes that should be applied to find the optimal solution and because global and regional environmental problems have emerged. The use of tool boxes and the more complex situation today is illustrated in Figure 1.2.

The tool box containing ecological engineering methods has been developed since the late 1970s. Ecological engineering is defined as a design of sustainable ecosystems that integrate human society with its natural environment for the benefit of both (Mitsch and Jørgensen, 2004). It is an engineering discipline, that operates in ecosystems, which implies that it is based on both design principles and ecology. The tool box contains four classes of tools:

1. Tools that are based on the use of natural ecosystems to solve environmental problems (for instance, the use of wetlands to treat agricultural drainage water)
2. Tools that are based on imitation of natural ecosystems (for instance, construction of wetlands to treat wastewater)

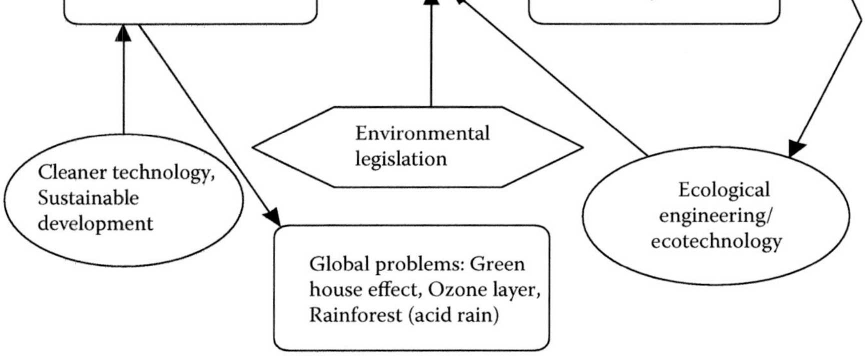

FIGURE 1.2
Conceptual diagram of the complex ecological–environmental management of present day, where there are various tool boxes available to solve the problems and where the problems are local, regional, and global.

3. Tools that are applied to restore ecosystems (for instance, restoration of lakes by the use of biomanipulation)
4. Ecological planning of the landscape (for instance, the use of agroforestry)

The introduction of ecological engineering has made it possible to solve many problems, that environmental technology could not solve, first of all nonpoint pollution problems, followed by a fast restoration of deteriorated ecosystems. Further details about ecological engineering are presented in Chapter 12.

As the environmental legislation has been tightened, it has been more and more expensive to treat industrial emissions, and the industry has of course considered whether it was possible to reduce the emission by other methods at a lower cost. That has led to the development of what is called cleaner production, which means the idea to produce the same product by way of a new method that would give a reduced emission and therefore less costs for the pollution treatment. New production methods have been developed by the use of innovative technology, that has created a completely new method to produce the same product with less environmental problems. Other emission reductions have been developed by the use of ecological principles in the industrial processes, for instance, recycling and reuse. In many cases, it has also been possible to achieve a reduction of the environmental problems by identification of unnecessary waste. Industrial ecology based on the author's opinion could be defined as the use of ecological principles in production such as recycling, reuse, and holistic solutions to achieve a high efficiency in the general use of resources. Industrial ecology today, however, is used to cover the use of waste from one production in another production. Some even use the word industrial ecology, when trees are planted in industrial areas, which of course is not very innovative. Chapters 13 and 14 focus on these methods from two different angles—from the point of view of pollution abatement and from that of longer duration of our resources.

Some environmental problems can, however, not be solved without a more strict environmental legislation and for some problems a global agreement may be needed to achieve a proper solution, for instance, by out-phasing the use of Freon to stop or reduce the destruction of the ozone layer. Note also that environmental legislation requires an ecological insight to assess the required reduction of emissions that is needed through the introduction of environmental legislation. This is the topic of discussion in Chapter 15.

Today, we have a possibility to solve any environmental problem and often at a moderate cost and sometimes even at a cost which makes it profitable to solve the problem properly by way of the four tool boxes with environmental management solutions, because of reduced use of resources. As is the case for the diagnostic tool boxes, also the tool boxes with solution tools to problems are rooted in recently developed ecological subdisciplines that are named after the tools: ecological engineering, cleaner technology, and environmental legislation.

1.4 Follow the Recovery Process

Environmental management is only complete if the environmental problem and the ecosystem are followed carefully after the tool boxes have been applied. It is usually not a problem because it is a question of providing the observations needed to follow the prognoses of the

1. Eventually developed ecological model
2. The selected ecological indicators
3. The recovery of the ecological services of the ecosystem (which can be done by focusing on a specific service or on the values of all the ecological services offered by the ecosystem)

1.5 Conclusions about Integrated Environmental Management

From the review of up-to-date integrated environmental management, it is possible to conclude the following:

1. Follow all the seven recommended steps, using a holistic approach. Integrate the knowledge about the problem(s), the ecosystems involved, and the sources with a good diagnosis and a combination of solution methods of reach an optimum environmental management. A good knowledge of systems ecology is important in this context; see Chapters 3–5.
2. It is recommended to consider the use of all three diagnostic tool boxes, but use at least the diagnostic tool box which fits into the problem, ecosystem, and the available observations. A good diagnosis is an indispensable step in the integrated management.
3. The three diagnostic tool boxes can also be applied to follow the development of an environmental–ecological problem—included in an eventually recovery process.
4. Eco-exergy—work energy including the work energy of information—is a useful indicator as it expresses sustainability and thereby the total amount of ecological services offered by an ecosystem.
5. Integrated environmental management based on the three diagnostic tool boxes and considering all sources of the problem may require the use of all four "problem-solving" tool boxes:
   a. Environmental technology
   b. Ecotechnology

c. Cleaner technology including industrial ecology
d. Environmental legislation

A combination (integrated use) of the four tool boxes will often be able to offer the best and most moderate-cost solution.

Integrated up-to-date environmental management requires the use of the seven presented tool boxes, which would not be possible if these tool boxes were not developed as a result of recently emergent ecological subdisciplines: ecological modeling, ecological engineering, application of ecological indicators, cleaner technology, and industrial ecology. These ecological subdisciplines are therefore crucial for the environmental management of today and they form an indispensable bridge between ecology and environmental management—between the basic science of ecology and its application in practical environmental management. This conceptual bridge symbolizing the close and integrated cooperation between the ecological subdisciplines and environmental management today is illustrated in Figure 1.3.

Integrated environmental management draws very much from the ecological disciplines shown in Figure 1.3. They are therefore presented in detail in various chapters of this book. The environmental problem is defined, which is the topic of discussion in Part I, Chapter 2; Systems Ecology is covered in

FIGURE 1.3
This conceptual bridge illustrates the close and integrated cooperation between the subdisciplines of ecology and environmental management, which is a prerequisite for an up-to-date and holistic solution to the environmental problem(s).

Part II, Chapters 3 through 5, where the important properties and processes of different ecosystems are covered. The sources and causes of environmental problems, Part III, Chapters 6 and 7 focus on the pollution sources and their effects on ecosystems and human health. To understand the effects of toxic substances, it is important to use ecotoxicology. It is therefore applied in Chapters 6 and 7. The diagnostic tools: ecological indicators, ecosystem services, and ecological modeling are presented with these headings in three chapters, Chapters 8–10. Environmental technology that was already integrated with environmental management to a high extent 50 years ago is the topic of Chapter 11, followed by the next four chapters, Chapters 12–15, with other problem-solving tools such as ecotechnology, cleaner production, resource management, and environmental legislation. The three latter tools are not ecological disciplines, but rooted in technological development and the development of environmental laws. They are the fruit of our efforts to solve the environmental problems after we have understood the root of the problems and developed a proper diagnosis by using the ecological disciplines. Chapter 16 discusses how to perform integrated, holistic ecological–environmental management on the basis of a synthesis of the seven steps. Chapter 17 is devoted to the seventh step: Follow the recovery process to ensure that the goals of ecological–environmental management have been achieved.

1.6 Classification of Environmental Problems

Traditional environmental management classification indicates the sphere that is touched by the environmental problem. Is it a water problem (is the hydrosphere involved)? Is it an air pollution problem (is the atmosphere involved)? Or is it a terrestrial problem (is the lithosphere involved)? This traditional classification is still widely used, but five new topics which have a particular emphasis today can be added:

- *Agriculture:* Many as-yet unsolved environmental problems are rooted in agriculture.
- *Energy:* Environmental problems associated with our use of energy are considered of particular importance for our future. Our present use of fossil fuel cannot continue. Oil and natural gas will be depleted in about 50 years and the use of fossil fuel implies massive, climate-changing emissions of greenhouse gases.
- *Toxicological/ecotoxicological problems:* Problems resulting from the lack of control over the emissions of toxic substances in the environment.

- *General:* Problems involving multiple spheres, often all three spheres.
- *Global:* Problems that have a global scope (the entire ecosphere is involved). The greenhouse effect or the deterioration of the ozone layer are typical examples.

It gives the following classification, which has also been applied in the recently published *Encyclopedia of Environmental Management* (Jørgensen, 2013); (the following abbreviations are those which are applied in the encyclopedia):

AGR: Agricultural issues

AIR: Air pollution

ENG: Energy use issues

GEN: General

GLO: Global

TER: Terrestrial pollution

TOX: Emission of toxic substances

WAT: Water pollution

The encyclopedia provides significant background knowledge to cover the seven presented steps to achieve an integrated, holistic environmental and ecological management of many of the actual environmental problems, from pollution problems rooted in the extensive use of fossil fuel and our water, air, and solid waste problems to erosion and nonpoint agricultural pollution of pesticides and fertilizers.

The articles (entries) have also been classified according to the presented seven-step procedure for the integrated, holistic environmental and ecological management. To repeat, the content has used the following topical classifications:

PSS covers entries that focus on a pollution problem and its sources (steps 1 and 3).

COV indicates that the articles are giving comparative overviews of important topics for environmental management or background knowledge that is important for the evaluation of environmental problems. The entries are of particular importance for step 2, but also touch upon steps 1 and 3, and also cover the follow-up—meaning step 7.

DIA means that the articles are about the diagnostic tools: monitoring, ecological modeling, ecological indicators, and ecological services. Refer to step 4.

CLT means that the solutions are based on cleaner technology, one of the tool boxes used in step 5.

ECT covers solutions of the problems based on ecotechnology, one of the four tool boxes used in step 5.

ENT refers to solutions of environmental problems with the use of environmental technology, one of the four tool boxes used in step 5.

ELE focuses on the use of environmental legislation to solve the environmental problems, one of the tool boxes used in step 5.

IMS are articles uncovering the possibilities to integrate the various tool boxes to derive an integrated and holistic management process. These entries focus on step 6. By using the two classifications simultaneously, $8 \times 8 = 64$ combinations of entries (articles) are presented.

References

Costanza, R., B.G. Norton, and B.D. Haskell. 1992. *Ecosystem Health, New Goals for Environmental Management*. Washington, DC: Island Press, 270 pp.

Costanza, R. et al. 1997. The value of the world's ecosystem services and natural capital. *Nature* 387, 252–260.

Jørgensen, S.E. 2002. *Integration of Ecosystem Theories: A Pattern*. Dordrecht: Kluwer, 386 pp.

Jørgensen, S.E. 2006. *Eco-Exergy as Sustainability*. Southampton, UK: WIT, 220 pp.

Jørgensen, S.E. 2010. Ecosystem services, sustainability and thermodynamic indicators. *Ecological Complexity* 7, 311–313.

Jørgensen, S.E. 2013. *Encyclopedia of Environmental Management*. Boca Raton, FL: CRC Press, 4 volumes, 3280 pp.

Jørgensen, S.E. and B. Fath. 2011. *Fundamentals of Ecological Modelling*. 4th edition. Amsterdam: Elsevier, 400 pp.

Jørgensen, S.E., N. Ladegaard, M. Debeljak, and J.C. Marques. 2005. Calculations of exergy for organisms. *Ecological Modelling* 185, 165–176.

Jørgensen, S.E. and S.N. Nielsen. 2012. Tool boxes for an integrated ecological and environmental management. *Ecological Indicators* 20, 104–109.

Marques, J.C. et al. 2009. Ecological sustainability trigon. *Marine Pollution Bulletin* 58, 1773–1779.

Mitsch, W.J. and S.E. Jørgensen. 2004. *Ecological Engineering and Ecosystem Restoration*. New York: John Wiley, 410 pp.

Part I

Defining the Problem

2

Conceptual Problems That Must Be Addressed to Create, Understand, and Test Environmental Management Scenarios

2.1 Sustainable Development and Ecological Sustainability

All types of environments worldwide are threatened as a consequence of pollution, overexploitation, and the impacts of climate change, which increased awareness of the intense impact of humans on ecosystem functioning. As a consequence, there is a well-identified need for approaches to sustain and, where necessary, restore ecosystems (Hughes et al., 2005). Approximately 25 years ago, the idea of *sustainable development* emerged (Pulselli et al., 2008), but there is no agreement regarding the precise understanding and definition of the term among researchers from different disciplines. "Development that satisfies present needs without compromising the possibility of future generations satisfying theirs" (Brundtland, 1987) is probably the most commonly adopted definition of sustainable development, but this definition is relatively vague and nonoperational. In fact, to apply the concept, additional quantification from the scientific, cultural, and socioeconomic points of view is required, and it is also necessary to take into account time, relationships, and biophysical limits (Pulselli et al., 2008).

More often, human society does not evolve in harmony with environment's capability to produce the assets necessary for its development, which turns time into a very important factor with regard to sustainability. But despite the fact that sustainable development became a crucial challenge, the way human society interacts with the various needed living and nonliving natural resources, which constitute what is called natural capital, is still controversial, to say the least. Essentially, there are two main indubitably contrary perspectives regarding the conceivable practical mean of sustainability: weak and strong sustainability. Weak sustainability admits that human well-being must continue through intergenerational timescales, considering that natural capital and man-made capital can be interconverted in the scope of specific production processes (Brand, 2009). Consequently, weak sustainability also accepts the option of depleting natural capital, unless its requirement tends to decline over time (Brand, 2009). Inversely, strong sustainability takes up a complementarity between natural capital and man-made capital, assuming that the whole stock of natural capital has to be preserved for present and future generations in the long run, and therefore that human society must keep each type of capital intact over time (Brand, 2009). Despite these conflicting perspectives, the fact that humans and their cultural diversity are recognized as an integral component of ecosystems, and awareness concerning predictable hazards arising from a serious worldwide environmental degradation, forced the entrance of ecological sustainability into international agendas.

Regardless of the conceptual approach adopted, resilience, the capacity to cope and adapt, and the conservation of sources of innovation and renewal constitute critical factors for sustainability when uncertainties and change are key questions of environmental and social organization (Lebel et al., 2006). Consequently, the sustainable management of natural systems implies attaining steadiness between delivering those economic goods and services provided by the environment, indispensable for societal health and functioning, and simultaneously continue and preserve ecological goods and services essential for natural systems' health and functioning.

Defining as clearly as possible the concepts involved in creating environmental management scenarios is indispensable. Jørgensen et al. (in prep), for instance, provided a set of useful ecological-oriented definitions, which we summarize in Table 2.1.

Nevertheless, such concepts are not always used with the very same meaning, obviously depending on the field where they are applied (e.g., ecology, economics, policy), as illustrated and discussed along with the present chapter.

The maintenance of environment and natural resources might be irreconcilable with our present economical paradigm, and therefore relationships become a crucial concern. Indeed, efficient allocation of resources is often not taken into account by economic instruments, which tend to consider only issues directly driven by the market (Pulselli et al., 2008), while relationships imply interdependencies and consequently the need to determine at what scale (regional, national, etc.) different aspects are interdependent.

Finally, at each place, human populations cannot easily fulfill their needs in terms of materials, energy, land, waste sinks, and information proceeding from local resources and therefore biophysical limits must be taken into account. The concept of "critical natural capital," developed between the "weak" and "strong" sustainability perspectives, reflects this evident fact. Critical natural capital may then be seen as part of the natural capital that perform significant and unique environmental functions, ensuring therefore ecosystem services that cannot be replaced by other types of capital (de Groot et al., 2003). Actually, especially in recent centuries, societal development has been driven mostly by finite, nonrenewable resources, and such perceptions have conveyed the general acceptance that the maintenance of

TABLE 2.1

Ecologically Oriented Definitions of Some Important Concepts

Concept	Definition
Resistance	The ability to resist changes of impacts or changes of forcing functions (external variable functions). It can be quantified as (change of forcing function)/(change of state variable). As there are many forcing functions and state variables in most cases, it implies that there are many quantifications of resistance. Note that the concept has a unit accordance with the definition.
Buffer capacity	Same definition as resistance.
Sensitivity (to external changes)	Inverse resistance or buffer capacity.
Resilience	The ability to return to normal (behavior, properties) after perturbations or disturbances. This concept focuses on the ability of systems to recover after disturbances. It can be quantified as (the change of forcing function or impact)/(time it takes for the system to recover = time it takes to return to normal). The quantification implies that normal behavior or normal properties have to be defined, for instance, as an interval of properties. Note that the unit is the unit of the forcing function/the applied time unit.
Sustainability	The resources of the system are maintained unchanged. For ecosystems, the ecosystem services offered to its living components or to mankind must be unchanged to denote the ecosystem sustainable. It is possible to distinguish between renewable and nonrenewable resources. As our society does not use 100% recycling, it is not possible to maintain the nonrenewable resources unchanged. In this case, society will be denoted sustainable if the nonrenewable resources are utilized at a rate that makes it possible to find substitutes for them. Unchanged renewable resources imply that the renewable resources cannot be used at a rate exceeding the rate at which they are created/renewed.

Source: Jørgensen, S.E., B. Fath, and S.N. Nielsen. (in prep). Definitions of important ecological concepts.

critical natural capital is required to attain environmental sustainability and sustainable development (Ekins et al., 2003).

The evaluation of natural capital is required in at least six domains: socio-cultural, ecological, sustainability, ethical, economic, and human survival (Brand, 2009). In addition, sustainable environmental management can only be attained if options and actions undertaken are environmentally and ecologically sustainable, economically realistic, technologically feasible, socially desirable or at least socially tolerable, administratively manageable, legally admissible, and politically opportune (e.g., Elliott et al., 2006; Bunce et al., 2008; Mee et al., 2008; Ojeda-Martínez et al., 2009). These are the essential conditions to achieve the main challenge for natural and social sciences, which is presently the maintenance or even the increase of economic goods and services demanded by a developing society and simultaneously maintaining and protecting ecological goods and services; these together represent environmental goods and services.

Natural and social aspects of ecosystems are intimately linked, which is reflected in "The Ecosystem Approach" in the strict sense (e.g., Convention for Biological Diversity; Kay et al., 1999). For instance, environmental restoration, crucial nowadays, involves dealing with complex problems like (a) loss of species diversity, habitats, and a reduction in habitat heterogeneity and size; (b) changes in dynamics and spatial distribution of many species, as well as diminution of their population size; (c) fragmentation of habitats and associated increase in the vulnerability of the remaining isolated pouches; and (d) reduction of economically significant services and goods naturally offered by ecosystems (e.g., Elliott et al., 2007).

The challenge of seeking for ecological sustainability must be met within this framework, in which some ecological concepts are well-understood or at least appropriately defined, for instance, the nature of ecosystem structure and functioning, whereas others, like resilience, carrying capacity, or ecosystem goods and services are not yet adequately quantified.

At present, from the ecological point of view, health of ecosystems is often assessed based on the abundances of few conspicuous (or even charismatic) species, namely birds, fish, and marine mammals. Weaknesses of such an approach are manifest because the mechanisms underlying temporal or spatial variations in abundance are frequently poorly understood and the way changes in these species influence ecosystems as a whole are rarely addressed (Hughes et al., 2005). Resilience-based management, for example, constitutes therefore an innovative and appropriate approach to marine and coastal ecosystems, especially if applied to fisheries, providing a shift in focus from the conservation of targeted species, frequently commercially important ones, to the active management of functional groups supporting critical processes and the maintenance of ecosystem services (Hughes et al., 2005). This focus on functional groups assumes the importance of species interactions and ecological roles (including that of humans) in sustaining ecosystems resilience through spatial and temporal scales, representing a clear change in perspective (Folke, 2006).

The resilience concept, despite its central position in sustainability science, has been suffering significant changes through the last three decades (Walker et al., 2004). In fact, there are confusions around the term, and several attempts have been undertaken to clarify it (Elliott et al., 2007), assuming the perspective that it refers to the intrinsic capacity of a system of coming back to a prior or similar state after a disturbance. In addition, Elliott et al. (2007) specified that resistance is the capacity of the system to tolerate stressors. Yet, there are at least two other meanings which can be found in the literature. One refers to the dynamics of close to equilibrium systems, being defined as the time necessary to their return to an equilibrium point after a disturbance, therefore addressing a system's recovery. This is commonly called "engineering resilience" (Holling, 1996; Folke, 2006), being to a large extent equivalent to the stability property "elasticity" (Grimm and Wissel, 1997), or "resistance to change" (Levin and Lubchenco, 2008). A second one is defined as the capacity to absorb stress and still maintain "function," referring to the dynamics of far from any equilibrium steady-state systems, and has been called "ecological resilience" (Gunderson and Holling, 2002; Folke, 2006). It essentially corresponds to the capacity to maintain functioning in spite of the multiple stressors which may affect an evolving system (Levin and Lubchenco, 2008). This second meaning addresses more the system's renewal, regeneration, or reorganization after a disturbance than postrecovery (Folke, 2006), which implies the assumption that disturbances and spatial heterogeneity determine the behavior of each system to be unique. As a consequence, recovery trajectories might be difficult or impossible to predict, due to the complexity of the system combined with unexpected compounded effects of disturbance, and a recovered system may eventually look identical to the previous one but is not the same system. As any other living system, it will be continuously developing (Folke, 2006). Differences in degradation and recovery trajectories can be called system hysteresis (Elliott et al., 2007), and in practical terms, ecological resilience can only be estimated by means of resilience proxies (Carpenter et al., 2005), which must be based on a broad resilience analysis, together with the identification of specific disturbance regimes and societal choices regarding desired ecosystem services (Brand, 2009).

What is the expected relationship between ecological resilience and critical natural capital? In simple terms, an ecosystem's degree of ecological resilience is assumed to be inversely related to its degree of threat (Brand, 2009), and therefore holding information on ecological resilience and system efficiency constitutes a prerequisite to assess whether an ecosystem will reach a critical state as a response to environmental stressors (Ulanowicz et al., 2009). Thus, a critical state regarding natural capital matches an extent in environmental degradation that exceeds a threshold beyond which the current level of social welfare cannot be supported. Such an ecological criticality appears therefore to be most important for the maintenance of ecosystem services and goods (Jax, 2005) and the sustainable use of their natural capital (Brand, 2009).

Ecosystems and their modification as a function of human stressors (e.g., overfishing or pollution) must be examined accounting to their carrying capacity and its loss next to stress. Carrying capacity is a concept originally defined as the number of individuals of a population that can be supported by a given environment without significant negative impacts to that population and that environment (Elliott et al., 2007). This definition is merely ecological and does not entirely capture the multilayered processes of human–environment relationships, which have a fluid and nonequilibrium nature, perhaps neglecting the influence of external forces on environmental change (Moore et al., 2009). Aiming to improve the concept of generality, Elliott et al. (2007) highlighted that the carrying capacity must also be related to the social and economic aspects of ecosystems, keeping in mind the level of human activities and anthropogenic pressure which an ecosystem can tolerate before undesirable changes occur. This means that it must be anticipated that humans' relationship with their environment are in principle more complex than those of other species with theirs (Pulselli et al., 2008). Actually, humans have the capacity to modify the type and degree of their impact on their environments, for instance, by increasing fisheries, by using more intensive farming techniques in aquaculture, when a defined local area is left free of productive activities or when marine products' consumption is scaled back. Humans can at a limit decrease environment productivity irreversibly and/or radically occupy or transform space. So, many species can modify the carrying capacity of their systems, but since only humans can manipulate ecosystems to such a drastic extent, the concept of carrying capacity must be expanded to include socioeconomic aspects (MacLeod and Cooper, 2005; Elliott et al., 2007).

A holistic approach to understand, manipulate, and manage the environment may be provided by linking and integrating these ecological concepts and the management framework. Ideas relating environmental management to the dynamics of ecosystems and the relations between biodiversity and ecosystem function seemed to be particularly crucial, leading to the perspective that ecosystems are complex adaptive systems, which are characterized by their historical dependency, as well as by nonlinear dynamics, threshold effects, various basins of attraction, and limited predictability (Folke et al., 2004; Duit and Galaz, 2008; Moore et al., 2009). For instance, it has been shown that estuarine ecosystems' temporal stability and biodiversity might follow different trends in relation to the systems' ecological quality status, exhibiting therefore a nonlinear relationship (Pinto et al., 2013b; 2014a). According to these findings, the optimal condition with regard, for instance, to estuarine stability (with higher values) might occur for average values of species richness, inducing that consequently this may influence ecosystem services provision. Furthermore, again in estuaries, this was confirmed through the analysis of food webs under distinct levels of eutrophication pressure, using mass-balanced models. In this case, a tendency for a decrease in food web connectivity was observed during a recovery process, from more eutrophic to less eutrophic conditions (Baeta et al., 2011), while the ecosystem's temporal stability tended to increase in the same period.

In which way can this panoply of theoretical concepts be applied to different environments, which management involves so many uses and users, sectorial interests, and governance regimes (Costanza et al., 1998; 1999; McLusky and Elliott, 2004)? How can we decide, for instance for the restoration or sustainable use of a specific ecosystem, which will be the best possible course of management in a multitude of driving forces which may be conflicting? In such multiuse/multiuser/multisectorial systems management solutions require that we are able to orientate ourselves among such complexity. Marques et al. (2009a) proposed the use a conceptual guidance tool—the Ecological Sustainability Trigon (EST)—which may possibly be useful as a sort of compass to provide orientation in the process of building and testing management scenarios to approach environmental problems. This proposal was mostly based on ideas resulting from the marine coastal and estuarine research, but seems to be also applicable to other domains.

2.2 Using EST in Building Management Scenarios toward Ecological Sustainability

It is generally accepted that there is an intrinsic complexity in environmental problems, and also that these are closely related to the development of human society. As a consequence, possible solutions to environmental problems always require taking into account different points of view, often expressing the conflicting perceptions of multiple sectors, uses, and users (Costanza et al., 1998; 1999). Different economic scenarios (Figure 2.1) (Turner et al., 1994; 2003; Turner, 2008) have been to a certain extent trying to address the uncertainty involved in dealing with environmental problems, and it is clear that in the way the natural or the social systems become favored will greatly depend on the different options and systems of decision (see Table 2.2). With regard to sustainability, environmental scenarios analysis

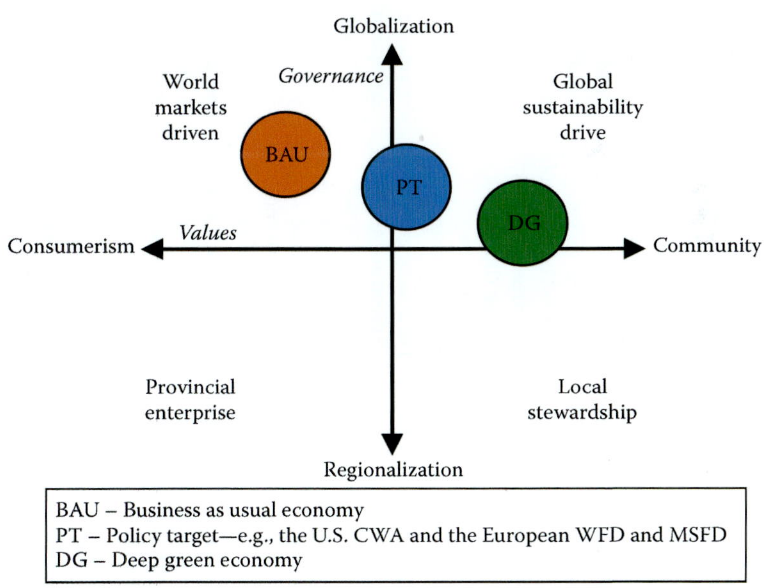

FIGURE 2.1
Different economic scenarios resulting from supporting the natural or to the social system, as a function of political decision systems. (From Turner, R.K., D. Pearce, and I. Bateman. 1994. *Environmental Economics: An Elementary Introduction*. London: Harvester Wheatsheaf, 328 pp. With permission.)

TABLE 2.2

Environmental Ideologies and Sustainability Perspectives

Very Weak Sustainability	Weak Sustainability	Strong Sustainability	Very Strong Sustainability	
Resource exploitative, growth maximization position	Resource conservationist and managerial position	Resource preservationist position	Extreme preservationist position	
Antigreen economy; Unfettered free markets; Widening income inequality not problematic; Free trade in international markets	Green economy, natural capitalism, and new industrial system green markets guided by economic incentive instruments (EIs) (e.g., pollution charges, etc.) in combination with voluntary agreements	Deep green economy, steady-state economy regulated by macro-environmental standards and supplemented by EIs and international agreements	Very deep green economy, heavily regulated to minimize resource take; National environmental duty of care formally regulated; Extensive and binding international agreements	Green labels
Primary economic policy objectives, maximize economic growth (max GNP); No formal policy integration processes	Modified economic growth (adjusted green accounting to measure GNP); Formal policy integration and review of institutional structures of growth and environmental quality	Zero economic growth; Zero population growth; Binding policy integration	Reduced scale of economy and population; Sustainability accounting, the primary approach	Type of economy
Taken as axiomatic that unfettered free markets in conjunction with technical progress will ensure infinite substitution possibilities capable of mitigating all local scarcity limits and constraints (environmental sources	Decoupling of growth and environmental quality, important but infinite substitution rejected; Sustainability rules: for example, constant natural capital rule; Use efficiency and productivity;	Decoupling plus no increase in scale; Systems perspective—health of whole ecosystems very important; Over-compliance with international environmental	Scale reduction imperative; At the extreme for some there is a literal interpretation of the Gaia hypothesis with moral obligations	Policies and management strategies

(*Continued*)

TABLE 2.2 (*Continued*)

Environmental Ideologies and Sustainability Perspectives

Very Weak Sustainability	Weak Sustainability	Strong Sustainability	Very Strong Sustainability	
and sinks); Voluntary approach to environmental regulation and intervention	Sustainability indicators and monitoring	agreements; Sustainability assessments and audits		
Support for traditional ethical reasoning; Rights and interests of contemporary individual humans; Instrumental value (e.g., recognized value to humans) in nature	Extension of ethical reasoning: caring for others, motive for intergenerational and intergenerational equity (i.e., contemporary poor and future people); Instrumental value in nature	Further extension of ethical reasoning: interests of the collective take precedence over those of individual primary value of ecosystems and secondary value of component functions and services	Acceptance of bioethics (i.e., moral rights/interests conferred on all nonhuman species and even the abiotic parts of the environment); Intrinsic value in nature (i.e., valuable in its own right regardless of human experience)	Ethics
Low level of environmental awareness in the public	Wider public education, establishment of stakeholder groups and of, for example, round tables to increase inclusion	Strong local/community awareness and action campaigns	Cultural shifts to the maintenance of local livelihoods and environmental stewardship	Degree of public inclusion

Sources: Nunneri et al. 2009a. *Mar. Pollut. Bull.* 58, 1773–1779. doi:10.1016/j.marpolbul.2009.08.020; Marques, J.C. et al. 2009a. *Mar. Pollut. Bull.* 58, 1773–1779. doi:10.1016/j.marpolbul.2009.08.020; Marques, J.C. et al. 2009b. *Ecological Indicators for Coastal and Estuarine Environmental Assessment—A User Guide*. Southampton, UK: WIT Press, 200 pp.

TABLE 2.3

Core Questions for Sustainability Science

1. How can the dynamic interactions between nature and human society—including time-lags due to inertia—be better incorporated into emerging models and conceptualizations that integrate the global systems, human development, and sustainability?
2. How are long-term trends and widely varying spatial scales in environment and development, including consumption and population change, reshaping nature–society interactions in ways relevant to sustainability?
3. What determines the vulnerability and resilience of the nature–society system in particular kinds of places and for particular types of ecosystem and human livelihoods?
4. Can scientifically meaningful limits or boundaries be defined that would provide effective warning thresholds beyond which the nature–society systems are at a significantly increased risk of serious degradation?
5. What systems of incentive structures—including markets, rules, norms, and scientific information—can most effectively improve social capacity to guide interactions between nature and society toward more sustainable trajectories?
6. How can today's operational systems for monitoring and reporting on environmental and social conditions be integrated or extended to provide more useful guidance for efforts to achieve sustainability?
7. How can today's relatively independent activities of research, planning, observation, assessment, and decision-support be better integrated into systems for adaptive engagement and societal learning?
8. How can future changes be determined and predicted in a creative, objective, rigorous, and policy-relevant manner that reflects sustainability and incorporates different perspectives?

Source: Modified from Kates, R.W. et al. 2001. *Science* 292(5517), 641–642. doi:10.1126/science.1059386; Swart, R., P. Raskin, and J. Robinson. 2004. *Global Environ. Chang.* 14, 137–146. doi:10.1016/j.gloenvcha.2003.10.002; Marques, J.C. et al. 2009a. *Mar. Pollut. Bull.* 58, 1773–1779. doi:10.1016/j.marpolbul.2009.08.020; Marques, J.C. et al., 2009b. *Ecological Indicators for Coastal and Estuarine Environmental Assessment—A User Guide.* Southampton, UK: WIT Press, 200 pp.

is therefore required as a tool to recognize and characterize central questions (Table 2.3), as well as to take decisions about priorities and solutions (Kontogianni et al., 2001; 2004; Swart et al., 2004).

Sustainability science is therefore complex and uncertain, but it is possible to encompass all the different viewpoints and concerns in the frame of three major driving forces (Marques et al., 2009a,b):

a. The search for human well-being and the maintenance of human health and safety;
b. The attempt of ecological sustainability and maintenance of natural environmental well-being;
c. Systems resilience in the face of increasing human pressure, resulting from population size and demand for wealth creation.

Certainly, it is necessary to assume that research should compulsorily cover these three views and simultaneously fulfill the goals of creating knowledge and wealth, and improving life quality. The search for human well-being, from the governance point of view, is often endorsed as corresponding to the Gross Domestic Product (GDP) or to stakeholder benefits, which could ultimately be expressed by some sort of metrics such as well-being indices (Diener et al., 1999; Diener, 2000) (Figure 2.2). The straightforward societal objective has been maximizing economic goods and services, which is usually reflected in governmental approaches, and at the same time, to a certain

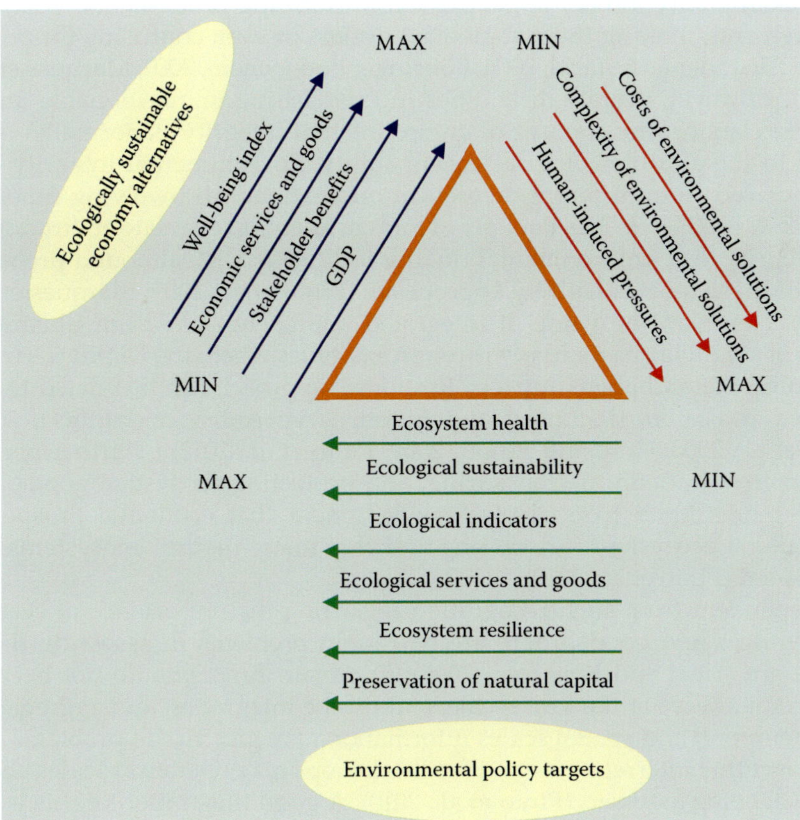

FIGURE 2.2
The EST (Marques et al., 2009a,b) illustrating the expected trends and relationships between variables assumed to be correlated with ecological sustainability, human well-being, and human-induced pressures. The bottom and left-hand axes of the EST indicate how governance and societal systems must be linked to environmental management, namely, through the definition of policy targets and the choice of ecologically sustainable economy alternatives, whereas the right-hand axis illustrates how increasing human pressures implies consequences on the other two axes. (Green—Good; Red—Bad.)

extent, protecting ecological goods and services, at least to avoid businesses of being accused of harming the latter ones, or preventing countries from being subjected to legal violation proceedings for not fulfilling rules settled in common legislation, as is the case with the European Union Directives (Marques et al., 2009a).

Sustainability indicators and composite indices have been progressively accepted as valued tools in terms of policy making and public communication with regard to the environment, economy, society, or technological improvement (Singh et al., 2009), as well as to the human quality of life (Diener and Tov, 2012). Nonetheless, in many cases, when applied in policy, sustainability indices appear to fail in fulfilling fundamental scientific requirements, making them somewhat useless or even confusing for policy advice (Brunner and Starkl, 2004; Böhringer and Jochem, 2007; Marques et al., 2009a). However, despite difficulties in establishing unquestionable quantitative relations, the concept of environmental integrity is normally associated to the aim of ecological sustainability. Environmental integrity can be perceived from different theoretical orientations, all involving inherent uncertainties (Figure 2.2). Panoply of tools available to evaluate environmental integrity (e.g., environmental quality indices) is vast, although probably none of those tools is entirely appropriate (Pinto et al., 2009; Marques et al., 2009b). However, the value of integrated approaches has been illustrated (Pinto et al., 2014a), particularly through examples where the DPSIR (Drivers, Pressures, State Change, Impacts, Response) approach can be related to the selected indicators (McLusky and Elliott, 2004; Aubry and Elliott, 2006; Borja et al., 2008; Gray and Elliott, 2009; Pinto et al., 2013b). Furthermore, a new interest in defining, measuring, and protecting ecosystem goods and services developed from the acknowledgment that economic prosperity depends on ecosystem functioning and that many natural ecosystems are threatened (Pinto et al., 2010).

Independent from such recognition regarding the importance of ecosystem services and goods, there are two main problems in assessing them, which are (Heal and Kristrom, 2005): (i) certain functions do not become important always at the same scale; and (ii) the integration and aggregation of all temporal and spatial scales information may give rise to problems due to the fact that interrelations and feedback loops may operate at scales above the level being assessed (Pinto et al., 2010). A good illustration of this is the dependency of economic goods and services in one area from the successful functioning elsewhere. For instance, estuarine fish nursery grounds in one area create the conditions for the existence of marine commercial stocks in another area (McLusky and Elliott, 2004; Pinto et al., 2010). As a consequence, scaling rules attempting to describe ecosystem services provision and delivery, especially for open and dynamic systems, still require to be quantified and defined (Limburg et al., 2002; Pinto et al., 2013a; 2014a).

The human population pressure is increasing as a consequence of population growth and the associated higher resource consumption and pollution related to the pursuit of satiating cumulative human needs. The complexity, difficulty, and intrinsic costs of solutions essential to deal with environmental problems created by such pressure have been increasing gradually (Figure 2.2). There is no reason to think that the needs of future generations will be different from ours in terms of quantity and quality, and therefore we may anticipate that two changes might occur as compared with today: (i) decrease in global carrying capacity and (ii) decrease in the number of choices available, meaning by both these the number of different resources and our capacity to exploit them. Possibly, both in terrestrial and aquatic ecosystems, shifts in regimes in relation to resilience and the functional roles of biological diversity will occur (Folke et al., 2004; Pinto et al., 2014a), and eventually future generations may have to adapt rapidly their ways in accordance to ecosystems sustainable management (Pinto et al., 2014b).

We assume here that ecological sustainability constitutes a major goal for human society, which reflects in international agreements adopted by most countries (e.g., the Convention on Biological Diversity) and national laws approved. Nevertheless, at the present levels of human population size and associated environmental pressure, it is not conceptually possible to maximize at the same time ecological sustainability and stakeholder benefits. Still, conceptually, it would eventually be possible to maximize ecological sustainability in a situation of extremely large human population size, but this would entail considerably low standards of human well-being. Of course, in the short term, we may consider maximizing stakeholders' benefits and environmental anthropogenic pressures, but only if we renounce to the goal of ecological sustainability, compromising the future in the longer run. It seems clear that in this "game of possibilities" there is no conceivable scenario that allows maximizing simultaneously the ecologically sustainable solutions for human development, the search for human well-being, and environmental human-driven pressures, although a trade-off between the three might be conceptually possible. Marques et al. (2009a) called such a trade-off the EST (Figure 2.3), in which the bottom and left-hand axes indicate how governance and societal systems must be linked to environmental management to increase the chances of moving both from minimum (MIN) in the direction of maximum (MAX), which constitutes the goal. The right-hand axis illustrates how increasing human pressures from MIN (best scenario) toward MAX (worth scenario) inevitably conditions the other two axes determining adverse consequences.

Different issues under discussion (indicators, objectives, pressures, etc.) can be arranged along the EST corresponding axis, which allows an immediate and intuitive rough integrated view of conceivable implications of policy or management decisions, even when processes behind relationships are not completely understood, as is often the case in estuarine and coastal ecosystems. Furthermore, while relationships are assumed to be nonlinear, all imaginable case studies roughly correspond to a position on the EST frame. At the country scale, Sweden for instance—a small, rich population, with

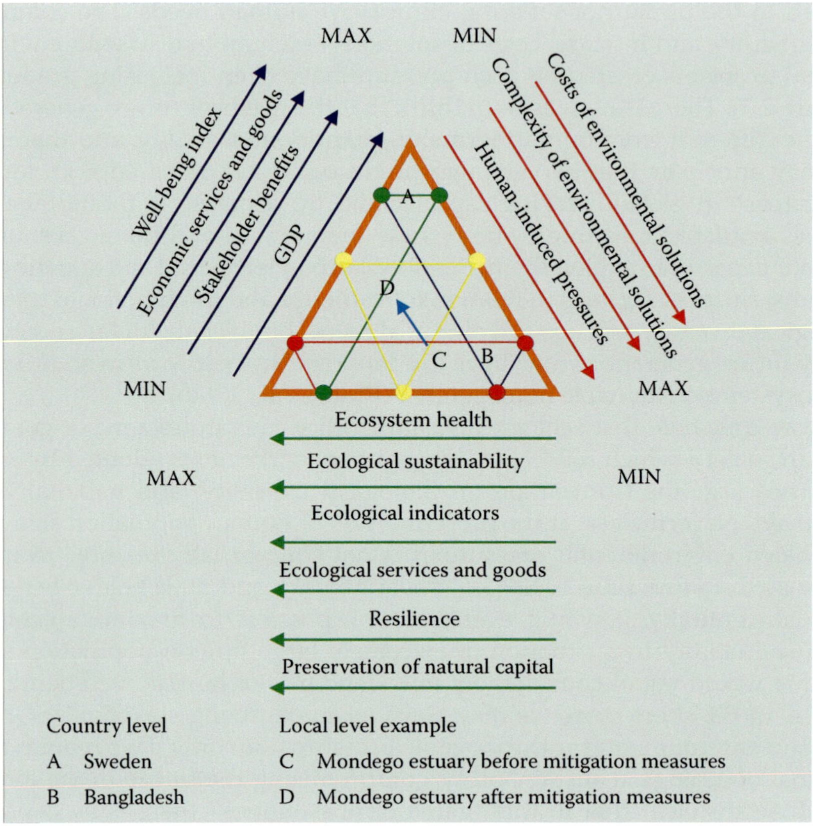

FIGURE 2.3
Proposed use of the EST in building and analyzing management scenarios: Analysis of the expected variations and relationships of different indicators and variables correlated with ecological sustainability, human well-being, and human-induced pressures (see legend in Figure 2.2 for details). (Green—Best scenario; Red—Worth scenario; Yellow—Acceptable scenario.)

good ecosystem governance, few environmental problems, and a high capability to tackle those problems—would be placed approximately in position A (Figure 2.3). However, Bangladesh, including the Sunderbans mangrove area, which has a large population and high human stress, poor funding and badly structured governance systems to implement resolutions, large pressures on ecosystems, and solutions almost ecologically insensible, which distresses its population well-being, could come approximately to position B. At the local scale, as a result of the application of some mitigation and recovery measures aimed at dealing with eutrophication problems, the Mondego estuary on the western coast of Portugal (which has been at long last studied) has shown an improvement in its ecological quality condition during the last decade, and a positive evolution of regulation and cultural ecosystem services, concomitant to a certain extent to an increase in stakeholders' benefits (Pinto et al., 2013b; 2014a). This corresponded roughly to a gradual evolution from position C to position D since the mid-1990s up to the present (Figure 2.3).

The two classic examples at the country scale plus the third one at the local scale, illustrate how the EST may help decision makers (a) in identifying the links between drivers and (b) in clarifying/deciding about the best management options to convey any system closer to an optimal condition in the trigon. Independent from the scale of the management scenario we may want to build, the EST may in principle be used as a conceptual guidance tool, which constitutes one of its most interesting features.

Obviously, we must accept that certain things can be managed and other things cannot, and consequently choices have to be made. It is with regard to such choices that the EST may constitute a useful intuitive tool. Bangladesh, for instance, at least in the short term, cannot alter its population size and concomitant human pressure, but governance and environmental management could eventually be improved. Similarly at a local environmental management scale, some pressures can be handled, for instance in the Mondego estuary, where mitigation measures included altering the discharge of point source polluting materials, allowing for a decrease in eutrophication symptoms in estuarine areas, although the general eutrophic situation in the river basin could not yet be solved (Veríssimo et al., 2013). However, climate change, as an "exogenic unmanaged pressure," is an example of something which cannot be managed. In fact, far from managing it, local managers can only respond to its consequences.

In general, it must be accepted that solving the increasing complexity of environmental problems cannot be the only condition to reconcile the difficulties in accounting for the three driving forces simultaneously (achieve and keep ecological sustainability human well-being goal, and dealing with human population or size pressure). Indeed, the benefits from pursuing a possible harmonization by implementing complex solutions will not increase linearly as a function of that complexity, as the intrinsic costs (energy and money) will plausibly become unsustainable in the long run (Figure 2.4).

2.3 What Might Be the Advantages of Using EST?

The EST proposal (Marques et al., 2009a) represents a tentative and intuitive view which requires further testing and debate (Pinto et al., 2013a; 2014a,b). In fact, the number of variables that must be taken into account when building management scenarios is often very high, as well as the uncertainties regarding their relationships and trends. The core problem resides in understanding to what extent interactions between given economic and ecological systems are sustainable, which involves cross scale, transcultural, and

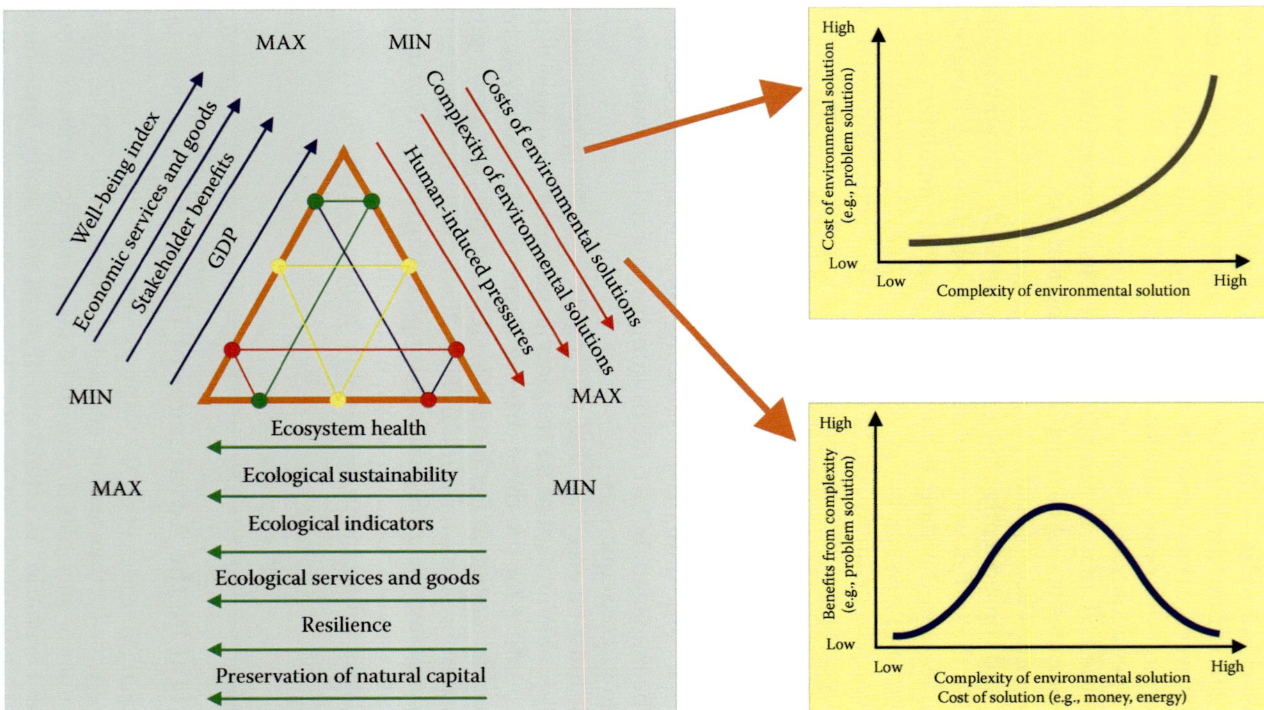

FIGURE 2.4
Proposed use of EST in building management scenarios: Expected trend in costs and benefits as a function of an increasing complexity of environmental problem solutions, namely to meet policy targets, forces the adoption of environmentally viable economic activities.

transdisciplinary fundamental questions, calling for innovative approaches in research, in policy, and to build social institutions (Costanza et al., 1998; 1999). EST as a conceptual framework has been tested with interesting results at local-level case studies (Pinto et al., 2013a; 2014a), as a kind of compass to provide orientation in building management scenarios ("In which direction do we want to go?" "How do we get there?"), alternatively to more conservative forecast scenarios ("Where are we going?"). In fact, besides addressing adaptations to current conditions and actions to be undertaken in the short term, management scenarios should also be built to accomplish transformations toward more sustainable development pathways (Folke, 2006).

As shown in Figures 2.2 through 2.4, different variables have the tendencies of variation of which are in principle correlated, can be arranged along the corresponding sides of the triangle, with their expected trends varying from MAX to MIN. Examples of variables are provided, but many other could of course be included, which may prove particularly useful, namely, when policy targeted approaches are pursued in exploring ecosystems' resources (e.g., implementing the European Water Framework Directive, or the European Marine Strategy Framework Directive) (Pinto et al., 2014b).

For instance, applying governance theory, which accounts for the legal and administrative aspects of policy implementation, implies establishing hypothesis on how different governance forms can be expected to handle processes of change, which are characterized by nonlinear dynamics, threshold effects, and cascades, being therefore poorly predictable (Duit and Galaz, 2008). Also, perhaps due to scientific uncertainty and to lack of consensus among scientists, linkages between ecological science and environmental policy are poor, which jeopardize the transfer of science into management (Moore et al., 2009; Pinto et al., 2010; Borja et al., 2011). The usefulness of the EST as an orientation tool may possibly be more explored in testing governance hypothesis. Moreover, EST can also be tested in analyzing and integrating, at least approximately (especially as correlations are far from being linear), the expected variations and interactions between different variables assumed to be correlated with ecological sustainability, human well-being, and anthropogenic pressures. Namely, for decision makers, these possibilities represent a great potential with regard to an intuitive clarification of what might be compliant and what is probably not compliant with the goal of sustainability, as well as in establishing possible safety margins in the "game of possibilities."

One of the advantages of the EST approach is being able to address and measure the different environmental components with a same species-specific currency, that is, the human society view and, at the same time, of describing in the light of ecological theory our behavior, energetics (economy), and dynamics. More than ever, the incorporation of our behavior, energetics, and dynamics into an ecosystem integrity framework represents a crucial challenge for the science of ecology, which requires measures of ecological status from the ecosystem organization and functioning points of view, rather than from pressures and vulnerability. Nevertheless, the EST approach allows making the evaluation criteria for environmental management scenarios more explicit, meaning that scales should match (time and space), interactions should match (relationships), and rates should match (biophysical limits), therefore constituting a promising tool for gap analysis (information concerning knowledge lacunae), as well as addressing new research questions.

References

Aubry, A. and M. Elliott. 2006. The use of environmental integrative indicators to assess seabed disturbance in estuaries and coasts: Application to the Humber Estuary. *Mar. Pollut. Bull.* 53(1–4), 175–185. doi:10.1016/j.marpolbul.2005.09.021.

Baeta, A., N. Niquil, J.C. Marques, and J. Patrício 2011. Modeling the effects of eutrophication, mitigation measures and an extreme flood event on estuarine benthic food webs. *Ecol. Modeling* 222, 1209–1221. doi:10.1016/j.ecolmodel.2010.12.010.

Böhringer, C. and P.E.P. Jochem. 2007. Measuring the immeasurable—A survey of sustainability indices. *Ecol. Econ.* 63, 1–8. doi:10.1016/j.ecolecon.2007.03.008.

Borja, A., E. Barbone, A. Basset, G. Borgersen, M. Brkljacic, M. Elliott, J.M. Garmendia et al. 2011. Response of single benthic metrics and multi-metric methods to anthropogenic pressure gradients, in five distinct European coastal and transitional ecosystems. *Mar. Pollut. Bull.* 62, 499–513. doi:10.1016/j.marpolbul.2010.12.009.

Borja, A., S.B. Bricker, D.M. Dauer, N.T. Demetriades, J.G. Ferreira, A.T. Forbes, P. Hutchings et al. 2008. Overview of integrative tools and methods in assessing ecological integrity in estuarine and coastal systems worldwide. *Mar. Pollut. Bull.* 56, 1519–1537. doi:10.1016/j.marpolbul.2008.07.005.

Brand, F. 2009. Critical natural capital revisited: Ecological resilience and sustainable development. *Ecol. Econ.* 68, 605–612. doi:10.1016/j.ecolecon.2008.09.013.

Brundtland, G.H. 1987. Report on the World Commission on Environment and Development: Our Common Future. Transmitted to the General Assembly as an Annex to Document A/42/427—Development and International Co-operation: Environment. UN Documents: Gathering a Body of Global Agreements, http://www.un-documents.net/wcedocf.htm.

Brunner, N. and M. Starkl. 2004. Decision aid systems for evaluating sustainability: A critical survey. *Environ. Impact Asses.* 24(4), 441–469. doi:10.1016/j.eiar.2003.12.001.

Bunce, M., L.D. Rodwell, R. Gibb, and L. Mee. 2008. Shifting baselines in fishers' perceptions of island reef fishery degradation. *Ocean Coast. Manage.* 51(4), 285–302. doi:10.1016/j.ocecoaman.2007.09.006.

Carpenter, S.R., F. Westley, and M.G. Turner. 2005. Surrogates for resilience of social-ecological systems. *Ecosystems* 8, 941–944. doi:10.1007/s10021-005-0170-y.

Costanza, R., F. Andrade, P. Antunes, M. Van den Belt, D. Boersma, D.F. Boesch, F. Catarino et al. 1998. Principles for sustainable governance of the oceans. *Science* 281, 198–199. doi:10.1126/science.281.5374.198.

Costanza, R., F. Andrade, P. Antunes, M. Van den Belt, D. Boesch, D. Boersma, F. Catarino et al. 1999. Ecological economics and sustainable governance of the oceans. *Ecol. Econ.* 31, 171–187. doi:10.1016/S0921-8009(99)00077-4.

de Groot, R., J. Van der Perk, A. Chiesura, and A. van Vliet. 2003. Importance and threat as determining factors for criticality of natural capital. *Ecol. Econ.* 44, 187–204. doi:10.1016/S0921-8009(02)00273-2.

Diener, E. 2000. Subjective well-being: The science of happiness and a proposal for a national index. *Am. Psychol.* 55(1), 34–43. doi:10.1037/0003-066X.55.1.34.

Diener, E., E.M. Such, R.E. Lucas, and H.L. Smith. 1999. Subjective well-being: Three decades of progress. *Psychol. Bull.* 125(2), 276–302.

Diener, E. and W. Tov. 2012. National accounts of well-being. In Land, K.C., Michalos, A.C., and Sirgy, M.J. (Eds.), *Handbook of Social Indicators and Quality of Life Research*. New York, NY: Springer, pp. 137–156.

Duit, A. and V. Galaz. 2008. Governance and complexity—Emerging issues for governance theory. *Governance* 21(3), 311–335. doi:10.1111/j.1468-0491.2008.00402.x.

Ekins, P., S. Simon, L. Deutsch, C. Folke, and R. de Groot 2003. A framework for the practical application of the concepts of critical natural capital and strong sustainability. *Ecol. Econ.* 44(2–3), 165–185. doi:10.1016/S0921-8009(02)00272-0.

Elliott, M., S.J. Boyes, and D. Burdon 2006. Integrated marine management and administration for an island state—The case for a new Marine Agency for the UK. *Mar. Pollut. Bull.* 52(5), 469–474. doi:10.1016/j.marpolbul.2006.03.017.

Elliott, M., D. Burdon, K.L. Hemingway, and S. Apitz. 2007. Estuarine, coastal and marine ecosystem restoration: Confusing management and science—A revision of concepts. *Estuar. Coast. Shelf Sci.* 74, 349–366. doi:10.1016/j.ecss.2007.05.034.

Folke, C. 2006. Resilience: The emergence of a perspective for social-ecological systems analyses. *Global Environ. Chang.* 16, 253–267. doi:10.1016/j.gloenvcha.2006.04.002.

Folke, C., S. Carpenter, B. Walker, M. Scheffer, T. Elmqvist, L. Gunderson, and C.S. Holling. 2004. Regime shifts, resilience, and biodiversity in ecosystem management. *Annu. Rev. Ecol. Evol. Syst.* 35, 557–581. doi:10.1146/annurev.ecolsys.35.021103.105711.

Gray, J.S. and M. Elliott. 2009. *Ecology of Marine Sediments: Science to Management*. Oxford: Oxford University Press, 240 pp.

Grimm, V. and C. Wissel. 1997. Babel, or the ecological stability discussions: An inventory and analysis of terminology and a guide for avoiding confusion. *Oecologia* 109, 323–334.

Gunderson, L.H. and C.S. Holling. 2002. *Panarchy: Understanding Transformations in Human and Natural Systems*. Washington, DC: Island Press, 536 pp.

Heal, G. and B. Kristrom. 2005. National income and the environment. In Maler, K.G. and Jeffrey, V. (Eds.), *Handbook of Environmental Economics*, vol. 3, Chapter 22. Amsterdam: Elsevier.

Holling, C.S. 1996. Engineering resilience versus ecological resilience. In Schulze, P.C. (Ed.), *Engineering within Ecological Constraints*. Washington, DC: National Academy Press, pp. 31–44.

Hughes, T.P., D.R. Bellwood, C. Folke, R.S. Steneck, and J. Wilson. 2005. New paradigms for supporting the resilience of marine ecosystems. *Trends Ecol. Evol.* 20(7), 380–386. doi:10.1016/j.tree.2005.03.022.

Jax, K. 2005. Function and "functioning" in ecology. What does it mean? *Oikos* 111, 641–648. doi:10.1111/j.1600-0706.2005.13851.x.

Jørgensen, S.E., B. Fath, and S.N. Nielsen. (in prep). Definition of important ecological concepts.

Kates, R.W., W.C. Clark, R. Corell, J.M. Hall, C.C. Jaeger, I. Lowe, J.J. McCarthy et al. 2001. Environment and development: Sustainability science. *Science* 292(5517), 641–642. doi:10.1126/science.1059386.

Kay, J.J., M. Boyle, H.A. Regier, and G. Francis. 1999. An ecosystem approach for sustainability: Addressing the challenge of complexity. *Futures* 31, 721–742. doi:10.1016/S0016-3287(99)00029-4.

Kontogianni, A., M.S. Skourtos, I.H. Langford, I.J. Bateman, and S. Georgiou. 2001. Integrating stakeholder analysis in non-market valuation of environmental assets. *Ecol. Econ.* 37(1), 123–138. doi:10.1016/S0921-8009(00)00270-6.

Kontogianni, A., M.S. Skourtos, I.H. Langford, I.J. Bateman, and S. Georgiou. 2004. Integrating stakeholder analysis in non-market valuation of environmental assets. In Turner, R.K., Georgiou, S., and Bateman, I.J. (Eds.), *Environmental Decision Making and Risk Management: Selected Essays by Ian H. Langford*. Cheltenham: Edward Elgar Publishing, pp. 246–263.

Lebel, L., J.M. Anderies, B. Campbell, C. Folke, S. Hatfield-Dodds, T.P. Hughes, and J. Wilson. 2006. Governance and the capacity to manage resilience in regional social-ecological systems. *Ecol. Soc.* 11(1), 19.

Levin, S.A. and J. Lubchenco. 2008. Resilience, robustness, and marine ecosystem-based management. *Bioscience* 58(1), 27–32. doi:10.1641/B580107.

Limburg, K.E., R.V. O'Neill, R. Costanza, and S. Farber. 2002. Complex systems and valuation. *Ecol. Econ.* 41, 409–420. doi:10.1016/S0921-8009(02)00090-3.

MacLeod, M. and J.A.G. Cooper. 2005. Carrying capacity in coastal areas. In Schwartz, M. (Ed.), *Encyclopedia of Coastal Science*. Heidelberg: Springer, 226 pp.

Marques, J.C., A. Basset, T. Brey, and M. Elliott. 2009a. The ecological sustainability trigon—A proposed conceptual framework for creating and testing management scenarios. *Mar. Pollut. Bull.* 58, 1773–1779. doi:10.1016/j.marpolbul.2009.08.020.

Marques, J.C., F. Salas, J. Patrício, J. Neto, and H. Teixeira. 2009b. *Ecological Indicators for Coastal and Estuarine Environmental Assessment—A User Guide*. Southampton, UK: WIT Press, 200 pp.

McLusky, D.S. and M. Elliott. 2004. *The Estuarine Ecosystem; Ecology, Threats and Management*. 3rd edition. Oxford: OUP, p. 216.

Mee, L.D., R.L. Jefferson, D.A. Laffoley, and M. Elliott. 2008. How good is good? Human values and Europe's proposed Marine Strategy Directive. *Mar. Pollut. Bull.* 56, 187–204. doi:10.1016/j.marpolbul.2007.09.038.

Moore, S.A., T.J. Wallington, R.J. Hobbs, P.R. Ehrlich, C.S. Holling, S. Levin, D. Lindenmayer et al. 2009. Diversity in current ecological thinking: Implications for environmental management. *Environ. Manage.* 43, 17–27. doi:10.1007/s00267-008-9187-2.

Nunneri, C., K. Turner, A. Cislak, A. Kannen, R. Klein, L. Ledoux, J. Marquenie et al. 2004. Integrated asssessment and future scenarios for the coast. In W. Salomons, Vermaat, J., and Turner, K. (Eds.), *Managing European Coasts: Past, Present and Future*. Berlin: Springer Press, pp. 271–289.

Ojeda-Martínez, C., F. Casalduero Giménez, J.T. Bayle-Sempere, C.B. Cebrián, C. Valle, J.L. Sanchez-Lizaso, A. Forcada et al. 2009. A conceptual framework for the integral management of marine protected areas. *Ocean Coast. Manage.* 52(2), 89–101. doi:10.1016/j.ocecoaman.2008.10.004.

Pinto, R., V.N. de Jonge, and J.C. Marques. 2014a. Linking biodiversity indicators, ecosystem functioning, provision of services, and human well-being in estuarine ecosystems: Application of a conceptual framework. *Ecol. Ind.* 36, 644–655. doi:10.1016/j.ecolind.2013.09.015.

Pinto, R., M.C. Cunha, C. Roseta-Palma, and J.C. Marques. 2014b. Mainstreaming sustainable decision making for ecosystems: Integrating ecological and socioeconomic targets within a decision support system. *Environ. Process.* 1, 7–19 doi:10.1007/s40710-014-0006-x.

Pinto, R., V.N. de Jonge, J.M. Neto, T. Domingos, J.C. Marques, and J. Patrício. 2013a. Towards a DPSIR driven investigation of ecological value, water uses, and ecosystem services for estuarine ecosystems. *Ocean Coast. Manage.* 72, 64–79. doi:10.1016/j.ocecoaman.2011.06.016.

Pinto, R., V.N. de Jonge, J.C. Marques, P. Chainho, J.P. Medeiros, and J. Patrício. 2013b. Temporal stability in estuarine systems: Implications for ecosystem services provision. *Ecol. Ind.* 24, 246–253. doi:10.1016/j.ecolind.2012.06.022.

Pinto, R., J. Patrício, A. Baeta, B.D. Fath, J.M. Neto, and J.C. Marques. 2009. Review and evaluation of estuarine biotic indices to assess benthic condition. *Ecol. Ind.* 9(1), 1–25. doi:10.1016/j.ecolind.2008.01.005.

Pinto, R., J. Patrício, J.M. Neto, Salas, F., and J.C. Marques. 2010. Assessing estuarine quality under the ecosystem services scope: Ecological and socioeconomic aspects. *Ecol. Complex.* 7, 389–402. doi:10.1016/j.ecocom.2010.05.001.

Pulselli, F.N., S. Bastianoni, N. Marchettini, and E. Tiezzi. 2008. *The Road to Sustainability. GDP and Future Generations*. Southampton, UK: WIT Press, 197 pp.

Singh, R.K., H.R. Murty, S.K. Gupta, and A.K. Dikshit. 2009. An overview of sustainability assessment methodologies. *Ecol. Ind.* 9, 189–212. doi:10.1016/j.ecolind.2011.01.007.

Swart, R., P. Raskin, and J. Robinson. 2004. The problem of the future: Sustainability science and scenario analysis. *Global Environ. Chang.* 14, 137–146. doi:10.1016/j.gloenvcha.2003.10.002.

Turner, G.M. 2008. A comparison of the limits to growth with 30 years of reality. *Global Environ. Chang.* 18, 397–411. doi:10.1016/j.gloenvcha.2008.05.001.

Turner, R.K., S. Georgiou, R. Brouwer, I.J. Bateman, and I.H. Langford. 2003. Towards an integrated environmental assessment for wetland and catchment management. *Geogr. J.* 169(2), 99–116. doi:10.1111/1475-4959.04953.

Turner, R.K., D. Pearce, and I. Bateman. 1994. *Environmental Economics: An Elementary Introduction*. London: Harvester Wheatsheaf, 328 pp.

Ulanowicz, R.E., S.J. Goerner, B. Lietaer, and R. Gomez. 2009. Quantifying sustainability: Resilience, efficiency, and the return of information theory. *Ecol. Complex.* 6, 27–36. doi:10.1016/j.ecocom.2008.10.005.

Veríssimo, H., M. Lane, J. Patrício, S. Gamito, and J.C. Marques. 2013. Trends in water quality and subtidalbenthic communities in a temperate estuary: Is the response to restoration efforts hidden by climate variability and the Estuarine Quality Paradox? *Ecol. Ind.* 24, 56–67. doi:10.1016/j.ecolind.2012.05.028.

Walker, B., C.S. Holling, S.R. Carpenter, and A. Kinzig. 2004. Resilience, adaptability and transformability in social-ecological systems. *Ecol. Soc.* 9(2), 5.

Part II

Understanding the Ecosystem

3

The Ecosystem

3.1 Introduction: Definition of Basic Concepts

Ecology is the scientific study of interactions among organisms and their environment, such as interactions organisms have with each other and with their abiotic environment. Ecology is an interdisciplinary field that includes biology and Earth science and both draw on chemistry and physics. Evolutionary concepts such as adaptation and natural selection are cornerstones of the modern ecological theory. The following are the core topics of ecology:

i. Life processes, interactions, adaptation, selection and survival, and their dependence on the conditions.
ii. Exchange of mass and energy between organisms and their environment. This implies a heavy use of thermodynamics.
iii. Ecosystems (see the definition as follows), and their reactions to changed conditions and development.
iv. Distribution and diversity of organisms in context of their environment.

An ecosystem is basically an organism community which interacts with one another and their environment in such a way that energy is transferred between them and system-level processes like the cycle of elements emerge. The ecosystem is the core concept in ecology and biology, and serves as the building block of biological organization where organisms interact with each other simultaneously and with the environment as well. Therefore, ecosystems in the hierarchical organization are a step above the ecological community level (in which organisms of different species interact with one another) but a step below the landscape, which is below regions and the ecosphere. All environmental problems involve ecosystems in its broadest sense. Development of an integrated ecological–environmental management requires therefore that we know the properties of ecosystems, which is the topic of this part "II" of the book—which includes this chapter and Chapters 4 and 5. All environmental management must include contemplated considerations of ecosystems and their properties.

Ecosystems include the living organisms, the dead organic matter and the abiotic environment which the organisms inhabit, for example, soils, water, the atmosphere, and so on, and the interactions with the components. Ecosystems follow the concept that the living organisms interact with one another and with their environment to create complex systems with different emergent properties. To cover these properties, we often use the following statements: "the whole is greater than the sum of its parts" and "everything is connected."

Hutchinson defined an ecological niche as an *n*-dimensional hyper volume, where the dimensions are environmental conditions and the resources that define the requirements of an individual or a species to practice "its" way of life.

An ecotone is a transition area between two biomes. Two communities meet and integrate in an ecotone exemplified by the transition between forest and grassland ecosystems. An ecotone may appear on the ground as a gradual blending of the two communities across a broad area or it may manifest itself as a sharp boundary line.

Ecosystems have 14 very fundamental properties that are able to explain the reactions of ecosystems to impacts and what is important to include in environmental management to alleviate the impacts and consequences of the impacts. These 14 properties are presented in the 14 next sections of this chapter. Further details are given of the core properties, the thermodynamic properties, and the network properties in the following two chapters.

3.2 Ecosystems as All Other Systems Conserve Matter and Energy

This means as for energy it is also possible for matter to use a bookkeeping of the exchange of matter with the environment. The amount of matter gained minus the amount of matter lost to the environment will tell us directly how much matter is accumulated—added to the system. The equation is sometimes expressed by the use of concentrations:

$$\frac{dC}{dt} = \text{accumulation} = (\text{inputs} - \text{outputs})/V \qquad (3.1)$$

where C is the concentration in the system and V is the volume of the system. In the case, the concentration is expressed per unit of area; V is replaced by the area A.

If the law of mass conservation is used for chemical compounds that can be transformed to other chemical compounds, then Equation 3.1 must be changed to

$$V\frac{dc}{dt} = \text{input} - \text{output} + \text{formation} - \text{transformation}\ (MT^{-1}) \qquad (3.2)$$

The principle of mass conservation is widely used in the class of ecological models called biogeochemical models. As we do not consider radioactive processes in ecology—or very rarely—the mass conservation principle can be used for all elements. The equation is set up for the relevant elements, for example, for eutrophication models for C, P, N, and perhaps Si (see Jørgensen, 1976; Jørgensen and Fath 2011; Jørgensen 2012).

3.3 Ecosystems Recycle All Matter and Partly All Energy: Liebig's Law of Minimum

The conservation laws have major environmental consequences. As we do not consider nuclear processes, we can use the bookkeeping principle for all 91 naturally occurring elements. This means that if an element needed for the growth of biomass is used up, it cannot be created inside the ecosystem but must be added from the environment. About 20–25 elements are needed for most organisms and the growth of an organism must stop, when the element that is present in the ecosystem in the smallest amount relatively to the needs for the considered organism is used up.

The needs are not necessarily a constant and fixed concentration. Table 3.1 provides the average concentration for freshwater plants (wet basis). In this table it is indicated, for instance, that the phosphorus concentration is 0.08% on a wet weight basis, which means that if the dry matter content is 10%, the average phosphorus concentration is 0.8%. Many plants have from about 0.4% to about 2.0% phosphorus on a dry weight basis. This implies that when there is no more phosphorus and a plant has reached the lowest feasible percentage of phosphorus (0.4%) the growth must inexorably stop.

The environment rarely has the exact chemical composition required for growth, which means the element in least supply compared with the needs determines the limits to growth. This is the classic Liebig's minimum law. The growth rate versus the nutrient concentration is illustrated in Figure 3.1, which is Michaelis–Menten's law. At low concentrations, the growth rate is proportional to the nutrient concentration, which therefore determines and limits the growth. If the nutrient concentration is high, the growth rate is unaffected or almost unaffected by the concentration—the needs are covered

TABLE 3.1

Average Elemental Composition of Freshwater Plants, Wet Weight Basis

Element	Plant Content (%)
Oxygen	80.5
Hydrogen	9.7
Carbon	6.5
Silicon	1.3
Nitrogen	0.7
Calcium	0.4
Potassium	0.3
Phosphorus	0.08
Magnesium	0.07
Sulfur	0.06
Chlorine	0.06
Sodium	0.04
Iron	0.02
Boron	0.001
Manganese	0.0007
Zinc	0.0003
Copper	0.0001
Molybdenum	0.00005
Cobalt	0.000002

Source: Wetzel, R.G. 1983. *Limnology*. Fort Worth, New York, and Orlando: Saunders College Publishing, 838 pp.

and it is another nutrient that is the limiting nutrient. Mathematically, the Michaelis–Menten equation is expressed by the following:

$$\text{Growth rate} = \mu_{max}\left(\frac{PS}{(PS + kp)}\right) \tag{3.3}$$

where μ_{max} is the maximum growth rate for the considered organisms and PS is the concentration of the nutrient (considered in this case of being phosphorus) and kp is a Michealis–Menten constant also denoted as the half-saturation constant, which is explained in Figure 3.2.

If multiple limiting factors are in play, further addition of only one of them will not influence the growth (Figure 3.1b). Liebig's law is a direct consequence of the conservation of matter.

Figure 3.2 shows the graph of the Michaelis–Menten equation (3.3). The nitrification rate is plotted versus the ammonium concentration. The microbiological oxidation of ammonium to nitrate follows as shown also the same equation.

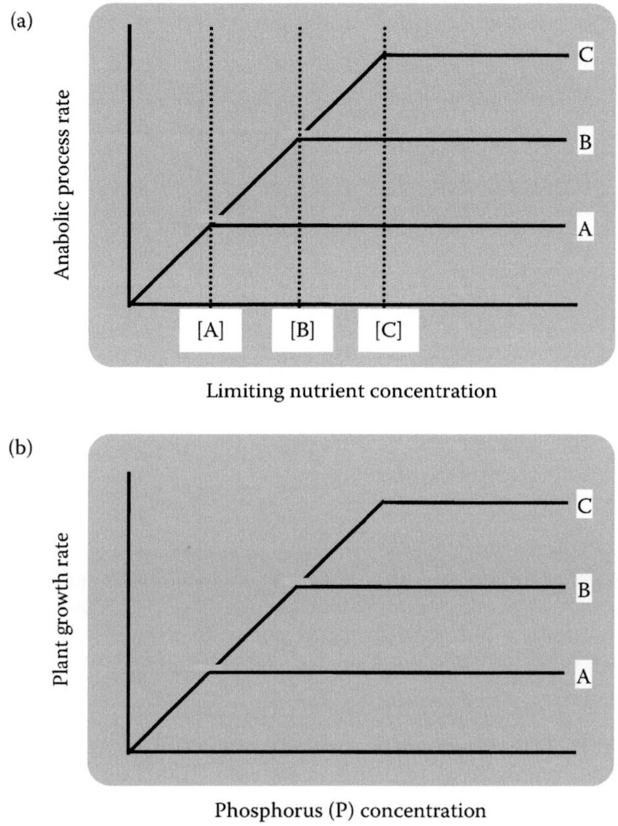

FIGURE 3.1
(a) General relationship between the rate of an anabolic process and the concentration of a limiting nutrient for that process. At limiting concentrations the process proceeds at the limited rates allowed. Three limiting concentrations ([A], [B], [C]) of a single nutrient are shown, and three correspondingly limited process rates (A, B, C). (b) Plant growth illustration of Liebig's law of the minimum. Phosphorus (P) concentration is plotted against growth rate. Under nonlimiting conditions, growth is linearly related to P concentration. At concentrations where elements other than phosphorus become limiting, higher P will not increase growth. The three levels A, B, and C in this case correspond to three different growth-limiting concentrations of elements other than phosphorus.

The chemical process is

$$NH_4^+ + 2O_2 \rightarrow NO_3^- + H_2O + H^+ \tag{3.4}$$

The renewable resources are used by man at a rate, which in many cases, is considerably higher than the rate at which the resources are regenerated. This implies that the renewable resources are decreasing. The nonrenewable resources are used by man at a rate which is higher than the rate at which

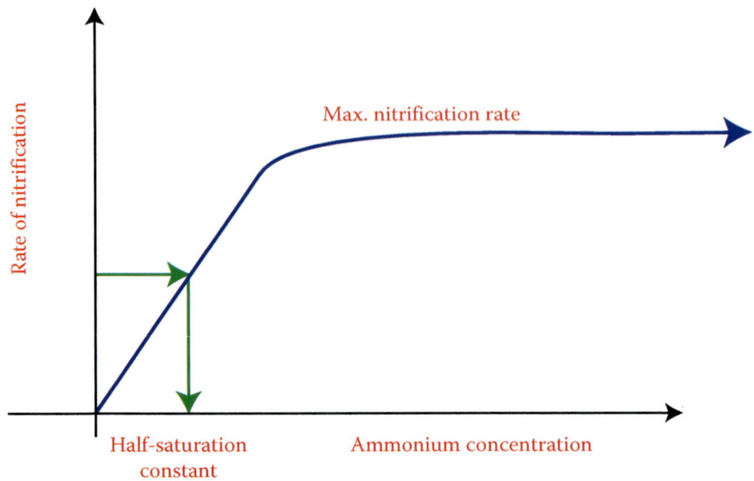

FIGURE 3.2
Graph of the Michaelis–Menten equation. In this case, the rate of nitrification versus the concentration of the substrate, ammonium, is shown. The same graph is obtained for the growth rate of phytoplankton versus the inorganic reactive phosphorus concentration, provided that phosphorus is the limiting nutrient. The figure shows that the half-saturation constant is corresponding to the ammonium concentration that yields half of the maximum rate.

alternatives to the nonrenewable resources are found. The decreasing renewable and nonrenewable resources demonstrate that the Earth is not in a sustainable development. Sustainable development is used in the same sense as in the *Bruntland Report*: a sustainable development means that we will hand over the Earth with the same possibilities for the next generation to plan and live their life as the previous generations have given us when the Earth was handed over to our generation.

EXAMPLE 3.1

It is often discussed in lake management which nutrient is limiting the phytoplankton growth and thereby the eutrophication (too high primary production of phytoplankton due to a too high nutrient concentration). Averagely the concentration of nitrogen in phytoplankton is seven times the concentration of phosphorus. For a considered lake, drainage water from agriculture contains 25 times as much nitrogen as phosphorus, while the wastewater discharged to the lake has only four times as much nitrogen as phosphorus. Above which ratio of wastewater to drainage water (R) would phosphorus not be the limiting nutrient?

Also, find R if the wastewater has only three times as much nitrogen as phosphorus.

Solution

Let us call the wastewater to drainage water where P is limiting RS.

The following R would correspond to RS:

$$R4 + 25 = 7(R + 1) \text{ or } 3R = 18; \; R = RS = 6$$

If the ratio is less than 6, P becomes limiting, while if the ratio is more than 6, nitrogen is the limiting element. For three times, we have the following equation:

$$R3 + 25 = 7(R + 1) \text{ or } R = RS = 4.5$$

It means that if the ratio is less than 4.5, P becomes limiting, and if the ratio is more than 4.5 nitrogen is the limiting element.

Some nonrenewable resources are recycled by man, for instance iron, but a 100% recycling is not possible in society and therefore the nonrenewable resources are declining, although at a lower rate corresponding to the recycling. The above 20–25 elements used by the organisms of ecosystems (see Table 3.1) are recycled in the ecosystems and the six elements that are considered absolutely necessary for life on the Earth—C, H, O, N, P, and S, are of course particularly important for nature to recycle. Figures 3.3 and 3.4 give two examples, phosphorus and nitrogen cycling in a lake. The two examples

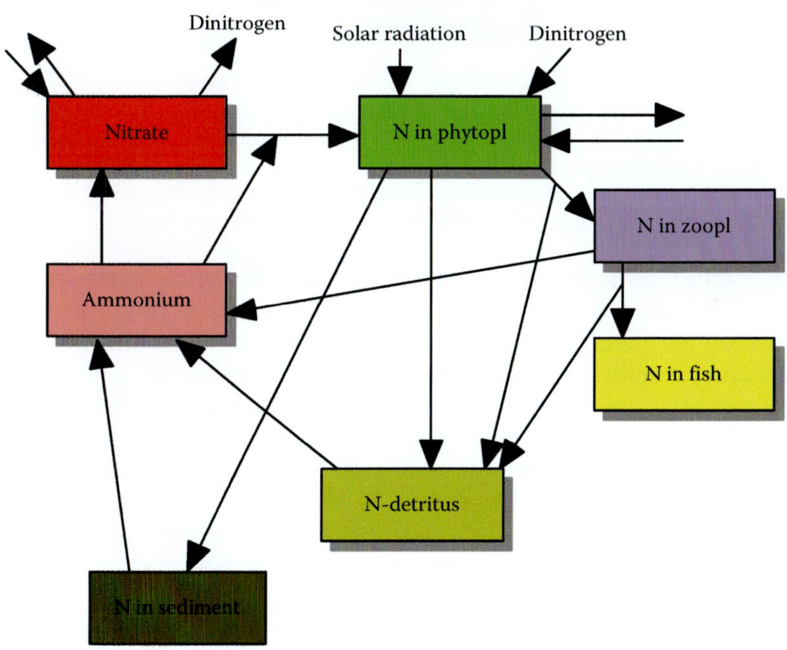

FIGURE 3.3
The nitrogen cycle in a lake is shown. A simple food chain nutrient—phytoplankton—zooplankton—fish is used to describe the process of the cycle. The cycle is closed by mineralization of detritus and by the release of nutrients by the sediment. (From Jørgensen, S.E. and G. Bendoricchio. 2001. *Fundamentals of Ecological Modelling*. 3rd edition. Amsterdam: Elsevier, 628 pp.)

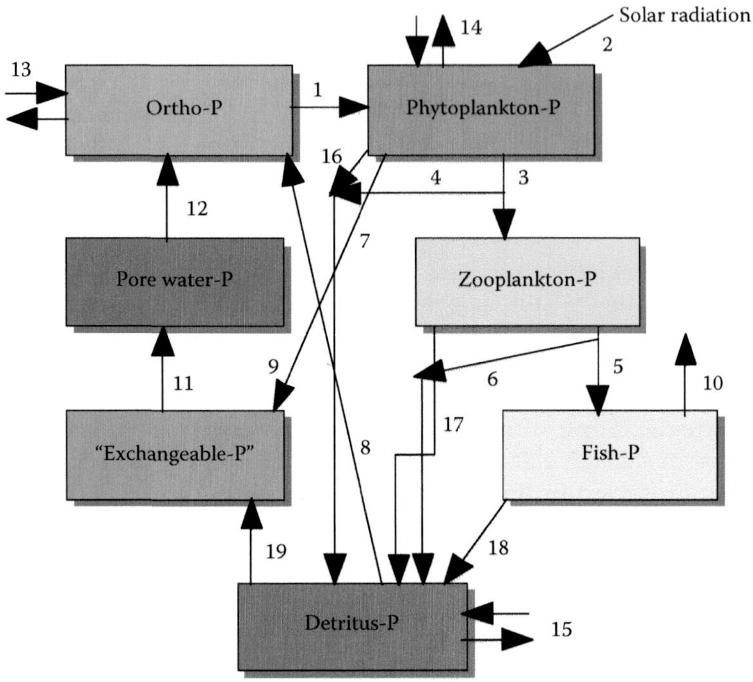

FIGURE 3.4
The phosphorus cycle. The processes are: (1) uptake of phosphorus by algae, (2) photosynthesis, (3) grazing with loss of undigested matter, (4, 5) predation with loss of undigested material, (6, 7, 9) settling of phytoplankton, (8) mineralization, (10) fishery, (11) mineralization of phosphorous organic compounds in the sediment, (12) diffusion of pore water P, (13, 14, 15) are inputs/outputs, (16, 17, 18) represent mortalities, and (19) is settling of detritus.

include the inorganic form either inorganic reactive phosphorus or ammonium and nitrate, the phosphorus and nitrogen in the phytoplankton, in zooplankton, in fish, in sediment and in detritus. The phosphorus cycle includes the phosphorus in the pore water of the sediment. The detritus is mineralized not only too close to the cycle, but also the sediment can release inorganic forms of phosphorus as phosphate and of nitrogen as ammonium. The cycles can include more or less details. In the case shown in Figure 3.3, the food chain is represented by nutrients, phytoplankton, zooplankton, and fish only. Phytoplankton could be represented by different groups of phytoplankton as for instance nitrogen-fixing species, diatoms, and so on. The possible nitrogen fixing is indicated in the diagram and so are the possibilities for denitrification. The two diagrams for the phosphorus and nitrogen cycle are almost parallel. The process numbers on the phosphorus refer to the processes, which are the same for the two cycles. The following processes are included: (1) uptake of nutrient, (2) photosynthesis (solar radiation converted to the organic matter in phytoplankton), (3) grazing, (4) loss of feces to detritus by the grazing process, (5) predation of zooplankton by fish, (6) loss of feces to detritus by predation, (7) settling of feces, (8) mineralization of detritus, (9) and (19) settling of phytoplankton and detritus, (10) fishery, (11) mineralization taking place in the sediment, (12) diffusion, (13)–(15) exchange of reactive phosphorus, phytoplankton, and detritus with the environment (inflows and outflows), and (16)–(18) mortality of phytoplankton, zooplankton, and fish. Notice that the nitrogen cycle includes nitrification, which is the oxidation of ammonium to nitrate.

The elements in the entire ecosphere are cycling too. It is important for a sustainable development of the Earth that there are no crucial changes of the concentrations of the elements in the atmosphere (hydrosphere and the lithosphere). It is, for instance, known that there is an unbalance in the global carbon cycle and that an accumulation of carbon dioxide is taking place in the atmosphere, because we are exhausting the storage of fossil fuel much faster than new fossil fuel can be formed. It entails what is known as the greenhouse effect, which can change the climate considerably, probably on the order of a 2–5°C increase of the temperature during this century, which will strongly influence the energy policy that we will implement in the coming decades. The global nitrogen cycle is unbalanced by a huge production mainly from atmospheric nitrogen of nitrogen fertilizers, which are used in the lithosphere by drainage water transported to the hydrosphere, where it is causing increased eutrophication (primary production is too high, causing several water quality problems). The global cycles have been modeled, see Jørgensen and Fath (2011).

The unbalancing of the global carbon cycle can only be solved effectively by cutting down on the consumption of fossil fuel by the development of alternative renewable energy sources such as wind and solar energy. The unbalancing of the nitrogen cycle can be solved by either a decreased production of fertilizers which is hardly possible in a world with an increasing population or by preventing the nitrogen to reach the hydrosphere by treatment of wastewater and drainage water. The use of constructed or natural wetland ecosystems (ecotechnology) offers a good cost-moderate solution particularly to treat drainage water. The nitrate and ammonium are by these treatment methods transformed into dinitrogen, which is transported to the atmosphere. A pattern of different types of wetlands in the landscape can considerably reduce the nitrogen loss to the hydrosphere; see Chapter 12 on ecological engineering methods.

All global cycle or ecosystem cycle unbalances require a solution, which respects the conservation principles. Matter or energy cannot be destroyed (or created) but only transformed, which implies that the energy sources used should be replaced by other energy forms that do not create an unbalance of the spheres. This means that fossil fuel, which causes accumulation of carbon dioxide in the atmosphere should be replaced by other energy forms and nitrogen in wastewater and drainage water should be transformed to dinitrogen, which would be harmless in the atmosphere, as it contains 98% dinitrogen.

The Appendix gives the composition of the spheres, which should be maintained on approximately the same level in the future to avoid global cycle unbalances.

3.4 All Processes in Ecosystems Are Irreversible and Consume Free Energy (Energy That Can Do Work)

It is important to understand this property of *all systems* including ecosystems, to distinguish between two forms of energy: energy that can do work and energy that cannot do work, but is "lost" as heat to the environment at the temperature of the environment. Energy that can do work has a gradient that is utilized to execute work, whereby the energy is transformed to other energy forms. The work energy for the different energy types is the gradient times an amount—or expressed differently an intensive times an extensive variable; see Table 3.2. The second law of thermodynamics tells us that whenever we do work (perform an activity) we will unfortunately lose a part of energy that can do work, to energy that cannot do work but is lost as heat to the environment at the temperature of the environment. This fundamental law is of course also valid for ecosystems.

If ecosystems were isolated, no energy or matter could be exchanged across their boundaries. The systems would stepwise degrade their initially stored work energy to heat energy at the temperature of the environment, corresponding to a loss of order and organization, and increase in the randomness of their constituents and microstates. This dissipation process would cease at thermodynamic equilibrium, where no further motion or change would be possible and all gradients would be eliminated. The physical manifestation would ultimately be a meltdown to the proverbial "inorganic soup" containing degradation products dispersed equi-probable through the entire volume of the system. Gradients of all kinds would be eliminated, and the system would be frozen in time in a stable, fixed configuration. The high-energy chemical compounds of biological systems, faced suddenly with isolation, would decompose stepwise to compounds with no work energy content. An isolated ecosystem could therefore, in the best case, sustain life for only a limited period of time, less than that required from the onset of isolation to reach a thermodynamic equilibrium.

TABLE 3.2

Different Forms of Energy and Their Intensive and Extensive Variables

Energy Form	Extensive Variable	Intensive Variable
Heat	Entropy (J/K)	Temperature (K)
Expansion	Volume (m^3)	Pressure (Pa = kg/s^2 m)
Chemical	Moles (M)	Chemical potential (J/moles)
Electrical	Charge (A s)	Voltage (V)
Potential	Mass (kg)	(Gravity) (Height) (m^2/s^2)
Kinetic	Mass (kg)	0.5(Velocity)^2 (m^2/s^2)

Note: Potential and kinetic energy are called *mechanical energy*.

Energy flow of about 10^{17} W by solar radiation ensures the maintenance of life on Earth. Ecosystems can maintain a certain concentration of compounds with work energy content against the second-law dissipation gradient, because they are not isolated. Ecosystems receive a continuous supply of work from the outside by solar radiation to compensate for the loss of work energy as a consequence of the second law of thermodynamics. On Earth, solar radiation is the main source of this input of work energy.

All ordered structures require work energy for maintenance. Ecosystems, in common with all real systems, have a global attractor state, thermodynamic equilibrium (see Jørgensen et al., 2000). Through their openness they avoid reaching this state by importing energy or matter carrying work energy. This anabolism combats and compensates for the catabolic deterioration of structures; the two processes operate against one another.

3.5 All Living Components in Ecosystems Have the Same Basic Biochemistry

The wet weight composition may be different because the water content varies more than the dry matter composition from organism to organism. A typical dry matter content of plants may be 12.5%. This means that as seen in Table 3.1, which gives the concentrations of various elements on a wet weight basis can be changed to a dry matter composition, see Table 3.3. The composition given in this table is with approximations that are applicable to all plants and even to animals.

Table 3.3 lists 19 elements that are generally found in all organisms. A handful of elements can be found in addition to the 19 listed at low concentrations, for instance, iodine and flour, and a few trace elements have been found to be characteristic in just a few species: selenium, nickel, vanadium, chromium, and even the toxic element cadmium may substitute into enzymes, that typically use zinc. In principle, all the applied elements can be a limiting factor for growth as discussed in Section 3.2. A limiting element must of course be absolutely necessary for the considered organism and may in some cases be present in a concentration which is lower than the other important element relative to its use for building a new biomass for the focal organism. The composition of the ecosphere reflects, however, the composition of the organisms, which means that the elements are often present in a ratio close to the concentrations shown in Table 3.3. Furthermore, the composition of organisms is not an exact unchangeable value but rather a range. For instance, most plants can manage to grow with a phosphorus content of 0.4%–2.0%. If the environment has a high phosphorus concentration relative to the need of the organism, the organism will accumulate more phosphorus, and if the phosphorus concentration of the environment is low, the plants are able to adapt to the

TABLE 3.3

Average Elemental Composition of Freshwater Plants, Dry Weight Basis

Element	Plant Content (%)
Oxygen	11.1
Hydrogen	1.4
Carbon	45.5
Silicon	9.1
Nitrogen	4.9
Calcium	2.8
Potassium	2.1
Phosphorus	0.56
Magnesium	0.49
Sulfur	0.42
Chlorine	0.42
Sodium	0.28
Iron	0.14
Boron	0.007
Manganese	0.0049
Zinc	0.0021
Copper	0.0007
Molybdenum	0.00035
Cobalt	0.000014

conditions and cope with less phosphorus. This adaptability of the plants to the composition of the environment is reflected in the range of concentrations in the organisms and implies that a limiting concentration of an element is not a sharp exact value, but is a range that gradually at lower concentrations will limit the growth more and more. Also see the discussion in Section 3.3.

Approximate ranges for C, N, and P are given for different organisms in Table 3.4 to show (1) the variability among species and (2) the possible ranges.

TABLE 3.4

Approximate Ranges of C, N, and P Concentrations as % Dry Matter for Different Organisms

Organisms	%C	%N	%P
Terrestrial plants	36–64	0.3–6.4	0.02–1.0
Benthic invertebrates	35–57	6–12	0.2–1.8
Terrestrial insects	36–61	7–12.5	0.5–2.5
Birds and mammals	32–60	6–12	0.7–3.7
Fish	38–52	7–12	1.5–4.5
Zooplankton	35–60	7–12.5	0.5–2.5
Phytoplankton	35–60	5–12	0.5–2.5

The differences among species are not very pronounced, while the ranges may be relatively wide for phosphorus, which are generally for elements with low biochemical concentrations.

3.6 Ecosystems Are Open Systems and Require an Input of Work Energy to Maintain Their Function

Ecosystems are open system in the sense that they are open for mass and energy transfers. Ecosystems receive energy from solar radiation and receive water from precipitation, dry deposition from the atmosphere, inputs by wind and inflows, and outflows of various types plus emigration or immigration of species. A system that is closed for inputs and outputs of energy and mass is called an *isolated system*, while a system that is close to inputs and outputs of mass, but open to energy transfers is denoted as a *closed system*. A nonisolated system is a closed or open system. If an ecosystem was isolated it would, as already underlined, inevitably move toward the thermodynamic equilibrium and become a dead system with no gradients to do the work.

The openness explains why an ecosystem can maintain life and stay far from thermodynamic equilibrium, because maintenance of life requires the input of work energy which of course is only possible if an ecosystem is nonisolated. For an exchange of matter between the environment and an ecosystem, the ecosystem must be open.

At first glance, it looks like ecosystems violate the first law of thermodynamics because they are moving away from thermodynamic equilibrium by formation of a biological structure, implying that they gain chemical energy. Ecosystems, however, receive work energy in the form of solar radiation, which is used for the formation of the biological structure and the work energy needed for maintenance of the system is far from thermodynamic equilibrium.

An energy balance equation for ecosystems might be written as follows in accordance with the principle of energy conservation:

$$\text{Ecap} = \text{Qevap} + \text{Qresp} + \cdots + \Delta\text{Ebio} \tag{3.5}$$

Here, Ecap is external energy captured per unit of time. A part of the incoming energy, solar radiation being the main source for the ecosystems on Earth, is captured and a part is reflected unused, determining the albedo of the globe. The more biological structure an ecosystem possesses, the more of the incoming energy will be captured, that is, the albedo is lowered. The structure functions as an umbrella capturing the incoming solar radiation.

In an ecosystem at steady states, the formation of biological compounds (anabolism) is in approximate balance with their decomposition (catabolism). That is, in energy terms:

$$\Delta Ebio \approx 0 \text{ and } Ecap \approx Qevap + Qresp + \cdots \quad (3.6)$$

The energy captured can, in principle, be any form of work energy (electromagnetic, electrical, magnetic, chemical, kinetic, etc.) but for the ecosystems on Earth, the shortwave energy of solar radiation (electromagnetic energy) plays the major role. The energy captured per unit of time is, however, according to Equation 3.5 used to pay the cost of maintenance per unit of time = Qevap + Qresp.

Energy inflows to ecosystems is high quality energy with high contents of work energy including information. The outflows, are in contrast, sinks for energy and matter and have a lower work energy capacity, and are closer to equilibrium. The work energy inflow is used to maintain the complex structure of the ecosystem and for growth, and it leads to the next property of ecosystems that they can use three growth forms to move away from thermodynamic equilibrium, whereby they develop more structure and gain work energy.

3.7 Ecosystems Apply Three Growth Forms: Growth of Biomass, Growth of the Network, and Growth of Information

The three growth forms are

I. Growth of biomass
II. Growth of the network
III. Growth of information

They are utilized to store more work energy in the ecosystem by using the surplus free energy that is available, when the maintenance is covered. The growth of the network is presented in Section 3.13 and in more detail in Chapter 5. Calculation of the contribution of information to work energy is revealed in Section 3.14. Development of ecological networks and information in the organisms entail that ecosystems use the energy and matter with higher efficiency. Furthermore, the network structure possesses work energy and information has work energy. Note in this context, that eco-exergy is work energy including the work energy of information embodied in the organisms.

To illustrate the three growth forms thermodynamically, the following thermodynamic variables will be used:

a. Work capacity stored in the ecosystem relative to thermodynamic equilibrium.
b. Power (throughflow of useful energy = the sum of all free-energy flows in the system).
c. Retention time.
d. Work energy used for maintenance of the system.
e. Specific work energy of the ecosystem (work energy stored divided by the biomass).

The usual description of ecosystem development illustrated for instance by the recovery of Yellow Stone Park after fire, an island born after a volcanic eruption, reclaimed land, and so forth, is well covered by Odum (1969, 1971): at first the biomass increases rapidly which implies that not only the percentage of captured incoming solar radiation increases but also the energy needed for the maintenance. The details of Odum's ecological attributes are presented in Chapter 12, which discusses the ecological indicators. The increase of biomass is mainly in the form of plants, which capture the solar radiation almost as an antenna captures radio waves. Growth form I is dominant in this first phase, where stored work energy increases (more biomass more physical structure to capture more solar radiation), but also the throughflow (of useful energy) is increasing because the flows are determined by the biomass and to maintain more biomass, more work energy is required. Furthermore, work energy consumption increases due to an increased need of energy for maintenance.

When the percentage of solar radiation captured reaches about 80%, it is not possible to increase the amount of captured solar radiation further (due in principle to the second law of thermodynamics). It is, of course, completely impossible to exceed 100%, but as work energy is lost by all energy transformations, about 80% is the upper limit in practice. Therefore, further growth of the physical structure (biomass) cannot improve the energy balance of the ecosystem. In addition, all or almost all the essential elements are in the form of dead or living organic matter and not as inorganic compounds ready to be used for growth. Growth form I will therefore not proceed; however, growth forms II and III can still operate and thereby increase the energy and matter efficiency. The ecosystem can still develop and improve the ecological network and can still change r-strategists with K-strategists, small animals and plants with bigger ones and less developed with more developed organisms with more information genes. A graphic representation of this description of

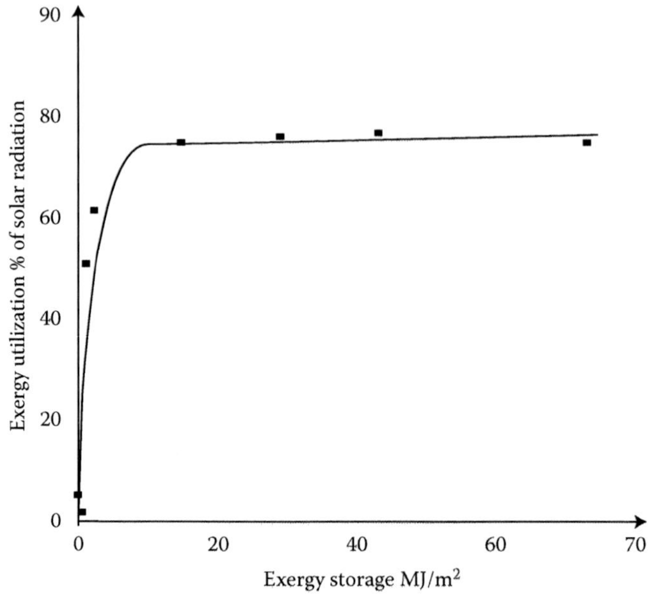

FIGURE 3.5
The work energy captured expressed as % of solar radiation, is plotted versus the exergy stored = work energy stored, including the work energy of information (unit J/m^2 or J/m^3), calculated from the characteristic compositions of the focal 8 ecosystems. The numbers from Table 3.5 are applied to construct this plot. Notice that exergy utilization for maintenance is parallel (proportional) to energy absorbed and the biomass of the ecosystem. (From Kay, J. and E.D. Schneider. 1992. *Proc. "Ecological Indicators,"* Amsterdam: Elsevier, pp. 159–182.)

ecosystem development is presented in Figure 3.5, where the solar radiation captured is plotted versus the work energy stored in the ecosystem in accordance with Table 3.5. As the increase of work energy is proportional to the increase of biomass, the specific work energy will in the first development phase not change, and as the free energy used for maintenance with good

TABLE 3.5
Exergy Utilization and Storage in a Comparative Set of Ecosystems

Ecosystem	Exergy Utilization (%)	Exergy Storage (MJ/m^2)
Quarry	6	0
Desert	2	0.073
Clear-cut forest	49	0.594
Grassland	59	0.940
Fir plantation	70	12.70
Natural forest	71	26.00
Old-growth deciduousforest	72	38.00
Tropical rainforest	70	64.00

TABLE 3.6

Accordance between Growth Forms and Thermodynamic Variables

Descriptor	Growth Form I	Growth Form II	Growth Form III
A. Work energy storage	up	up	up
B. Power/throughflow	up	up	up
C. Retention time	equal	up	up
D. Work energy consump.	up	equal	equal
E. Specific work energy	equal	up	up

Source: Jørgensen, S.E. 2002. *Integration of Ecosystem Theories: A Pattern.* 3rd edition. Dordrecht: Kluwer Academic Publisher, 428 pp. (1st edition 1992, 2nd edition 1997).

approximation is proportional to the biomass, the specific work energy is also unchanged.

In accordance with the five descriptors, A to E, listed and the three growth forms based on this description of ecosystem development is shown in Table 3.6. The table clearly shows that work energy storage and power are the two thermodynamic descriptors that increase with all three growth forms.

3.8 If the Ecosystems Receive More Work Energy Than They Need to Maintain Their Function, the Surplus Work Energy Will Be Applied to Move the System Further Away from Thermodynamic Equilibrium; It Means for Growth and Development

If ecosystems receive more work energy than needed for the maintenance, the surplus work energy is used to move the system further away from thermodynamic equilibrium by the use of three growth forms. Either the biomass can grow, but it will stop, when the element which is less abundant relatively to the needs is exhausted. It is however possible for an ecosystem to increase the amount of information which also contains work energy (Boltzmann, 1905) and increase the network (number of components and number of linkages) which will make the ecosystem more effective in its use of mass and energy, because the recycling is increased. There are, however, many possible pathways to move away from the thermodynamic equilibrium, because ecosystems have many components and they are linked in a complex ecological network. It implies that the three growth forms can be used by the ecosystem to move away from thermodynamic equilibrium in an almost astronomically high number of possibilities—so, which one of these numerous possibilities will be selected? The next section gives the answer.

3.9 Ecological Law of Thermodynamics Gives Information about the Development of Ecosystems

Darwin has given the answer: as "survival of the fittest," meaning that the species which have properties that are best fitted to the prevailing conditions determined by the forcing functions or constraints, will survive and grow. Survival is represented by biomass and information. Survival is in this context a question about the survival of the genes—this means the information is important, because the survival of the genes ensures that the processes, the biomass and the functions of the ecosystem are conserved for the next generation of organisms and therefore on a longer-term basis. The more biomass the more difficult it will be to eliminate the biomass (Svirezhev, 1990); and more information means that the resources (necessary elements to build the biomass and the use of the solar radiation) are used more effectively. The work energy of biomass and information are expressed by eco-exergy and as all the species in an ecosystem are interrelated and dependent on each other through a complex ecological network, the work energy of the ecosystem can be used to account for the survival of the entire ecosystem. These holistic considerations are included in the translation of Darwin's theory to thermodynamics by the ecological law of thermodynamics (ELT). Darwin's theory focuses on the survival of the organisms, but all the organisms are dependent on all other organisms because they are linked in a network and affect and influence each other. It is therefore absolutely necessary to consider the survival of the entire ecosystem with all the living and interdependent components. To summarize: ELT is a translation of Darwin's theory to thermodynamics and at the same time an expansion from the organisms—or according to the neo-Darwinian interpretation, the genes of the organisms—to the entire ecosystem.

ELT is formulated as follows: *A system that receives a throughflow of work energy will try to utilize the work energy flow to move away from thermodynamic equilibrium (more biomass, more structure, and more information), and if more combinations of components and processes are offered to utilize the work energy flow, the system will select the organization that gives the system as much work energy content (storage) as possible.*

Without the genes, or rather a heritage system, there would have been no development or evolution. The interplay between the three growth forms plays also an important role, because the growth of the biomass and networks may create constraints, which inevitably will influence the third growth form—the information that is embodied in the genes. The neo-Darwinian theory has taught us that adaptation occurs through natural selection of changed genetic variations. It is a part of the story, but not the full story. Let us denote it as heredity system number one. It includes the so-called *Hox genes* that play a pivotal role in specifying regional identity in the body plans. It has been suggested that increasing the complexity of body plans during evolution might be causally correlated with increasing complexity of the Hox gene complexes (Barbieri, 2001).

The support for the validity of the tentative law in its present formulation is strong and may be summarized in the following four points:

1. It may be considered a translation of Darwin's theory to thermodynamics and is consistent with the basic thermodynamic laws. The selected organization is the one which offers the most "survival" that can be measured as work energy. The selection is in accordance with the latest formulations of Darwin's theory still taking place on the levels of species. The species are surviving, growing, and fighting for the resources. All the species are, however, connected in ecological, cooperative, and synergistic networks and are dependent on each other. The survival is under the prevailing conditions, which includes the presence of all the components in the ecological network. All the species in the ecological network are influencing all the other species. The result is therefore that the entire ecological network gets as much survival and therefore eco-exergy is possible under the prevailing conditions.
2. The application of the hypothetical law in models gives (many) results, which are consistent with ecological observations, see Jørgensen (2002) and Jørgensen and Svirezhev (2004). It includes the use of structurally dynamic models (SDMs).
3. Many ecological observations, including our description (image) of the evolution, can be explained by the presented hypotheses (see Jørgensen, 2002; Jørgensen et al., 2000; Jørgensen and Svirezhev, 2004; Jørgensen et al., 2007). One example is shown in the following as an illustration: the sequence of organic matter oxidation (see Table 3.7). Many more examples can be found in Jørgensen (2012).
4. Proteins carry out all the biochemical reactions in the cells and their physicochemical properties are the prerequisite for the existence of cells. Their amino acid sequence, which is fundamental for enzymatic properties, is coded by the information embodied in the genome.

The sequence of biological organic matter oxidation (e.g., Schlesinger, 1997) takes place in the following order: by oxygen, by nitrate, by manganese dioxide, by iron (III), by sulfate, and by carbon dioxide. This means that oxygen, if present, will always outcompete nitrate which will outcompete manganese dioxide, and so on. The amount of work energy stored as a result of an oxidation process is measured by the available kJ/mole of electrons which determines the number of adenosine triphosphate molecules (ATPs) formed. ATP represents exergy storage of 42 kJ/mole. Usable energy as exergy in ATPs decreases in the same sequence as indicated above. This is as expected if the

TABLE 3.7

Yields of kJ and ATPs per Mole of Electrons, Corresponding to 0.25 Moles of CH_2O Oxidized (Carbohydrates)

Reaction	kJ/mole e-	ATP's/mole e-
$CH_2O + O_2 = CO_2 + H_2O$	125	2.98
$CH_2O + 0.8NO_3^- + 0.8\,H^+ = CO_2 + 0.4N_2 + 1.4H_2$	119	2.83
$CH_2O + 2MnO_2 + H^+ = CO_2 + 2Mn^{2+} + 3H_2O$	85	2.02
$CH_2O + 4FeOOH + 8H^+ = CO_2 + 7H_2O + Fe^{2+}$	27	0.64
$CH_2O + 0.5\,SO_4^{2-} + 0.5H^+ = CO_2 + 0.5HS^- + H_2O$	26	0.62
$CH_2O + 0.5\,CO_2 = CO_2 + 0.5CH_4$	23	0.55

Note: The released energy is available to build ATP for various oxidation processes of organic matter at pH = 7.0 and 25°C.

exergy-storage hypothesis were valid (Table 3.7). If more oxidizing agents are offered to a system, the one giving the highest storage of free energy will be selected.

EXAMPLE 3.2

Explain why

1. Denitrification requires anaerobic conditions.
2. Phosphorus is released from the sediment in aquatic ecosystems with thermocline.

Solution

1. Nitrate can only be used for oxidation when oxygen is not present.
2. The bottom water in aquatic ecosystems with thermocline has a low redox potential because the sediment is usually rich in organic matter and oxygen cannot be transferred to the bottom water due to the thermocline. It implies that iron is in the form of Fe(II), and iron(II) phosphate is considerably more soluble than iron(III) phosphate.

3.10 Ecosystems Are Hierarchically Organized

The biological hierarchy is easy to observe. The biochemical processes take place in the cells, which have molecular components and structures to control the processes and protect the genome. In vertebrates, there are different types of cells, which are specialized to carry out the biochemical processes

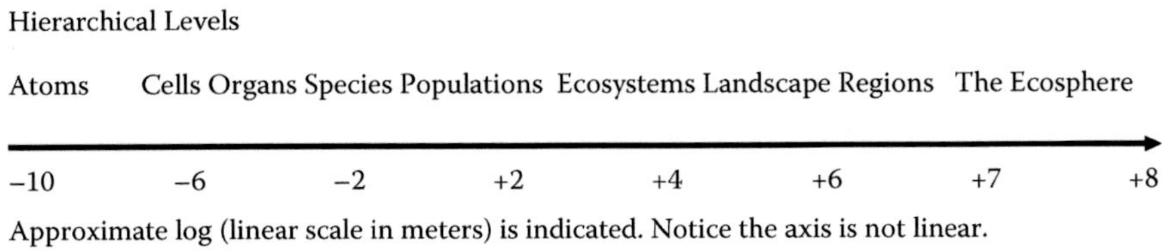

FIGURE 3.6
The presented hierarchical levels are shown. Log to the linear scale in meter is indicated, but notice that the scale of the axis is not linear.

which are characteristic for different organs: the liver, the muscles, the kidneys, the heart, and so on. The cells that carry out the processes that take place in the liver make up the liver, and so on. It is a proper and effective hierarchical solution that the cells which have certain biochemical functions are working together to ensure the function of the organs.

The next hierarchical level after the organs is the species. They are working together in populations that have numerous methods to ensure survival and growth for the individuals. The grazers form, for instance, a herd that makes it more difficult for the predators to attack the individuals of the herd. On the other side, the predators hunt together to obtain by cooperation a higher probability for successful hunting. Populations are also using communication among the individuals to increase the probability for survival. Populations are interacting in a network and make up together with the nonbiological component of the environment, the ecosystem. Landscapes are formed by interactions among several ecosystems, and regions comprise of many landscapes. The entire living matter on the Earth makes up the biosphere and the biosphere plus the nonbiological components are denoted the ecosphere.

The entire hierarchy is illustrated in Figure 3.6, which indicates the corresponding spatial scale. The hierarchy is well fitted to the constraints that biological systems strive to move as far away as possible from thermodynamic equilibrium by the conditions of the thermodynamic laws, including ELT and the biochemical characteristics for life on Earth.

EXAMPLE 3.3

Give an example that illustrates that the population determines the framework of the individuals and that the individuals determine the properties and characteristics of the population.

Solution

A population of deer, for instance, determines in which direction the deer should move to get food and avoid predators. The individuals determine which type of food the population would prefer (could eat). The population sticks together to be better protected, which is particularly important for the young and weak deer. The individuals are carrying the genes that determine the properties and characteristics of the deer.

3.11 Ecosystems Have a High Diversity in All Levels of the Hierarchy

The best solution to move as far away from thermodynamic equilibrium as possible under the prevailing conditions is continuously selected. The prevailing conditions, determined by the constraints, impacts or forcing functions on the ecosystems, vary, however, enormously in time and space. Therefore, there are continuous interactions, on the one side, attempts to find the best solution for the most possible growth; and on the other side, the prevailing conditions, which vary in time and space. As a result and due to the possibilities by the use of the genes to remember good solutions, the selection leads inevitably to differentiation on all levels of the hierarchy. The process of finding better and better solutions by selection is denoted by development, when the focus is on the short- or medium-term changes, while we would use the term evolution when we are focusing on the long-term results. The diversity in the hierarchical levels is presented: (a) in biochemistry—the molecular level, (b) in information carried by the cells—the genetic diversity, (c) of cells, (d) of organs, (e) of individuals in a population, (f) of species, (g) of communities and ecological networks, and (h) of ecosystems.

All known life on Earth resides in the thin layer enveloping the globe known as the ecosphere. This region extends from sea level about 10 km into the ocean depths and approximately the same distance up into the atmosphere (Jørgensen et al., 2007). It is so thin that if an apple was enlarged to the size of the Earth the ecosphere would be thinner than the peel. Yet a vast and complex biodiversity has arisen in the ecosphere. However, the conditions in the ecosphere for living organisms vary enormously in time and space, but the ecosphere has been able to meet these enormous variations by the development of an enormous differentiation.

The climatic conditions that are extremely important for life (see Chapter 4) vary enormously:

1. The temperature can vary from about –70°C to about 55°C.
2. The wind speed can vary from 0 km/h to several hundred km/h.
3. The humidity may vary from almost 0% to 100%.
4. The precipitation from a few millimeters in average per year to several meters per year, which may or may not be seasonally aligned.
5. Annual variation in day length according to longitude from 0 to 24 h.
6. Unpredictable extreme events such as tornadoes, hurricanes, earthquakes, tsunamis, and volcanoes can also change the climate very significantly.

The physical–chemical environmental conditions:

1. Nutrient concentrations (C, P, N, S, Si, etc.).
2. Salt concentrations (it is important both for terrestrial and aquatic ecosystems).
3. Presence or absence of toxic compounds, whether they are natural or anthropogenic in origin.
4. Rate of currents in aquatic ecosystems and hydraulic conductivity for soil.
5. Space requirements and availability.

The biological conditions:

1. The concentration of food for herbivore, omnivore, and carnivore organisms.
2. The density of organisms.
3. The density of competitors for resources (food, space, etc.).
4. The concentration of pollinators, *symbionts*, and mutualists.
5. The density of decomposers.

The human impact on natural ecosystems today add to this complexity.

The list of factors determining the life conditions is even much longer, as we have only mentioned the most important forcing functions on ecosystems. In addition, ecosystems have history or path dependency, meaning that the initial conditions offer the possibilities of development. If we modestly assume that 100 factors define the life conditions and each of these 100 factors could be on 100 different levels, then at least 10^{200} different life conditions are possible, which can be compared with the number of elementary particles in the universe 10^{81} or the number of seconds since the big bang 5×10^{17}. The confluence of path dependency and an astronomical number of combinations affirms that the ecosphere could not experience the entire range of possible states. Furthermore, its irreversibility ensures that it cannot track back to other possible configurations.

EXAMPLE 3.4

Approximately how many amino acids are in 100 mg proteins (enzymes)? How much more information is contained in the amino acid sequence of the amino acids in 100 mg proteins than in the human nucleotides? What are the possibilities to increase the information of living organisms by a continuous evolution?

Solution

Let us presume an average molecular weight of amino acids of 200 g, which means that 100 mg will contain 1/2000 moles of amino acids or A (Avogadro's number)/2000 = $6.2 \times 10^{23}/2000 = 3.1 \times 10^{20}$ amino acids or about 10^{11} times as much as the coded amino acids in the genomes. This means that there are almost no limits to how much the information of organisms can increase.

EXAMPLE 3.5

The species diversity is low toward the poles and high toward the tropics? Give at least a partial explanation to this gradient in the species diversity.

Solution

A detailed and full explanation would require inclusion of many factors. Two factors seem to be particular important:

1. The productivity gradient associated with the latitude is very similar to the species gradient as a function of the latitude.
2. The development of species diversity as all biological processes is generally increasing with increasing temperature.

3.12 Ecosystems Have a High Buffer Capacity toward Changes

Are ecosystems generally stable? What makes them unstable? These two questions are among the core problems of system ecology and environmental management, but we can easily prolong the list of relevant questions: Can ecosystems persist in the course of sufficiently long time in spite of perturbations, changing forcing functions, constraints or impacts? What will happen when the perturbations stop? Or the forcing functions return to normal values valid before the major impacts or disturbances started? We will try to answer these questions in this section, but before the discussion of the questions, it is necessary to clarify what we mean by stability and give definitions of the most important stability concepts:

a. *Resilience* is generally in science defined as the capability of a strained body to recover its size and shape after deformation has been caused especially by compressive stress. It has been introduced in ecology by Holling (1986) as "the capacity of a system to absorb disturbances and reorganize while undergoing change so as to still retain essentially the same function, structure, identity, and feedbacks," but a slightly more quantitative definition can also be found in the ecological literature: the maximum amount a system can be changed before losing its ability to recover. It has also been defined quantitatively as the change of the system by disturbances divided with the time it requires to recover. Some ecological systems display, however, several possible stable states. They may also show a hysteresis effect in which, even after a long time, the state of the system may be partly determined by its history. The concept of resilience depends therefore upon our objectives, the types of disturbances that we anticipate, control measures that are available and desirable, and the timescale of interest. Ecosystems are very complex and it makes it difficult to apply the concept of resilience, at least quantitatively. It is necessary, if the concept should be applied in environment management to define the exact meaning of "the same function and structure" and which disturbances are actual. An ecosystem will never return to the same conditions and state again, because of the complexity of the ecosystem and the enormous variability of the forcing functions.

b. *Resistance* is understood as the ability to resist changes, when the impact on the ecosystem or its forcing functions is changed or perturbations are introduced. It can easily be applied quantitatively as the ratio changes in the ecosystem of forcing functions to the changes of state variables or process rates. The concept could be used multidimensionally in sense that there are many different ratios that could be used corresponding to all relevant combinations of forcing functions and state variables. The stability information that we gain by the use of this concept is more quantitative than resilience. On the other hand, resilience looks into to the ability of the ecosystem to recover, if the forcing functions are changed more significantly. Therefore, both the concepts are relevant in environmental management and ecosystem conservation, and they supplement each other. Resistance gives a good quantitative answer to the stability questions, but it would still be relevant to discuss: Can the ecosystem recover—particularly, can it recover its functions and main structure? This question should be answered as detailed and quantitatively as possible.

c. *Buffer capacity*

$$\text{Buffer capacity} = \frac{\text{change in forcing functions}}{\text{change in state variables}} \tag{3.8}$$

Forcing functions are the external variables that are driving the system, such as discharge of wastewater, precipitation, wind, and so on, while state variables are the internal variables that determine the system, for instance the concentration of soluble phosphorus, the concentration of zooplankton, of certain species, and so on. As seen, the concept of buffer capacity has a definition which allows us to quantify, for instance, in modeling and it is furthermore applicable to real ecosystems, as it acknowledges that some changes will always take place in an ecosystem as a response to changed

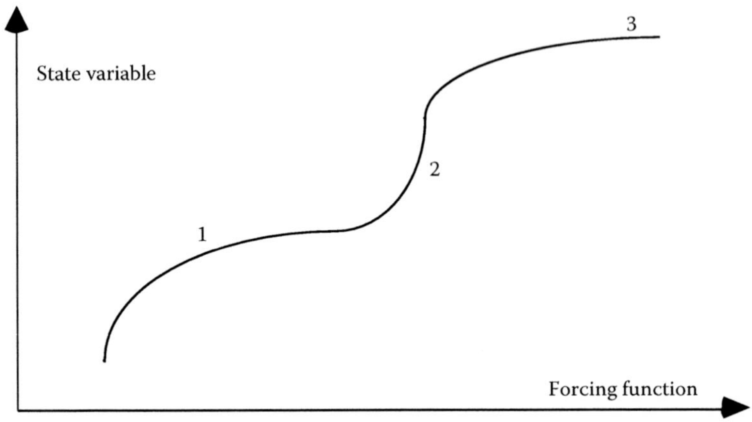

FIGURE 3.7
The relation between state variables and forcing functions is shown. At points 1 and 3, the buffer capacity is high; at point 2, it is low.

forcing functions. The question is how large these changes are relative to changes in the conditions (the external variables or forcing functions). The concept should be applied multidimensionally, as we may consider a number of relevant combinations of state variables and forcing functions. It implies that even for one type of change there are many buffer capacities corresponding to each of the state variables.

The relation between forcing functions (impacts on the system) and state variables indicating the conditions of the system are rarely linear, and buffer capacities are therefore not constant, but strongly dependent on the state of the ecosystem. It may therefore in environmental management be important to reveal the relationships between forcing functions and state variables to observe under which conditions buffer capacities are small or large, compare with Figure 3.7.

Unfortunately, it is possible in ecological and environmental literature to find wrong use and a misinterpretation of the concepts. Resistance is sometimes called resilience, but the three definitions given above will consequently be used throughout the book.

High diversity gives a higher probability for the ecosystem to meet new and sometimes unexpected impacts and it may sometimes at least imply higher resistance or buffer capacity.

To summarize, buffer capacity and resistance seem to be applicable stability concepts in environmental management, as they are based:

1. On an acceptance of the ecological complexity—it is a multidimensional concept, as there is a buffer capacity to each combination of forcing functions and state variables.

2. On reality, that is, that an ecosystem will never return to exactly the same situation again.

Environmental management could be supplemented by such questions as:

- Are there possibilities that the ecosystem will be able to return to a normal function?
- Can the ecosystem recover and offer the ecosystem services that we are dependent on?

This means that it is still relevant to have the concept of resilience in mind in environmental management, because ecosystems may have a high buffer capacity which is not easily recovered after major disturbances or may have a low buffer capacity and still be able to recover due to very effective feedback and adaptation properties.

EXAMPLE 3.6

Do the observed buffer capacities for eutrophication and for toxic substances comply with ELT and the before mentioned statistical analysis that show that eco-exergy is a good measure of the sum of buffer capacities?

Solution

The answer is "Yes." A lake that has a high concentration of phytoplankton has a very high buffer capacity for changes of the eutrophication level. Some other buffer capacities may be low, but the eco-exergy is at least relatively high due to the high biomass concentration. When a toxic substance has a high concentration and several species may be extinguished, the buffer capacity is relatively high for changes in concentration of the toxic substances, because the resistant species have survived. The eco-exergy has most probably decreased due to the reduced biomass, but ELT emphasizes that the system tries to obtain highest possible eco-exergy under the prevailing conditions, which are of course less favorable when a toxic substance is discharged into an ecosystem.

3.13 Components of Ecosystems Work Together in a Cooperative Network

Ecosystems recycle the about 20 elements that are used to build biomass; see Section 3.3. Due to the mass conservation principles, all growth and development would stop without the recycling. Ecological networks are the prerequisite for the indispensable recycling of matter, energy, and information

(Patten, 1985). The networks reinforce at the same time the cooperation of the components in the ecological network (Patten, 1978, 1981, 1982, 1991). It is therefore beneficial for the ecosystems to develop ecological networks and as presented in Section 3.7 ecosystem grows and develops by

1. Increasing the biomass
2. Increasing the information
3. Development of the ecological networks and their efficiencies

One might say that ecological networks have a synergistic effect in the sense that all the components in the network benefit from the formation of the network (Patten, 1978, 1981, 1982, 1985, 1991). The properties of networks are very important for the reactions of ecosystems to disturbances and to consider in integrated environmental management. Chapter 5 will therefore present these properties in more detail.

3.14 Ecosystems Embody a Lot of Information in the Genes

The work energy density can be found from the concentrations of the ecosystem components and their β-values, that express the information content (see Table 3.7)

$$\text{Work energy density} = \sum_{i=1}^{i=n} \beta_i c_i$$

as g detritus equivalents at the temperature T = 300 K

The work energy density in kilojoules is found by multiplication by 18.7, because 1 g of detritus contains 18.7 kJ. The β-values are found by considering the information in the genome about determination of the amino acid sequence. The work energy of information can be expressed by an equation proposed by Boltzmann (1905):

$$k\,T\,\ln M,$$

where k is Boltzmann's constant 1.38×10^{-23} J/(molecule K), T is the absolute temperature (310 K in the human body), and M is the number of microstates, that we do know. Let us use this equation to calculate the work energy of information embodied in the human genome to illustrate the method.

The number of amino acids coded in the human body is according to the human genome project, 1 billion equal to one-third of the number of amino bases in the genes, as three amino bases are needed to code one amino acid in the enzymes. The sequence of amino acids is very important for enzyme function, which implies that each amino acid can be placed in 10^9 different positions and in each position there is a possibility of selecting 20 different amino acids. Therefore, each amino acid molecule has $20^{1,000,000,000}$ different microstates.

We obtain the following by the use of Boltzmann's equation on the case of amino acids in the human body:

$$1.38 \times 10^{-23}\ \text{J/(molecule K)}\ 310\ \text{K}\ \ln(20^{1000,000,000}) = 1.38 \times 10^{-23}\ \text{J/(molecule K)}\ 310 \times 10^9 \ln 20 = 1.283 \times 10^{-11}\ \text{J/(molecule)}$$

The human body has 10^{14} cells and each cell contains about 10^9 molecules of amino acids. This means that the total amount of work energy due to information in the human body is

$$1.283 \times 10^{-11}\ \text{J/(molecule)} \times 10^{23} = 1.283 \times 10^{+12}\ \text{J}$$

The β-value 1.00 is defined as the ratio between this amount of information work energy and the chemical work energy of the biomass (indicated earlier as 18.7 J/mg or 18.7 kJ/g. A human has on average a biomass of 32 kg biomass dry weight and it yields a chemical work energy of the biomass = 32,000,000 × 18.7 J = 5.98×10^8 J. The β-value minus 1.00 is therefore = (12.83/5.98) 10^3 = 2145 or β = 2146. The previously found value is 2173 or 1.2% higher.

The same calculation methods have been applied to find the β-value for the field mice and for yeast (*Saccharomyces cerevisiae)* to illustrate the method. It means that the following equation has been applied:

$$\beta\text{-value} = 1 + 1.38 \times 10^{-23}\ T \ln(20^{NA}) NC\ NM/(W\ 18{,}700)$$

where T is the absolute body temperature (here 310 K is used), NA is the number of coded amino acids, NC is number of cells, NM is number of molecules per cell, and W is the weight in gram. The results of the calculations and the relevant information used as presented in Table 3.9. The values in Table 3.8 are found by a slightly different method (see Jørgensen et al., 2005 and Jørgensen, 2012). The previous values (Table 3.8) are included in Table 3.9 for comparison. Notice that the values in Table 3.8 are based on information about the genomes for many of the values, while a few values are found indirectly from the number of cell types, the evolutionary tree, or other relevant information; see Jørgensen et al. (2005).

Specific work energy is work energy relatively to the biomass and for the *i*th component: Sp. $ex._i = Ex_i/c_i$. It implies that the total specific exergy per unit of area or per unit of volume of the ecosystem is equal to the β-value.

TABLE 3.8

β-Values = Exergy Content Relatively to the Exergy of Detritus

Organisms	Plants	Animals
Detritus	1.00	
Viroids	1.0004	
Virus	1.01	
Minimal cell	5.0	
Bacteria	8.5	
Archaea	13.8	
Protists (Algae)	20	
Yeast	17.8	
	33	Mesozoa, Placozoa
	39	Protozoa, amoeba
	43	Phasmida (stick insects)
Fungi, molds	61	
	76	Nemertinea
	91	Cnidaria (corals, sea anemones, jelly fish)
Rhodophyta	92	
	97	Gastrotricha
Prolifera, sponges	98	
	109	Brachiopoda
	120	Platyhelminthes (flatworms)
	133	Nematoda (round worms)
	133	Annelida (leeches)
	143	Gnathostomulida
Mustard weed	143	
	165	Kinorhyncha
Seedless vascular plants	158	
	163	Rotifera (wheel animals)
	164	Entoprocta
Moss	174	
	167	Insecta (beetles, flies, bees, wasps, bugs, ants)
	191	Coleoidea (Sea squirt)
	221	Lepidoptera (butterflies)
	232	Crustaceans
	246	Chordata
Rice	275	
Gymnosperms (incl. pinus)	314	
	310	Mollusca, bivalvia, gastropodea
	322	Mosquito
Flowering plants	393	
	499	Fish
	688	Amphibia
	833	Reptilia

(Continued)

TABLE 3.8 (*Continued*)

β-Values = Exergy Content Relatively to the Exergy of Detritus

Organisms	Plants	Animals
	980	Aves (Birds)
	2127	Mammalia
	2138	Monkeys
	2145	Anthropoid apes
	2173	Homo sapiens

Source: From Jørgensen, S.E. et al., 2005. *Ecol. Model.* 185, 165–176.
Note: β-Values = eco-exergy content relatively to the eco-exergy of detritus.

TABLE 3.9

Results and Calculation of β-Values for Humans, Field Mice, and Yeast

	NA	NC	NM	W	β-Value	Previous β-Value[a]
Human	10^9	10^{14}	10^9	32 kg	2146	2172
Field mouse	833×10^6	10^{11}	10^9	25 g	1907	2127
Yeast	4×10^6	1	10^9	200 pg	14	16

Source: Jørgensen, S.E., B. et al., 2007. *A New Ecology. Systems Perspectives.* Amsterdam: Elsevier, 288 pp; Jørgensen, S.E. and B. Fath. 2011. *Fundamentals of Ecological Modelling.* 4th edition. Amsterdam: Elsevier, 400 pp; Jørgensen, S.E. 2012. *Introduction to Systems Ecology.* CRC: Boca Raton, FL, 320 pp. Chinese edition 2013.

[a] Previous applied value.

EXAMPLE 3.7

The state variables of the P-cycle in Figure 3.4 have the following concentrations:

Ortho P: 0.1 mg/L; Phyt-P: 0.1 mg/L; Zoopl-P: 0.02 mg/L; Fish-P: 0.0025 mg/L; Det-P: 0.025 mg/L; Exch. P: 0.022; Por-P: 0.4 mg/L.

Calculate the eco-exergy in kJ per liter of lake water, when it is presumed that P for phytoplankton, zooplankton, detritus, fish, and exchangeable P is 1%.

It is presumed that organic matter has 18.7 kJ/g free energy.

Solution

The equation for calculation of eco-exergy (work energy including work energy of information) shown above can be used directly as we know the β-values for the state variables.

$(0.1 \times 0 + 0.1 \times 100 \times 20 + 0.02 \times 100 \times 163$ (Rotifera represents zooplankton) $+ 0.0025 \times 100 \times 499 + 0.025 \times 100 \times 1 + 0.022 \times 100 \times 1 + 0.4 \times 0) \times 18.7\ \text{J/l} = 12.252\ \text{kJ/l}.$

As the average β-value for many ecosystems is typically between 50 and 250, the amount of information carried by ecosystems corresponds to 50–250 times as much as the energy of the biomass. The biomass per m^2 is for most ecosystems between 1 kg and 1000 kg which would correspond to about 18.7 MJ and 18.7 GJ per m^2. This implies that the information content will correspond to in the order of between 1.87 GJ and 1870 GJ per m^2 if we use an average information content of 100. This means that the number of amino acids coded in an ecosystem is enormous, but it is the information which is controlling all the life processes of the ecosystem.

An ecosystem is a typical middle number system, which implies that most ecosystems will have between 10^{15} and 10^{20} individuals with different properties, different genomes, and different reactions to the prevailing conditions of the ecosystem. The individuals may be presented by 1000–100,000 species, which implies that the ecosystem would have 1000–100,000 completely different genomes. The species and the individuals embody with other words another set of information, that is crucial for the ecosystem and its reaction to the enormous variability of the forcing functions, determining the prevailing conditions.

The evolution has increased the amount of information. Both the biodiversity and the genomes have increased, when we exclude the last few hundred years where human beings have reduced the area of nature and the biodiversity. The work energy of an ecosystem is very big due to the information, which means that the ecosystems should be conserved. Particularly, of course, because ecosystems offer us ecosystem services, which can be expressed by the total work energy content; see Chapter 1.

3.15 Ecosystems Have Emergent Properties

A human body consists of a number of chemical compounds: lipids, proteins, calcium compounds, and so on. The total value of these chemical compounds is maybe in the order of 100 dollars at the most. Hence, it is the not the value of the chemical compounds that makes the human being (or any other mammal) a unique creature or we may say system. It is the cooperation of these compounds in various synergistic networks based on an enormous organizational knowledge that makes it possible for a human being to coordinate and direct his movements, to come up with new ideas, solve problems, speak, sense, read, write, and express emotions. A human being is much more than just a collection of chemical compounds. He is a self-organizing system with many surprising, and advantageous properties.

The ecosystems work similarly as unique systems with surprising and advantageous properties that have been presented in the 13 other sections of the chapter. Ecosystems are complex, adaptable, hierarchical systems with an enormous amount of feedbacks and regulations that are a result of the evolution, which has learned the ecosystems to deal with a huge number of impacts and environmental problems.

Nielsen (2007) and Nielsen and Müller (2000, 2009) have as observers listed and discussed several ecosystem properties that they applied in industrial ecology and for the presentation of ecosystem services. They have listed the following holistic properties of ecosystems, which are supplementary to the 14 basic properties or rather a consequence of the 14 properties:

1. *Complexity*: It is of course a consequence of the information content and the diversity of all the levels in all the ecological hierarchical levels. The complexity of ecosystems is determined by their openness far from thermodynamic equilibrium and rooted furthermore in the many flows and processes that are continuously operating in all ecosystems. Ecosystems are very dynamic systems due to these flows and processes, which are functions of the time, as also all the components in all the hierarchical levels are also functions of the time. Living systems as ecosystems are many magnitudes more complex than physical systems.
2. *Evolution*: It is a consequence of the thermodynamic properties of ecosystems that will be presented in Chapter 4. The components of ecosystems are struggling under the prevailing conditions to move away from thermodynamic equilibrium. At the same time, the prevailing conditions determined by the forcing functions vary in time and space, and are continuously changing. The ecosystems generate simultaneously new possibilities to meet the prevailing conditions on all levels of the ecological hierarchy. Of particular importance is the generation of new possibilities on the species level, by mutations, sexual recombinations, and general transfer of genes among organisms. Ecosystems evolve on all the levels of the ecological hierarchy and the evolution is both vertical and horizontal, which implies that the information embodied in the genes is increasing and the complexity and efficiency of the networks are increasing.
3. *Compartmentalization*: The description of the hierarchical organization requires that on each level we have different compartments, which of course have a clear definition and boundaries.
4. *Flow and processes*: These are the transfers among the compartments in the ecological network. It is clear that a more quantitative description of ecosystems, for instance, in ecological models will require that mathematical expressions are applied for the various flows and processes. It should be emphasized that ecological networks consist not only of compartment, that are connected in a network, but that the connections are flows and processes with a causality, which we want to capture when we are constructing ecological models.

5. *Cybernetics, feedback, and controls*: Ecosystems possess an enormous amount of self-organization and self-regulation. All processes in an ecosystem are practically regulated by enzymes directly or indirectly. The concentrations of the enzymes and the conditions of the processes such as pH and temperature are to a high extent regulated by the process. Because of the wide spectrum of possibilities and components on all levels of the ecological hierarchy and the continuous selection, there is an enormous ability to adapt to the prevailing conditions.
6. *Cycling*: Matter, all the biological important elements, energy and information cycle in ecosystems, which is the immediate result of the network. The cycling is crucial for ecosystems, because all further growth would stop when the most limiting element was used up in the form applied, if the elements were not returned or recovered by the cycling. Energy cycling is important, too, because it makes it possible to increase the efficiency of the free-energy utilization. Information cycling is important because a dissemination of the information to all corners of the ecosystem is thereby made possible and can be done by the network with no costs or almost no costs. Without the cycling, no evolution would have taken place, because as already stressed, the growth would stop and the struggle for growth is the prerequisite for the evolution.
7. *Network properties* have been presented shortly in Section 3.13 and more details will be presented in Chapter 5. They are crucial for the ability of ecosystems to follow and utilize all possibilities to move away from thermodynamic equilibrium, it means to utilize the three growth forms, which is significant for ecosystems and the evolution.
8. *Organization*: The hierarchical organization is very beneficial for the ecosystems—not only because the openness of the different levels makes it possible to obtain the right dynamics on all levels, but also because disturbances are absorbed very effectively. Furthermore, the diversity on the different levels fits to the size, processes, dynamics, and the components of the levels.
9. *Diversity*: In the light of the enormous variability of the forcing functions in time and space, it is very beneficial for ecosystems to meet this variability of the challenges to move as far away as possible from thermodynamic equilibrium by the availability of as many solutions as possible.
10. *Openness and dissipation*: Because of the dissipation of free energy, ecosystems must be open or at least nonisolated. The huge complexity of ecosystems is making ecosystems ontic open in the sense that it is impossible to make certain predictions as it is possible for physical systems. It is, however, possible to give high probabilities or

propensities for possible changes of ecosystems as a result of given changes of forcing functions or impacts on ecosystems, which is important in environmental management context.

EXERCISES/PROBLEMS

1. A lake receives 300,000 m^3 agricultural drainage water with 11 mg/L nitrogen and 0.2 mg/L phosphorus. The lake also receives 150,000 m^3 wastewater with 30 mg/L nitrogen and 10 mg/L phosphorus. The lake is eutrophic and the phytoplankton has seven times as high a nitrogen concentration as phosphorus concentration. How effective does the removal of phosphorus from the wastewater have to be to make phosphorus the liming element?
2. What are the extensive variables for the following energy forms: electrical energy, chemical energy, and heat energy? What are the intensive variables for the following energy forms: pressure energy, heat energy, and kinetic energy?
3. Explain how ecosystems are able to recycle trace metals?
4. Explain why the formation of organic matter with a molecular weight of 10,000 cannot be formed spontaneously? We find, however, high molecular organic matter in ecosystems. How is it possible?
5. Explain why eco-exergy measures the distance from thermodynamic equilibrium.
6. Give at least three different formulations of the second law of thermodynamics that could be useful in ecology.
7. A wetland has a per square meter: 2500 g of wild rice, 400 g of detritus, 0.2 g of crabs, 2 g of insects, 0.2 g of fish, and 0.1 g of frogs. Calculate the eco-exergy density of the wetland in megajoules per square meter.
8. The ratio N:P of waste released by bodies with increasing N:P is decreasing. The following observations have been made:

N:P Molar Ratio in Body	N/P Ratio in Waste
2	80
3	50
8	20
9	16
20	6

9. If iodine is the limiting factor for brown algae, what is the approximate maximum concentration of brown algae as dry weight per liter in water with a concentration of 0.01 mg iodine/liter. Brown algae contains 1.5 g iodine/kg dry matter.

10. How much faster is a marine ecosystem with an area of 5 ha to recover after a tsunami, than a 50,000 ha marine system?
11. Which region is most vulnerable to pests?
    a. A region consisting of only wheat fields with an area of 5000 km^2.
    b. A region consisting of 5 different landscapes and 88 different ecosystems.

    The answer must be explained.
12. The information embodied in the genes is far from its limits. Is the information embodied in the ecological network close or far from its limits? Give a detailed explanation of the answer.

References

Barbieri, M. 2003. *The Organic Codes*. Cambridge: Cambridge University Press, 301 pp.

Boltzmann, L. 1905. The Second Law of Thermodynamics (Populare Schriften. Essay No. 3 [Address to the Imperial Academy of Science in 1886]). Reprinted in English in: Theoretical Physics and Philosophical Problems, Selected Writings of L. Boltzmann. D. Riedel: Dordrecht.

Holling, C.S. 1986. The resilience of terrestrial ecosystems: Local surprise and global change. In Clark, W.C. and Munn, R.E. (Eds.), *Sustainable Development of the Biosphere*. Cambridge: Cambridge University Press, pp. 292–317.

Jørgensen, S.E. 1976. A eutrophication model for a lake. *J. Ecol. Model.* 2, 147–165.

Jørgensen, S.E. 2002. *Integration of Ecosystem Theories: A Pattern*. 3rd edition. Dordrecht: Kluwer Academic Publisher, 428 pp. (1st edition 1992, 2nd edition 1997).

Jørgensen, S.E. 2012. *Introduction to Systems Ecology*. Boca Raton, FL: CRC Press, 320 pp. (Chinese edition 2013.)

Jørgensen, S.E. and G. Bendoricchio. 2001. *Fundamentals of Ecological Modelling*. 3rd edition. Amsterdam: Elsevier, 628 pp.

Jørgensen, S.E. and B. Fath. 2011. *Fundamentals of Ecological Modelling*. 4th edition. Amsterdam: Elsevier, 400 pp.

Jørgensen, S.E. and Y. Svirezhev. 2004. *Toward a Thermodynamic Theory for Ecological Systems*. Amsterdam: Elsevier, 366 pp.

Jørgensen, S.E., B. Fath, S. Bastiononi, M. Marques, F. Müller, S.N. Nielsen, B.C. Patten, E. Tiezzi, and R.E. Ulanowicz. 2007. *A New Ecology. Systems Perspectives*. Amsterdam: Elsevier, 288 pp.

Jørgensen, S.E., N. Ladegaard, M. Debeljak, and J.C. Marques. 2005. Calculations of exergy for organisms. *Ecol. Model.* 185, 165–176.

Jørgensen, S.-E., B.C. Patten, and M. Straškraba. 2000. Ecosystems emerging: 4. *Growth. Ecol. Model.* 126, 249–284.

Jørgensen, S.E. and Y. Svirezhev. 2004. *Toward a Thermodynamic Theory for Ecological Systems*. Amsterdam: Elsevier, 366 pp.

Kay, J. and E.D. Schneider. 1992. Thermodynamics and measures of ecological integrity. *Proc. "Ecological Indicators."* Amsterdam: Elsevier, pp. 159–182.

Nielsen, S.N. 2007. What has modern ecosystem theory to offer to cleaner production, industrial ecology and society? The views of an ecologist. *J. Clean. Prod.* 15, 1639–1653.

Nielsen, S.N. and F. Müller. 2000. Emergent properties of ecosystems. In Jørgensen, S.E. and Müller, F. (Eds.), *Handbook of Ecosystem Theories and Management*. Boca Raton, FL: Lewis Publishers, pp. 195–216.

Nielsen, S.N. and F. Müller. 2009. Understanding the functional principles of nature: Proposing another type of ecosystem services. *Ecol. Model.* 220, 1913–1925.

Odum, E.P. 1969. The strategy of ecosystem development. *Science* 164, 262–270.

Odum, E.P. 1971. *Fundamentals of Ecology*. 3rd edition. Philadelphia, PA: W.B. Saunders Co., 360 pp.

Patten, B.C. 1978. Systems approach to the concept of environment. *Ohio J. Sci.* 78, 206–222.

Patten, B.C. 1981. Environs: The super-niches of ecosystems. *Am. Zool.* 21, 845–852.

Patten, B.C. 1982. Environs: Relativistic elementary particles or ecology. *Am. Nat.* 119, 179–219.

Patten, B.C. 1985. Energy cycling in the ecosystem. *Ecol. Model.* 28, 1–71.

Patten, B.C. 1991. Network ecology: Indirect determination of the life–environment relationship in ecosystems. In Higashi, M. and Burns, T.P. (Eds.), *Theoretical Ecosystem Ecology: The Network Perspective*. London: Cambridge University Press, pp. 288–351.

Schlesinger, W.H. 1997. *Biogeochemistry: An Analysis of Global Change*. 2nd edition. San Diego, London, Boston, New York, Sydney, Tokyo, Toronto: Academic Press, 680 pp.

Svirezhev, Yu. M. 1990. Entropy as a measure of environmental degradation. *Proc. Int. Conf. on Contaminated Soils*, Karlsruhe, Germany.

Wetzel, R.G. 1983. *Limnology*. Fort Worth, New York, and Orlando: Saunders College Publishing, 838 pp.

4

Thermodynamic Properties of Ecosystems

4.1 How Ecosystems React to a Work Energy Flow

Holling (1986), see Figure 4.1, suggested that an ecosystem progresses through the sequential phases of renewal (mainly growth form I), exploitation (mainly growth form II), conservation (dominant growth form III), and creative destruction. The creative destruction phase also fits into the three growth forms but will require a further explanation. It is a result of either external or internal factors. In the first case (for instance, hurricanes and volcanic activity), further explanation is not needed, as an ecosystem has to use the growth forms under the prevailing conditions which are determined by the external factors. If the destructive phase is a result of internal factors, the question is: "Why would a system be self-destructive?" A possible explanation is that a result of the conservation phase is that almost all nutrients will be contained in organisms, which implies that there are no nutrients available to test new and possibly better solutions to move further away from thermodynamic equilibrium or, expressed in Darwinian terms, to increase the probability of survival. This is also implicitly indicated by Holling, as he talks about creative destruction. Therefore, when new solutions are available, it would, in the long run, be beneficial for the ecosystem to decompose the organic nutrients into inorganic components which can be utilized to test the new solutions. The creative destruction phase can be considered a method to utilize the three other phases and the three growth forms more effectively in the long run.

Odum described the development of ecosystems from the initial stage to the mature stage as a result of continuous use of the self-design ability (Odum, 1969, 1971). See the significant differences between the two types of systems listed in Table 4.1 and notice that the major differences are on the level of information. The content of information increases in the course of an ecological development, because an ecosystem encompasses an integration of all the modifications that is imposed on the environment. Thus, it is on the background of genetic information that systems develop which allow interaction of information with the environment. Herein lies the importance in the feedback from the organism to the environment, which means that an organism can only evolve in an evolving environment. The differences

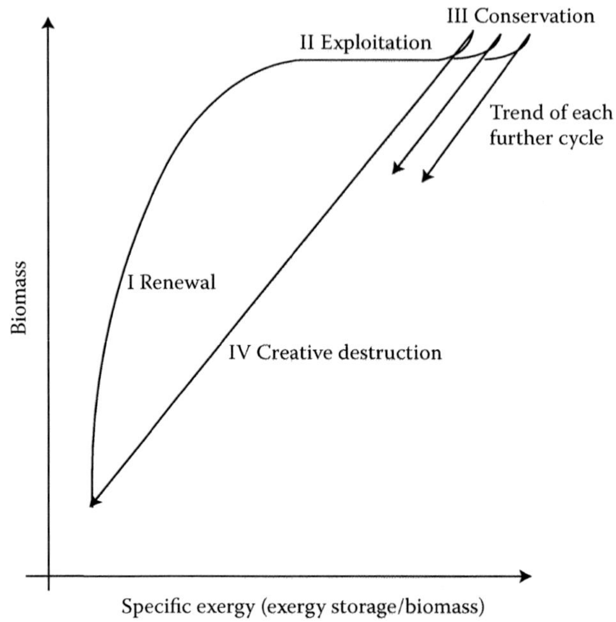

FIGURE 4.1
Holling's four stages are expressed in terms of biomass and specific exergy. Notice that the trend of each further cycle is toward a higher exergy storage.

between the two stages include entropy and exergy (work energy). The latter concept has been discussed in Chapter 3 and entropy has previously be defined as the extensive variable for the energy form heat (see Table 3.2).

Most of the free energy (work energy) of the solar radiation captured by the vegetation is used for maintenance—it means respiration. The development of the work energy captured and the work energy applied for maintenance may vary considerably from ecosystem to ecosystem and from season to season. If we follow Odum's attributes in our description of the ecosystem development, Figure 4.2 gives a typical average picture. The vegetation at an intermediate stage yields the relative biggest contribution to the storage of work energy.

Work energy (exergy, eco-exergy if information is included as work energy) storage and utilization patterns follow the seasonal trends in biomass, throughflow (power), and informational characteristics. In winter, biomass and information content are at seasonal lows. In the spring, the flush of new growth (dominantly form I) in an ecosystem at the early stage produces, rather quickly, a significant biomass component of work energy (Figure 4.2), but the information component remains low because most active flora, fauna, and microbiota of this nascent period tend to be lower phylogenetic forms.

These lower forms rapidly develop biomass but make relatively low informational contributions to the stored work energy. Growth form I (biomass)

TABLE 4.1

Differences between the Initial Stages and Mature Stages Are Indicated

Properties	Early Stage	Late or Mature Stage
A Energetic		
P/R	≫1 ≪1	Close to 1
P/B	High	Low
Yield	High	Low
Specific entropy	High	Low
Entropy production per unit of time	Low	High
Exergy (work energy)	Low	High
Information	Low	High
B Structure		
Total biomass	Small	Large
Inorganic nutrients	Extrabiotic	Intrabiotic
Diversity, ecological	Low	High
Diversity, biological	Low	High
Patterns	Poorly organized	Well organized
Niche specialization	Broad	Narrow
Size of organisms	Small	Large
Life cycles	Simple	Complex
Mineral cycles	Open	Closed
Nutrient exchange rate	Rapid	Slow
Lifespan	Short	Long
C Selection and Homeostatis		
Internal symbiosis	Undeveloped	Developed
Stability (resistance to external perturbations)	Poor	Good
Ecological buffer capacity	Low	High
Feedback control	Poor	Good
Growth form	Rapid growth	Feedback controlled growth
Types	r-Strategists	k-Strategists

Source: Odum, E.P. 1969. The strategy of ecosystem development. *Science* 164, 262–270.

is dominant during the spring. As the growing season advances, in summer, growth forms II and III become successively dominant. Following the expansion of system organization that this represents, involving proliferation of food webs and interactive networks of all kinds, and all that this implies, waves of progressively more advanced taxonomic forms can now be supported to pass through their phenomenology and life cycles. Albedo and reflection are reduced, dissipation increases to seasonal maxima following developing biomass, and as seasonal maxima are reached, further increments tend to negligible amounts (Figure 4.3). The biotic production of advancing summer reflects more and more advanced systemic organization,

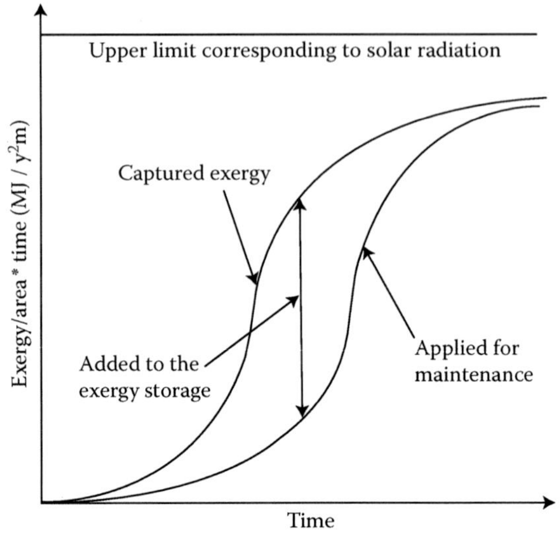

FIGURE 4.2
Work energy (exergy) utilization of an ecosystem under development is shown versus time. Notice that the consequence of the growth in exergy is increased utilization of exergy for maintenance.

manifested as increasing accumulations of both biomass and information to the exergy stores. In autumn, the whole system begins to unravel and shut down in preadaptation to winter, which is the phenological equivalent of senescence. Networks shrink, and with this all attributes of exergy storage, throughflow, and information transfer declines as the system slowly degrades to its winter condition. Biological activity is returned mainly to the more primitive life forms as the ecosystem itself returns to more "primitive" states of exergy organization required for adaptation to winter. The suggestion from phenology is that the exergetic (work energetic) principles of organization also apply to the seasonal dynamics of ecosystems. The work energy of the ecosystem continues in the Northern Hemisphere to increase until about 1st of August, because the free energy captured is still bigger than the free energy needed for maintenance.

4.2 New Ecosystems

It would be interesting to follow the development of the work energy storage when new ecosystems are formed. If the eco-exergy actually increases for new ecosystems, it would be a strong support for the proposition presented in

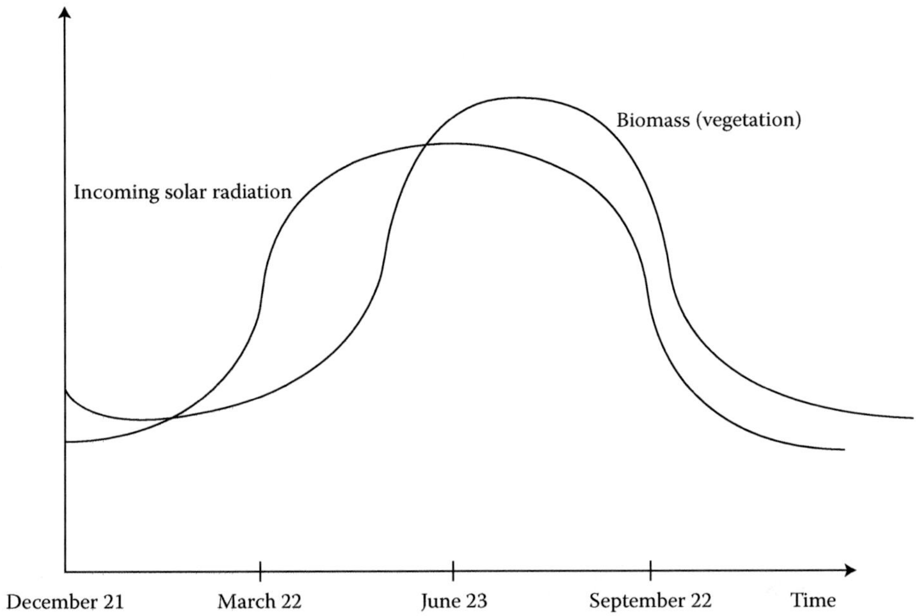

FIGURE 4.3
The seasonal changes in incoming solar radiation and biomass (vegetation) are shown for a typical temperate ecosystem. The x-axis gives the time of the year, while the y-axis shows either the free energy of the incoming solar radiation or the biomass per unit of area or volume. The slope of the curve for biomass indicates the increase in work energy due to growth form I. Growth form I can continue as long as the captured solar radiation is bigger than the eco-exergy applied for maintenance or as long as there is enough inorganic matter to build new biomass. Therefore, the biomass is often at its maximum around August 1st. The biomass is at its minimum around February 1st, because at that time the captured work energy and the work energy applied for maintenance (although it is low) are in balance.

Sections 3.7 and 3.8 about the use of the surplus free energy to move further away from thermodynamic equilibrium—the so-called ELT hypothesis. To illustrate the development of eco-exergy for new ecosystems we have chosen Surtsey, a volcanic island located in the south of Iceland. The island is 150 ha and was formed in 1962 by a volcanic eruption; the ecological development on the island has been followed by Reykjavik University since 1964. In the first year and a half, it was impossible to start investigation on the island due to the heat. The development of work energy of plants and birds is shown in Figure 4.4. As shown, the work energy (expressed as eco-exergy), means that the work energy of information is included, has increased since 1964 and up to year 2000 (indicated on the graph as 100) approximately exponentially, which is supported by a log graph in Figure 4.5. It is expected that the eco-exergy is growing as a first-order reaction and therefore exponentially,

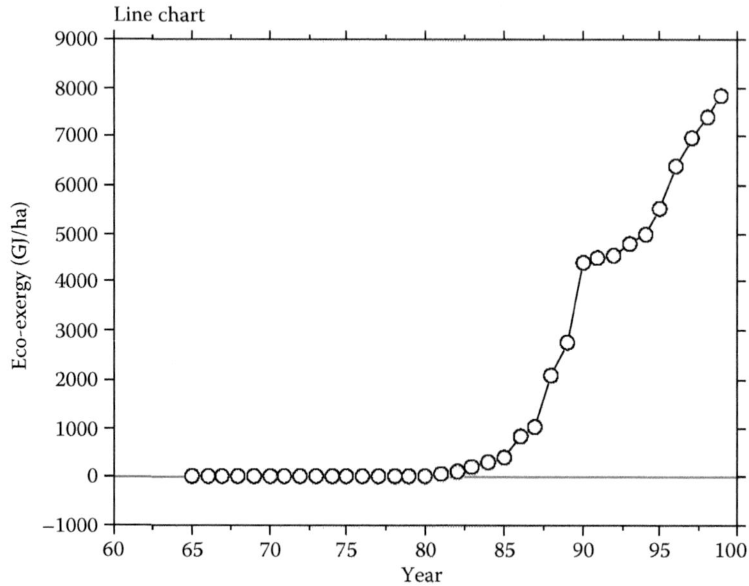

FIGURE 4.4
The development of eco-exergy on the island Surtsey, a volcanic island formed in the south of Iceland by a volcanic eruption in 1962. The island is 150 ha and the eco-exergy for plants and birds has been found in GJ/ha from 1964 to the year 2000 (indicated as 100 on the graph). The eco-exergy (work energy including information) is approximately increased exponentially—see also Figure 4.5 and Jørgensen et al. (2007).

because the growth is proportional to the growth already achieved by the three growth forms.

Examinations of the development of other new ecosystems have shown similar trends, and although the observations for the development of Surtsey island were particularly dense and detailed, it seems possible to conclude that new ecosystems with good approximation will show an exponential increase of eco-exergy, completely in accordance with the presented ecosystem properties in Sections 3.6 through 3.9.

Generally, there are many ecological observations that support the ELT hypothesis. A few were presented in Section 3.9. Two additional examples, Examples 4.1 and 4.2, are given below, and further examples can be found in Jørgensen (2012). The evolutionary theory furthermore gives support to ELT.

EXAMPLE 4.1

Certain parasites are able to attack and damage one ear of insects, but never attack the second ear. How can these observations be explained by ELT?

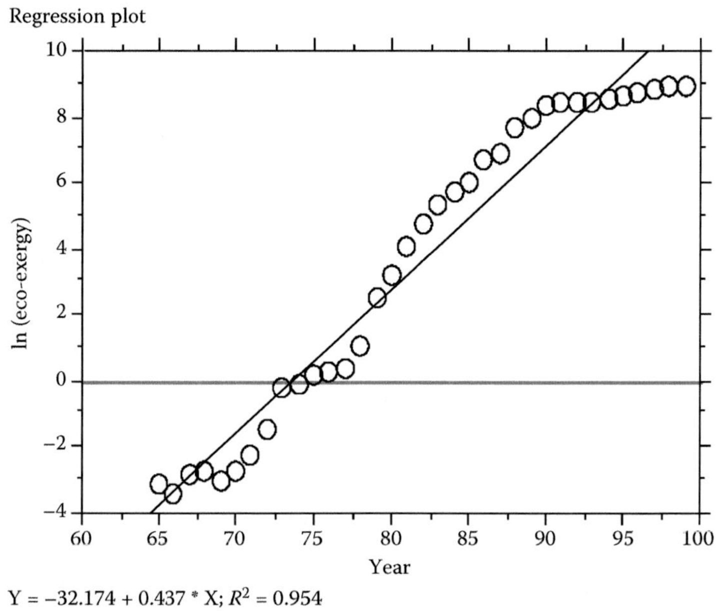

FIGURE 4.5
A regression plot of log eco-exergy in gigajoule per hectare (see Figure 6.8) versus the years 1964–2000 (indicated on the graph as 100). As the regression plot with good approximation is linear with $R^2 = 0.954$, it is possible to conclude that the eco-exergy on the island with good approximations has increased exponentially. For further details, see Jørgensen et al. (2007).

Solution

If the parasites damage both ears, the insects would be very easy victims for bats, because the insects could not hear them. It is therefore not beneficial for the parasites to attack and damage both ears, because it would imply that their hosts would certainly be eaten by bats. The result is that the parasites and the insects survive and therefore contribute more to the eco-exergy of the system.

EXAMPLE 4.2

The horse was only 35 kg about 45 million years ago.

Today the horse is several hundred kilograms, that is, it could be 500 kg.

There has obviously been an evolution toward a bigger size. The hypothesis is that the bigger size utilizes the food more effectively.

Find the eco-exergy efficiency of grass as food used for metabolism and growth as a function of the weight of horses. A model could be applied to answer the question.

Solution

A STELLA model has been developed to show the hypothesis; see the STELLA diagram below.

Note: org is the horse that is growing and covering the maintenance of the complex structure far from thermodynamic equilibrium by respiration. The total food consumption covers the respiration and growth. A logistic growth should be applied to describe the growth, the growth should be proportional to the weight with the exponent 2/3, and respiration should be proportional to the weight with the exponent 3/4. The β-values applied are for horse 2127 and for grass 200. The eco-exergy efficiency is found for the lifespan of the horse from 1 kg to the final weight. The following equations have been applied, using the STELLA format for the differential equations:

```
org(t) = org(t - dt) + (growth - respiration) * dt
INIT org = 1
INFLOWS:
growth = 3*org^(0.67)*(1-org/690)
OUTFLOWS:
respiration = 0.5*org^(3/4)
total_food(t) = total_food(t - dt) + (consumption) * dt
INIT total_food = 1
INFLOWS:
consumption = growth + respiration
food_eff = 2127*100*org/(200*total_food)
```

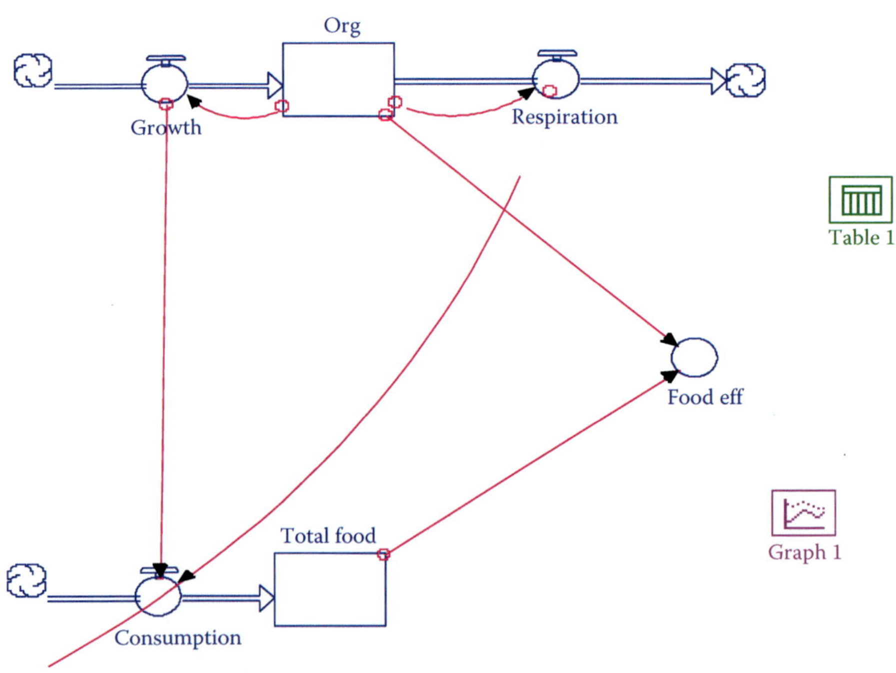

The results of running the model to different maximum values are as follows:

Final Weight (kg)	Efficiency of Food (%)
500	2.75
400	2.60
300	2.42
200	2.18
100	1.85
50	1.55
35	1.41

As shown, the efficiency clearly increases with an increase in body weight, which implies that it is beneficial for horses to increase their body weight.

Figure 4.6 illustrates the difference of energy utilization between the early stage and the mature stage. The biomass is smaller in the early stage, which implies that it captures less solar radiation, while on the other hand it also requires less energy for maintenance (respiration). In contrast, the mature stage

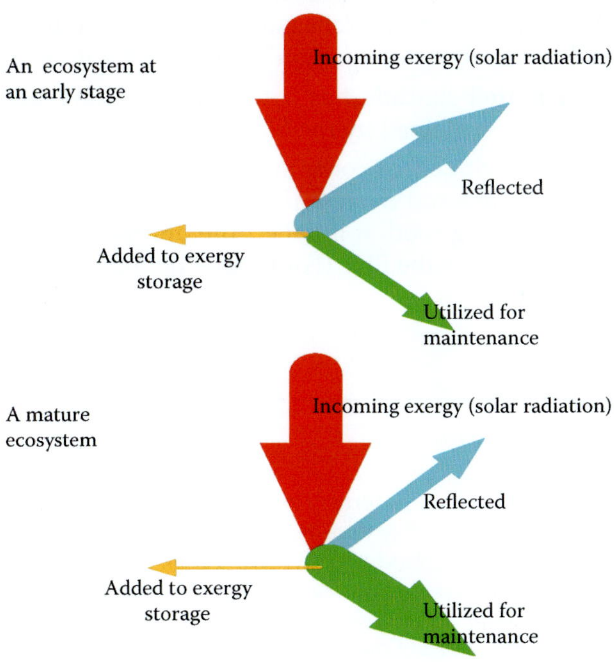

FIGURE 4.6
The difference of energy utilization between the early stage and the mature stage.

captures more solar radiation but also requires more energy for maintenance. In both cases, part of the solar radiation will be reflected (Jørgensen, 2012).

EXERCISES

1. Wallace hypothesized that insects that resemble in color of the trunks on which they reside will be the best survivors. The peppered moth (*Biston betularia*) has two color forms, white moths peppered with black spots and a black form produced by a single allele. Between 1850 and 1920, the black form became dominant in England as response to the air pollution, because the dark form was better protected from predation by birds. Between 1950 and 1995, the white form again became dominant as a result of the air pollution abatement. Explain these observations by the use of the ELT.
2. Is the prey–predator coevolution, that is sometimes called the *arms race*, consistent with the ELT?
3. Number of bird species versus the latitude with good approximations has the following equation: Log number of bird species = Latitude/80 + (80 – latitude) 2/70.
4. A higher species diversity is presumed to give a better possibility to exploit all the ecological niches and therefore the better natural resources. Explain that the equation is consistent with ELT.
5. Species at a higher latitude has a tendency to have bigger body weight. Can it be explained by the ELT?
6. Prey abundance influences consumers' selectivity. Diets are broad when prey are scare and narrow if food is abundant, as a function of time for searching food. It is called the optimal foraging theory. Explain this theory by the ELT. (For further details about this theory see Chapter 15.)

References

Holling, C.S. 1986. The resilience of terrestrial ecosystems: Local surprise and global change. In Clark, W.C. and Munn, R.E. (Eds.), *Sustainable Development of the Biosphere*. Cambridge: Cambridge University Press, pp. 292–317.

Jørgensen, S.E. 2012. *Introduction to Systems Ecology*. (Chinese edition.) Boca Raton, FL: CRC Press, 320 pp.

Jørgensen, S.E. et al. 2007. *A New Ecology. Systems Perspectives*. Amsterdam: Elsevier, 288 pp.

Odum, E.P. 1969. The strategy of ecosystem development. *Science* 164, 262–270.

Odum, E.P. 1971. *Fundamentals of Ecology*. 3rd edition. Philadelphia, PA: W.B. Saunders Co., 360 pp.

5

Network Properties of Ecosystems

5.1 Ecological Networks Increase Utilization Efficiency of Matter and Energy

The components of ecosystems work together in a network. Two examples are shown in Figures 5.1 and 5.2. The properties of ecological network are very important for the proper function of ecosystems. Ecological networks make it possible for ecosystems to move away from thermodynamic equilibrium and thereby gain work energy storage by the prevailing conditions for the ecosystems including the limitations of the resources available for the ecosystems to grow and develop.

The advantages of network aggradation (Patten, 1991) are illustrated in Figure 5.3. Steady state is presumed for this simple example or input = output and the flows are donor determined by a first-order reaction. The flows and storage could represent the work energy (exergy; if the work energy of information is included it would be eco-exergy) of the simple system. The outflows from the compartments represent the respiration or the energy needed for maintenance of the compartment (the organisms). The retention time is five time units, which entails that the work energy (exergy) stored in the compartments is five times the inputs. As seen, the throughflow and the work energy increase because of the coupling; reference to Figure 5.3.

The assumptions that are the basis for the illustration, are often applied in ecological modeling and give, in many cases, a close reality image of the flows and storage.

Figures 5.4 through 5.7 show similar results for slightly more complex networks. The examples in Figures 5.4 through 5.7 show that the formation of networks clearly give a better utilization. It is presumed in the examples that eco-exergy as input is driving the cycling, but it would also be possible to follow matter or total energy or even information. From the examples it can be seen that more couplings or more recycling by additional connections in the network increases the utilization efficiency. Network formation offers a great advantage for the energy and matter utilization of resources by ecosystems.

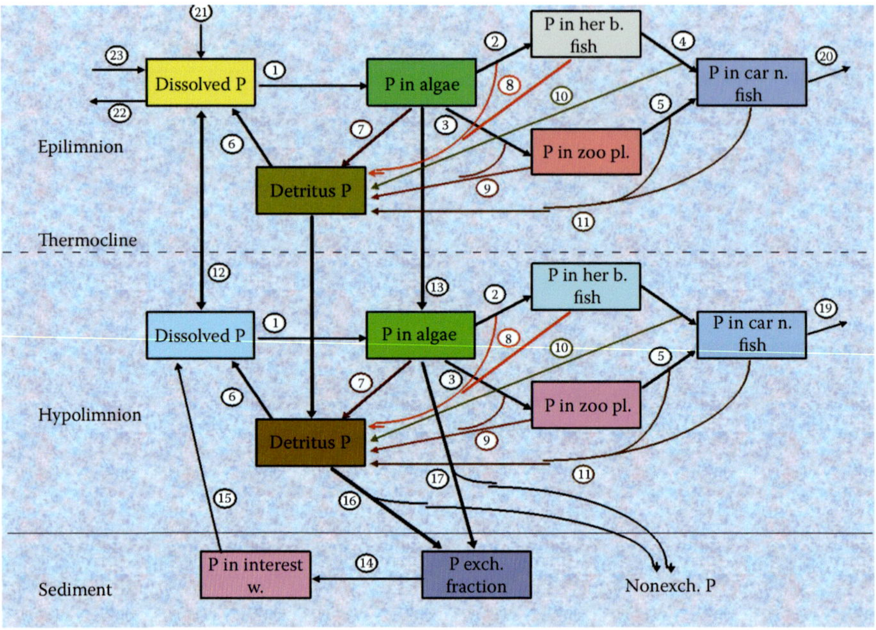

FIGURE 5.1
Deeper lakes have comprehensive networks, because the lake is divided by the thermocline into epilimnion and hypolimnion.

It should, however, be remembered that the network examples are all based on a steady state, but ecosystems *are* often at steady state or close to steady state. First-order reactions, which are also assumed, are also often valid for ecological processes (see for instance, Jørgensen and Fath, 2011), at least in a narrow range. The Michaelis–Menten equation which is often used to represent ecological processes corresponds for small concentrations to a first-order reaction. The retention time in Figure 5.3 is five time units, which implies that the sum of the compartments is 20% of the total throughflow. The retention time in Figures 5.4 through 5.7 is one time unit, which means that the sum of the compartments becomes equal to the total throughflow. In any case, observations on how ecological networks are working in ecosystems confirm that the networks can explain recycling, reuse, and the relative high utilization efficiency that is observed in ecosystems (Jørgensen et al., 2007; Jørgensen, 2012). The network idea has due to the high utilization efficiency been adopted in industries and agriculture (see for instance Jørgensen, 2006). The sum of the compartments is correlated strongly with the throughflow. It is in accordance with ELT: ecosystems move away from thermodynamic equilibrium and obtain thereby most possible eco-exergy stored in the components and the maximum throughflow (maximum power).

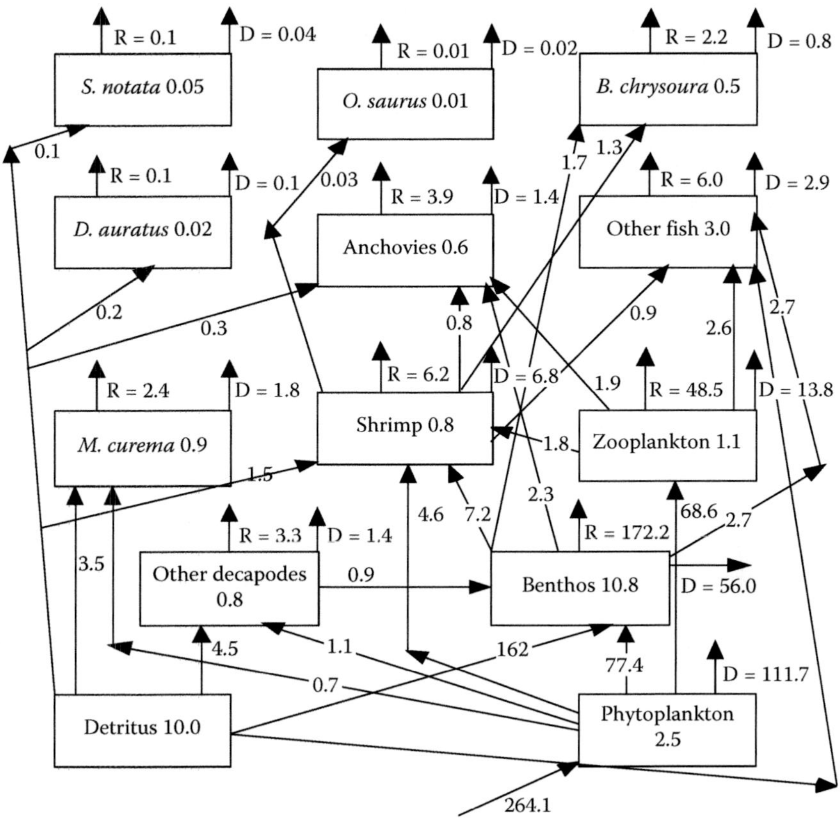

FIGURE 5.2
Example of a marine Ecopath model, taken from Christensen and Pauly (1993). R means respiration and D the transfer to detritus. Notice that all the components are in a steady state: input = output. The unit for the state variables is g/m^2 and for the processes it is $g/(m^2y)$.

5.2 Network Analyses

The network diagrams (Figures 5.3 through 5.7) are denoted diagraphs and to illustrate the most simple network analyses, we will use the diagraph in Figure 5.7. The first applied matrix analysis is the adjacency matrix, which indicates the realized pathways. The realized pathways are marked by "1" in the matrix and the nonrealized pathways by "0." The columns indicate "from" compartment 1, 2, 3, and 4 and the rows indicate to compartment 1, 2, 3, and 4. If a direct arc from j to i exists, then $a_{ij} = 1$. Direct arcs in Figure 5.7 are from 1 to 2, from 2 to 3, from 3 to 1 and 4, and from 4 to 1.

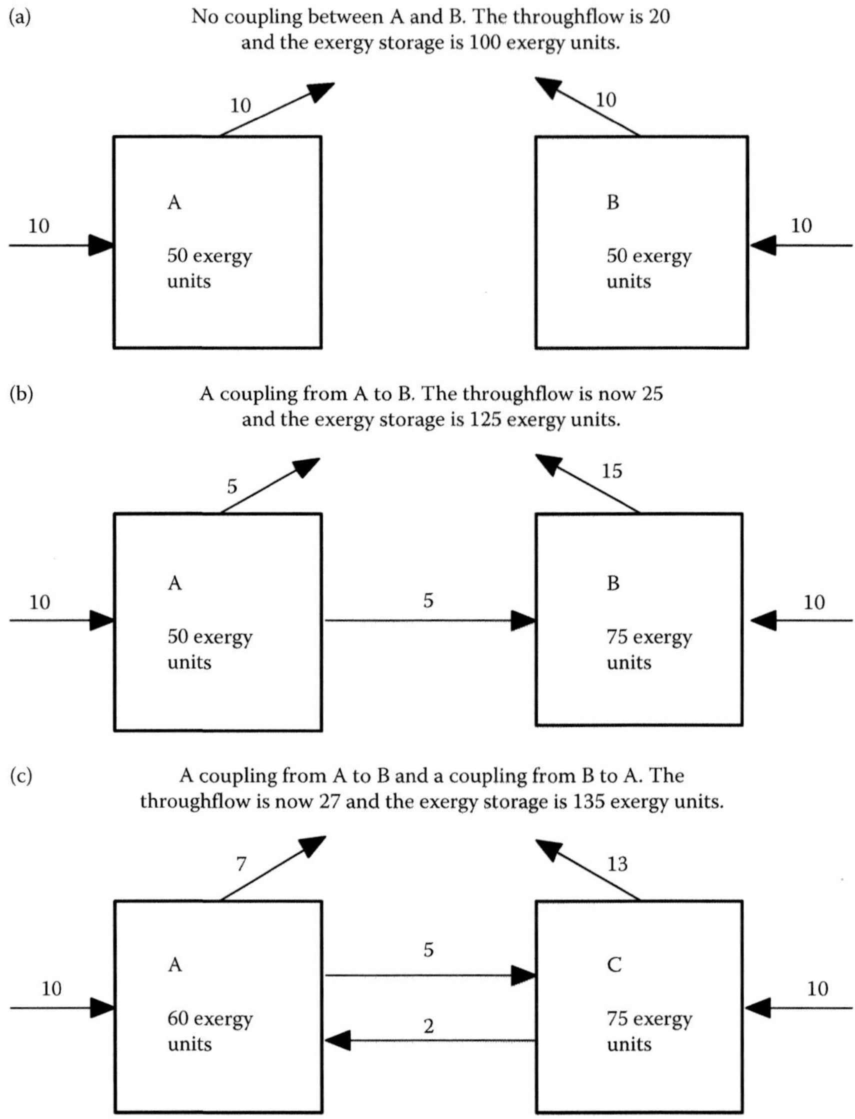

FIGURE 5.3
The two compartment system illustrates the advantages of network aggradation. The eco-exergy and throughflow increase from no coupling, to one coupling, to cyclic coupling. (a) Total eco-exergy is 100 and the throughflow is 20. Inputs and outputs are 10 for both A and B. They are usually denoted as input environ and output environ to indicate that they are the exchange between the system and its environment. (b) Total eco-exergy is 125 and the throughflow is 25. (c) Total eco-exergy is 135 and the throughflow is 27.

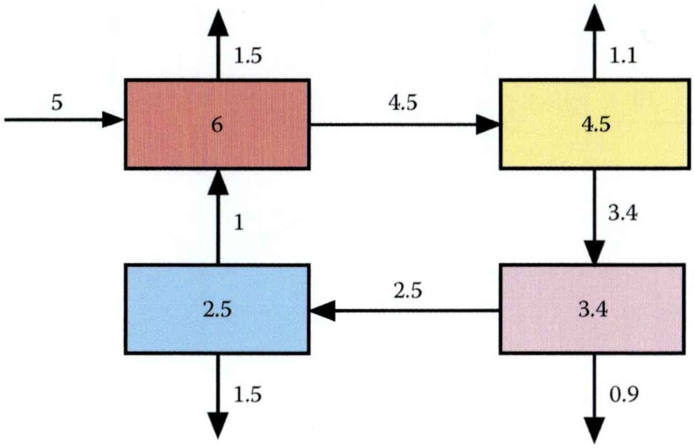

FIGURE 5.4
The network is at a steady state and the flows are first-order donor determined. The double input of eco-exergy or energy is able to support twice as much storage in all four compartments.

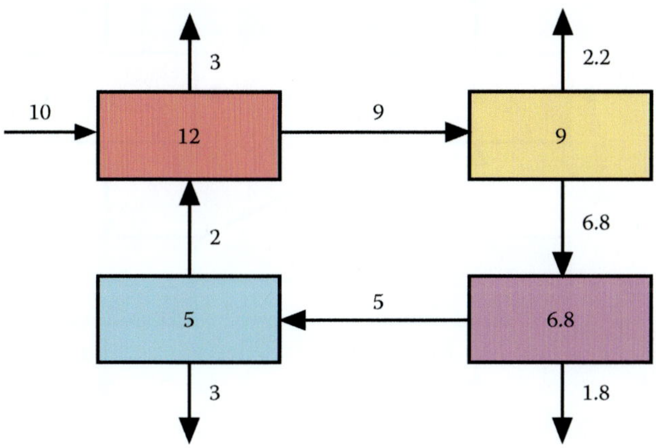

FIGURE 5.5
The difference between Figure 5.4 and this figure is only the input which has been changed from 5 to 10.

The adjacency matrix for Figure 5.7, A, is shown in Figure 5.8. This matrix allows us to calculate the connectivity, which is the fraction of all possible pathways that are realized. The possible pathways $4 * 4 - 4 = 12$ and 5 are realized which means the connectivity is $5/12 = 0.41$. A connectivity >0.5 usually entails a rigid network, while a connectivity of <0.25 or 0.3 makes a too loose network. A connectivity between 0.3 and 0.5 is preferred. The adjacency matrix indicates the presence of a direct connection by "1," and matrix A^2 gives the number of pathways that take two steps between two compartments. Correspondingly, A^3 gives the number of pathways that take

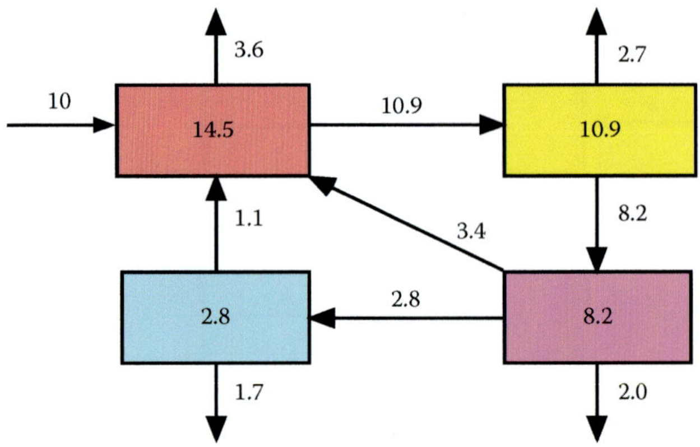

FIGURE 5.6
Figure 5.6 has introduced an extra cycling from compartment 3 to compartment 1. The result is that the eco-exergy, which is 32.8 in Figure 5.5, increases to 36.4.

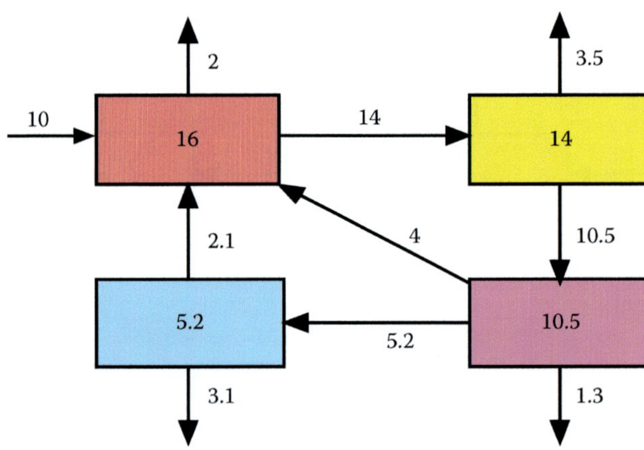

FIGURE 5.7
This diagram presumes that the first compartment represents bigger organisms than in Figure 5.6, which implies that the relative loss by respiration is reduced. Therefore, more eco-exergy is available to support the other compartment and recycle. The result is a total eco-exergy of 45.7.

exactly three steps and so on. Therefore, A^m gives the number of pathways of the length, m.

F is the flow matrix. 1s in A are replaced by the actual flows. The reuse of matter or energy in the network is important for the network properties. The yellow compartment in Figure 5.7 receives 14 units but only uses 3.5 for maintenance, while the remaining 10.5 units are reused in the network. This means that 10.5/14 = 0.75 or 75% of the matter or energy will be reused and will eventually return to the yellow compartment. We distinguish between direct (where the causes and the effects are adjacent) and indirect effects. Any transfer of energy or matter propagates and cycles many times in the

$$A = \begin{vmatrix} 0 & 0 & 1 & 1 \\ 1 & 0 & 0 & 0 \\ 0 & 1 & 0 & 0 \\ 0 & 0 & 1 & 0 \end{vmatrix}$$

$$F = \begin{vmatrix} 0 & 0 & 4 & 2.1 \\ 14 & 0 & 0 & 0 \\ 0 & 10.5 & 0 & 0 \\ 0 & 0 & 5.2 & 0 \end{vmatrix}$$

$$T = \begin{vmatrix} 16.1 & 14 & 10.5 & 5.2 \end{vmatrix}$$

$$G = \begin{vmatrix} 0 & 0 & 0.38 & 0.31 \\ 0.87 & 0 & 0 & 0 \\ 0 & 0.75 & 0 & 0 \\ 0 & 0 & 0.49 & 0 \end{vmatrix}$$

FIGURE 5.8
The adjacency matrix—A, the flow matrix—F, the throughflow vector—T, and the nondimensional or normalized direct flow matrix—G for the diagraph in Figure 5.7 are shown.

network before it is finally dissipated as heat used for maintenance for the environment. It is therefore not surprising that relatively complex calculations are needed to express the role of indirect effects compared with the direct ones. These calculations can be carried out by EcoNet, which is accessible on internet. Let us however make some simple calculations to illustrate the indirect effect. If we consider for the yellow compartment in Figure 5.7 that 75% of the matter or energy is reused in the second cycle, then 75% of the 75% or 56.25% will be reused in the third cycle, and 75% of 56.25% in the fourth cycle, or 42.2% and so on. Notice, however, that we have simplified the effect by not considering other inputs or outputs. It means that

The indirect effect after 8 cycles direct effect = direct effect (5.1)
$(0.75 + 0.5625 + 0.422 + 0.3165 + 0.237 + 0.178 + 0.141) = 2.6 * \text{direct effect}$

The indirect effects calculated by EcoNet consider an infinite number of cycles and will therefore of course indicate a higher ratio of indirect to direct effects, but these simple calculations show clearly that indirect effects in many cases will be several times the direct effects. The calculations are of course more complicated particularly when we consider many inputs and outputs, but it is here where the software EcoNet can help us to make the calculations considering these complications. The mathematics behind these calculations will not be presented in this textbook, because they are complex and not needed to understand how ecosystems are working and which benefits the ecosystems gain by the network organization.

T gives the throughflow of the four compartments. Also note that the throughflow is 45.8 and the storage (= the sum of the compartments) is 45.7. The retention time is one unit of time. The difference is due to rounding off errors. With the first-order expression applied for the respiration and the steady state with the retention time one unit, the storage (for instance, eco-exergy stored) and the power (= the total throughflows) should be the same. Correlations between eco-exergy or work energy storage and the maximum power (= the sum of the throughflows) are in accordance with ELT hypothesis: the ecosystem strives toward highest possible eco-exergy storage under the prevailing conditions, but highest possible eco-exergy storage also implies the maximum power is reached (see Chapter 3).

The information contained in the F- and G-matrices could be applied as inputs to EcoNet, which is programmed to carry out the complicated calculations needed to reveal all the relevant ecological network properties. The F-matrix gives the direct flows, from compartment 1 in the first column and for compartment 2 in the second column and so on and to compartment 1 in the first row and so on. The arc from j to i is indicated as a_{ij}. The G-matrix indicates the flow relative to the donor storage. It means that for instance the flow 14 in Figure 5.7 is in the G-matrix indicated as 14/16 or 0.87; see Figure 5.8.

The software does have extended results which include control analysis and an eco-exergy calculator. The analysis data can be downloaded in MATLAB format (.m file) or in spreadsheet format (.csv file). Overall, the EcoNet software provides a very easy to use application for researchers or students interested in conducting network analysis. The results of using EcoNet on the network in Figure 5.7 and with the corresponding matrices in are shown in Table 5.1. The software is also able to calculate the eco-exergy (work energy including information), but it would of course require that the β-values have to be indicated for the compartment. The software also explains the background knowledge (press the buttons "Learn More") about all the results and particularly the system-wide properties are important to read and understand in this context. An explanation of the system properties is provided in Table 5.2. For more information about the software, refer to Kazanci (2007, 2009).

Notice particularly the following results in Table 5.1:

1. The indirect effects are more than 3 times the direct ones.
2. The aggradation index is 4.58, meaning that the total system throughflow (TST) and the work energy storage are 4.58 times the inflow.
3. The synergism index is 6.73.
4. The mutualism index is 2.2.

TABLE 5.1

The Results Obtained by the Application of EcoNet on Figure 5.7

Network diagram

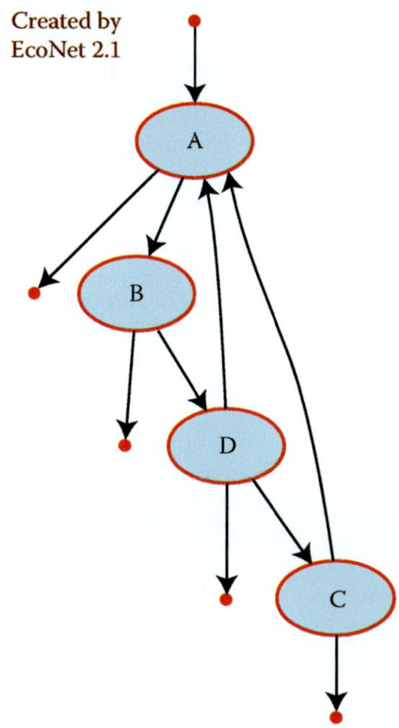

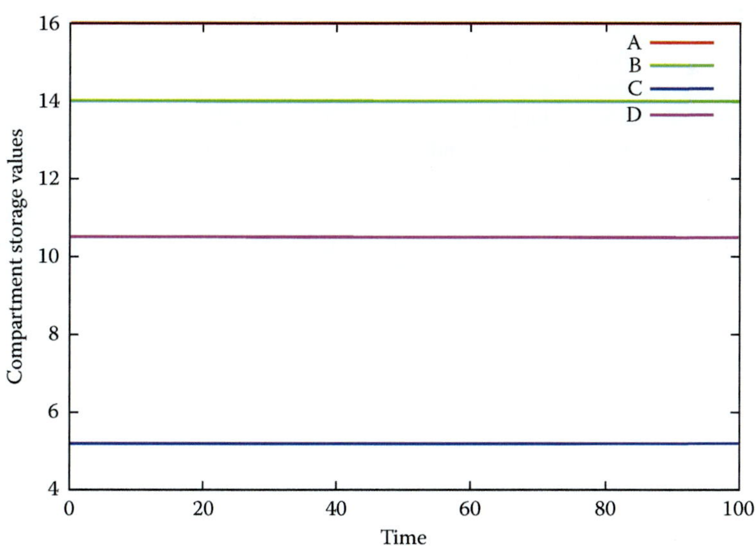

Data for time-course figure

(Continued)

TABLE 5.1 (*Continued*)

The Results Obtained by the Application of EcoNet on Figure 5.7

Compartmental Properties

–	Initial Storage	Final Storage (x)	Environmental Input (z)	Environmental Output (y)
A	16	16	10	2.1
B	14	14	0	3.5
C	5.2	5.2	0	3.1
D	10.5	10.5	0	1.3

–	Input Throughflow	Output Throughflow	Residence Time
A	16.1	16.1	0.993789
B	14	14	1
C	5.2	5.2	1
D	10.5	10.5	1

System-Wide Properties

Link Density	1.25
Connectance	0.3125
Total System Throughflow	45.8
Finn's Cycling Index	0.355569
Indirect Effects Index	3.03594
Ascendency	58.5963
Development Capacity	86.4408
Aggradation Index	4.58
Synergism Index	6.73482
Mutualism Index	2.2
Homogenization Index	3.92417

Adjacency Matrix (A)

–	A	B	C	D
A	0	0	1	1
B	1	0	0	0
C	0	0	0	1
D	0	1	0	0

Stoichiometric Matrix (SM)

–	* → 1	1 → 2	2 → 4	4 → 3	4 → 1	3 → 1	1 → *	2 → *	3 → *	4 → *
A (1)	1	–1	0	0	1	1	–1	0	0	0
B (2)	0	1	–1	0	0	0	0	–1	0	0
C (3)	0	0	0	1	0	–1	0	0	–1	0
D (4)	0	0	1	–1	–1	0	0	0	0	–1

(*Continued*)

TABLE 5.1 (*Continued*)

The Results Obtained by the Application of EcoNet on Figure 5.7

Flow Matrix (F)

–	A	B	C	D
A	0	0	2.1	3.99999
B	14	0	0	0
C	0	0	0	5.2
D	0	10.5	0	0

Throughflow Analysis

Normalized Flow Matrix (G)

–	A	B	C	D
A	0	0	0.403846	0.380952
B	0.869565	0	0	0
C	0	0	0	0.495238
D	0	0.75	0	0

Throughflow Analysis (N)

–	A	B	C	D
A	1.61	0.701499	0.650192	0.935332
B	1.4	1.61	0.565384	0.813332
C	0.52	0.598	1.21	0.797333
D	1.05	1.2075	0.424038	1.61

Normalized Flow Matrix (G′) Output Oriented

–	A	B	C	D
A	0	0	0.130435	0.248447
B	1	0	0	0
C	0	0	0	1
D	0	1	0	0

Throughflow Analysis (N′) Output Oriented

–	A	B	C	D
A	1.61	0.609999	0.21	0.609999
B	1.61	1.61	0.21	0.609999
C	1.61	1.61	1.21	1.61
D	1.61	1.61	0.21	1.61

(*Continued*)

TABLE 5.1 (*Continued*)

The Results Obtained by the Application of EcoNet on Figure 5.7

Storage Analysis

Partial Turnover Rate Matrix (C)

–	A	B	C	D
A	–1.00625	0	0.403846	0.380952
B	0.875	–1	0	0
C	0	0	–1	0.495238
D	0	0.75	0	–1

Storage Analysis (S)

–	A	B	C	D
A	1.6	0.697142	0.646153	0.929523
B	1.4	1.61	0.565384	0.813332
C	0.52	0.598	1.21	0.797333
D	1.05	1.2075	0.424038	1.61

Partial Turnover Rate Matrix (C′) Output Oriented

–	A	B	C	D
A	–1.00625	0	0.13125	0.25
B	1	–1	0	0
C	0	0	–1	1
D	0	1	0	–1

Storage Analysis (S′) Output Oriented

–	A	B	C	D
A	1.6	0.609999	0.21	0.609999
B	1.6	1.61	0.21	0.609999
C	1.6	1.61	1.21	1.61
D	1.6	1.61	0.21	1.61

Utility Analysis

Mutual Relations

–	A	B	C	D
A	+	–	–	+
B	+	+	+	–
C	–	+	+	+
D	+	+	–	+

(*Continued*)

TABLE 5.1 (*Continued*)

The Results Obtained by the Application of EcoNet on Figure 5.7

Utility Series Matrix (D)

–	A	B	C	D
A	0	–0.869565	0.130435	0.248447
B	1	0	0	–0.75
C	–0.403846	0	0	1
D	–0.380952	1	–0.495238	0

Utility Analysis (U)

–	A	B	C	D
A	0.646781	–0.265404	–0.0627299	0.297014
B	0.469825	0.473169	0.119862	–0.118288
C	–0.0252585	0.455752	0.781877	0.433788
D	0.235941	0.348569	–0.243456	0.553736

Control Analysis

Control Analysis I (CD)

–	A	B	C	D
A	0	0.564286	0.596154	0.419048
B	0	1.11022e-16	0.64883	0.494824
C	0	0	0	0
D	0	0	0.736622	0

Control Analysis II (CR)

–	A	B	C	D
A	0	–0.0564286	–0.0596154	–0.0419048
B	0.0564286	0	–0.0746154	–0.0569048
C	0.0596154	0.0746154	0	0.112949
D	0.0419048	0.0569048	–0.112949	0

In other words, despite Figure 5.7 being a simple ecological network, it provides significant advantages to the participants of the ecological network. It is very beneficial for an ecosystem to enhance the growth of the ecological network and increase the efficiencies, expressed by the various indices, of the network. By growth form III (see Section 3.7), there are enormous possibilities for the ecosystem to move away from the thermodynamic equilibrium.

Adjacency Matrix: Adjacency matrix A consists of 0 and 1 entries and indicates if there exists a direct flow between two stocks. See Figure 5.8 and the explanation in the text for this figure.

TABLE 5.2

Explanation of the Results Obtained by EcoNet Applied on Figure 5.7 (EcoNet is also given, the explanations with indication of relevant references.)

System-Wide Properties

Some of these indexes depend on the analysis matrices defined below.

1. *Link Density:* Number of intercompartmental links (d) per compartment: d/(Number of compartments).
2. *Connectance:* Ratio of the number of actual inter-compartmental links (d) to the number of possible intercompartmental links: d/(Number of compartments).[2]
3. *Total System Throughflow:* Also called Total System Throughput, it is the sum of throughflows of all compartments: $TST = T_1 + T_1 + \cdots + T_n$.
4. *Finn's Cycling Index:* Measures the amount of cycling in the system by computing the fraction of total system throughflow that is recycled.
5. *Indirect Effects Index:* Measures the amount of flow that occurs over indirect connections versus direct connections. When the ratio is greater than 1, indirect flows are greater than direct flows.
6. *Ascendency:* It quantifies both the level of system activity and the degree of organization (constraint) with which the material is being processed in ecosystems.
7. *Aggradation Index:* Measures the average path length. In other words, it is the average number of compartments a unit flow quantity (e.g., an N atom, unit biomass, energy quanta, etc.) passes through before exiting the system: $TST/(z_1 + z_2 + \ldots + z_n)$.
8. *Synergism Index:* Based on utility analysis, it provides a system-wide index for pairwise compartment relations. Values larger than 1 indicate a shift toward positive interactions (mutualism). It is computed as the ratio of sum of positive entries over the sum of negative entries in the utility analysis matrix U.
9. *Mutualism Index:* Similar to the synergism index, mutualism index provides a system-wide index for pairwise compartment relations. Values larger than 1 indicate a shift toward positive interactions (mutualism). It is computed as the ratio of number of positive entries over the number of negative entries in the mutual relations matrix.
10. *Homogenization Index:* Quantifies the action of the network making the flow distribution more uniform. Higher values indicate that resources become well mixed by cycling in the network, giving rise to a more homogeneous distribution of flow.

Stoichiometric Matrix: Entries of the stoichiometric matrix can be 0, –1, and 1. The entry (i, j) indicates how the storage value of compartment i change when flow j occurs.

Note that the adjacency matrix contains only inter-compartmental relationships. Stoichiometric matrix contains the flows between the system and the environment as well. Therefore the stoichiometric matrix alone is enough to define network topology.

PROBLEMS

1. Mention, in addition to ecosystems, other systems which use the advantages of networks.
2. Explain how industrial systems could benefit from networks.
3. Explain how agricultural systems could benefit from networks.

4. Develop an adjacency matrix for Figure 5.1.
5. Explain under which circumstances TST and eco-exergy are closely correlated. Explain why TST or power as denoted by Lotka and Odum generally under all circumstances should be more or less correlated.
6. Explain why networks which have larger biological components result in higher eco-exergy and power.

References

Jørgensen, S.E. 2006. *Eco-Exergy as Sustainability*. Southampton: WIT Press, 220 pp.

Jørgensen, S.E. 2012. *Introduction to Systems Ecology*. (Chinese edition.) Boca Raton, FL: CRC Press, 320 pp.

Jørgensen, S.E. et al., 2007. *A New Ecology. Systems Perspectives*. Amsterdam: Elsevier, 288 pp.

Jørgensen, S.E. and B. Fath. 2011. *Fundamentals of Ecological Modelling*. 4th edition. Amsterdam: Elsevier, 400 pp.

Kazanci, C. 2007. EcoNet: A new software for ecological modeling, simulation and network analysis, *Ecol. Model.* 208(1), 3–8.

Kazanci, C. 2009. Network calculations II: A user's manual for EcoNet. In Jørgensen, S.E., Chon, T.S., and Recknagel, F. (Eds.), *Handbook of Ecological Modelling and Informatics*, Chapter 18. Southampton: WIT Press, 432 pp.

Patten BC. 1991. Network ecology: Indirect determination of the life–environment relationship in ecosystems. In Higashi, M. and Burns, T.P. (Eds.), *Theoretical Ecosystem Ecology: The Network Perspective*. London, Cambridge: University Press, pp. 288–351.

Part III

Sources and Causes

6

The Pollution Sources and Effects

6.1 Pollution Sources

This section provides an overview of pollution forms. We distinguish between point sources and nonpoint sources and can identify 13 forms of pollution. The impact on our environment is mainly due to three factors: the increasing population, the increasing consumption and production by us, and our use of technology, which can increase or decrease the impact depending on the technology.

Pollution is defined as the introduction of contaminants into the natural environment that cause adverse and undesirable changes in the ecosystems. Pollution can either be matter such as chemical substances or energy, such as noise, heat, or light. Pollutants, the components of pollution, can either be human-controlled—it means as a result of our activities—or naturally occurring contaminants resulting from natural events, for instance a volcanic eruption. It is very important to distinguish between point source and nonpoint source pollution, because the two sources will in most cases require different tool boxes to be eliminated. Environmental technology is developed to solve point sources pollution and will in far most cases not be applicable to nonpoint pollution, whereas ecological engineering is better able to cope with nonpoint pollution. Therefore, it is important to consider the entire spectrum of possible solutions for each of the sources. Point sources are, for instance, discharge from wastewater treatment plants, discharge of wastewater from industries, smoke from a chimney, or ash from a solid waste incineration plant. Nonpoint sources are diffuse and can be exemplified by the agricultural pollution by drainage water containing pesticides and nutrients or by storm water. Typical analyses of storm water and agricultural drainage water are shown in Table 6.1.

The pollution is often classified by the sphere that is polluted. When the hydrosphere is contaminated, it is denoted as water pollution. When the atmosphere is contaminated, we call it air pollution. Terrestrial pollution or sometimes called soil pollution is the pollution of the lithosphere. There are, however, often pollutants that are contaminants in all the three spheres and in addition in the biosphere, where the effect often is most radical.

TABLE 6.1

Pollution from Urban and Agricultural Runoff

Constituent	Storm Water	Agricultural Drainage Water
Suspended solids (mg/L)	5–1200	5–5000
Chemical oxygen demand (COD)[a] (mg/L)	20–700	5–1000
Biological oxygen demand (BOD5)[b] (mg/L)	1–200	1–300
Total phosphorus (mg/L)	0.02–8	0.1–2.0
Total nitrogen (mg/L)	0.3–8	0.5–10
Chlorides (mg/L)	1–35	—

[a] COD = Oxygen demand is determined by the use of chromate.
[b] BOD5 = Oxygen used by microbiological decomposition over a period of 5 days.

The major forms of pollution are covered with the following list:

1. Common gaseous pollutants include carbon monoxide, sulfur dioxide, chlorofluorocarbons (CFCs), and nitrogen oxides produced by industries, power plants, and motor vehicles.
2. Photochemical ozone and smog are caused as nitrogen oxides and hydrocarbons react to sunlight. The sources are industry and vehicles.
3. Particulate matter or fine dust is characterized by the size, contaminating the atmosphere and deteriorating the air quality.
4. Light pollution includes light trespass, over-illumination, use of laser, and astronomical interference.
5. Noise pollution encompasses roadway noise, aircraft noise, industrial noise, as well as high-intensity sonar.
6. Radioactive contamination, resulting from nuclear power generation and nuclear weapons research, manufacture and deployment of the weapons, and from the use of radioactive isotopes.
7. Thermal pollution is a temperature change in natural water bodies caused mainly by use of water as a coolant in a power plant.
8. Soil contamination occurs when chemicals are released by spill or underground leakage. The possible contaminants are listed below as toxic substance pollution.
9. Visual pollution, which can refer to the presence of overhead power lines, motorway billboards, scarred landforms (as from strip mining), open storage of trash, littering, municipal solid waste, or space debris. It could also be named as landscape pollution.
10. Pollution by organic matter of water causing oxygen depletion due to the use of oxygen mainly for biological decomposition of the

organic matter. Note that this pollution is a result of the two sources: nonpoint sources (agricultural drainage water and storm water) and insufficiently treated wastewater.

11. Pollution of aquatic ecosystems (surface waters) by nutrients causing eutrophication. The nutrients cause an extensive growth of algae (or other plants), which reduce the transparency of the water, cause oxygen depletion in the bottom water, and/or in the fall and winter. Note that this pollution is a result of mainly two sources: nonpoint sources (agricultural drainage water and storm water) or insufficiently treated wastewater.
12. Groundwater pollution covering the disposal and leaching of nutrients and toxic substances into groundwater. Note that this pollution is often the result of an uncontrolled use of nutrients and pesticides in agricultural production.
13. Toxic substance pollution which can cause adverse effects in all spheres include both the soil (see point 8) and the biosphere. Later in this chapter, a short overview and classification of the toxic substances will be presented.

In accordance to the proposed procedure in Chapter 1, it is important to quantify *all* possible pollution sources. Quantification is the key, because only by having the knowledge of the quantities of all the sources, we can with the help of diagnostic tools (see the next part, Part IV) find the sources necessary to be eliminated or reduced to obtain the needed and desired impact reduction. The mass conservation principles are often used to determine the quantities of the various pollution sources. From emissions at the discharge point to the impact on the ecosystem, some processes may have changed the concentrations and the amounts of the polluting components. For instance, from the emission of gases from a chimney to the dry deposition in an ecosystem, chemical transformations and dilution processes may have occurred.

All these pollution sources and problems are rooted in urbanization, industrialization, inappropriate environmental planning or application of available solutions, and natural disasters. It is, however, due to human activities, that most pollution problems occur. The increased impacts of environmental problems are predominantly due to increased urbanization and industrialization. During the last few decades, we have observed a distinct increase in pollution. Many examples illustrate these observations. The concentrations of many toxic substances have increased in soil, water, and the biosphere. The concentrations of carbon dioxide, sulfur dioxide, and other gaseous pollutants have increased drastically in the atmosphere. Oxygen depletion has been recorded in many rivers and many aquatic ecosystems are suffering from eutrophication with high concentrations of nitrogen and phosphorus. What has caused this almost sudden increase in pollution? The answer is not simple, but Ehrlich and Raven (1969) have proposed a crude but useful model. They claim that the total environmental impact, I, is mainly determined by three multiplicative factors: the population size, P, the consumption per capita, C, and the environmental impact per unit of consumption, which is dependent on the technology, T:

$$I = P * C * T \tag{6.1}$$

It is understandable that the impact is proportional to P, as twice as many people with the same consumption and technology of course will cause twice as large an impact. Similarly, if we consume and produce twice as much, the impact will inevitably be twice as large. The factor T is more difficult to find and manage because on one side, for instance, a technological development leading to production of more harmful chemicals increase the impact, whereas development of more recycling and reuse will lead to less impact. Development of, for instance, renewable energy will decrease the carbon dioxide emission and thereby decrease the total impact. P has increased by a factor of 2 from year 1960 to year 2000—from 3 to 6 billion. The global GNP/capita—gross national product per capita—which reflects the consumption and the production has also increased by a factor of about 2 from 1960 to 2000. It means that the global impact has increased four times from 1960 to 2000, presuming that T has not changed. We produce more harmful chemicals today than 50 years ago, but we have also developed technology which emits less pollution and applies more recycling. However, an increased impact of four times (400%) in 40 years can only be considered to be a very rapid increase, calling for pollution abatement.

The next three sections are devoted to the effects of the various forms of pollution sources. Section 6.2 gives an overview of the possible effects. Sections 6.3 and 6.4 present the effects of inorganic and organic compounds, respectively.

6.2 Overview of Pollution Effects

Pollution has many adverse effects, and it is the task of the environmental management to eliminate and reduce the same. By the use of the diagnostic tools, we can find the relationship between the environmental quality and the impact, or expressed differently, between the reduction of the various sources of pollution and the environmental quality. Consequently, it is crucial to understand the relationships among the quantity of pollution and the effect on ecosystems as well as human health.

It is not easy to achieve a full understanding of all the effects of pollution, because they are complex and most pollution problems have several possible effects that are interacting and valid, simultaneously. The diagnostic tools can help in this context to overview the pollution problems and their effects. First, a list of effects can be used as a checklist. The possible effects are as follows:

1. Effect on the human health. This effect is mainly observed by pollution forms: 1 + 6 + 8 + 12 + 13.
2. Oxygen depletion in aquatic ecosystems. It has a very significant effect as the aquatic organisms are solely dependent on sufficient oxygen level, corresponding to their needs. For instance, Salmon fish species require at least 6–7 mg oxygen/L. The sources of pollution are mentioned in point 10. Oxygen depletion is a very clear indicator of deterioration of the environmental quality of aquatic ecosystems.
3. Eutrophication caused by high nutrient concentrations. Plant growth may be limited in accordance with Liebig's law, which implies that discharge of nutrients into aquatic ecosystems inevitably causes extensive plant growth which is named eutrophication. Particularly for lakes and coastal areas it is a major problem, because the extensive algal growth entails reduced transparency and the plant biomass as well as organic matter is decomposed microbiologically, which reduces the oxygen concentration significantly, in the bottom region of the water source and/or during the fall and winter. The amount of oxygen produced by photosynthesis is approximately equal to the amount of oxygen used for decomposition of the organic matter; but the problem is that the production and the consumption of oxygen occur in different places at different times. The pollution sources are mentioned as point 11. Eutrophication causes an unacceptable deterioration of aquatic ecosystems.

Example 6.1 illustrates how to curb deterioration of the aquatic ecosystems: Which nutrient is limiting the eutrophication in a focal aquatic ecosystem and which pollution abatement methods should therefore be used in combination to achieve the best results? It is a matter of understanding the ecosystem properties. Phosphorus can be removed from wastewater at a cost which is roughly 10 times lower than that for the removal of nitrogen; for further details, see Chapter 11. Nutrient removal from drainage water is possible with the use of natural or constructed wetlands. For further details, see Chapter 12.

EXAMPLE 6.1

Example 3.1 shows how to determine the limiting nutrient. The question is, however, not which nutrient is limiting, but which nutrient can easily (most cost-moderate) be made limiting.

If R is for instance 3, the ratio of nitrogen to phosphorus is 4 in the wastewater and 25 in the drainage water, which limiting nutrient would it be *beneficial* to control to reduce the eutrophication? Phytoplankton has 7 times more nitrogen than phosphorus.

Solution

1 kg P in the drainage water would correspond to 25 kg N, and 3 × 1 kg P and 3 × 4 = 12 kg N in the wastewater. Therefore, without further treatment of the wastewater, 4 kg P and 37 kg N indicate the ratio of these two nutrients and the ratio is 9.25. P is therefore the limiting nutrient already, but if we remove 90% for instance of the P in the wastewater, the ratio would be 1.3 kg P to 37 kg N and the reduction in the discharge of P will limit the eutrophication correspondingly. If the peak phytoplankton concentration is proportional to the total P concentration, it would be reduced by a factor of 4/1.3 = 3.08, which is a significant reduction and the cost to obtain this result would be limited. Consequently, it would be beneficial to try to limit the P concentration in the lake further. It is possible to reduce the phosphorus concentration in wastewater further for instance 99%, but it would mean that the P concentration in the lake is reduced relatively from 1.3 to 1.03. The phosphorus concentration in the drainage water can be reduced by ecotechnology and a reduction by 90% is possible; see Chapter 12.

4. Contamination of soil by toxic substances. It implies that the land cannot be utilized for any related activities or it requires a very expensive purification process. The plants and animals living in the area will inevitably be affected by the toxic substance. Particularly high concentrations may occur in some organisms depending on the processes in the ecosystem and the ecological network; see Section 3.2. The pollution source is mentioned as point 8.
5. Toxic substances have furthermore a pronounced effect on aquatic ecosystems, of course dependent on the resulting concentration. The plants and animals living in the area will inevitably be affected by the toxic substance. Particularly high concentrations may occur in some organisms, which is dependent on the processes in the ecosystem and the ecological network; see Section 3.2. The pollution source is mentioned as point 13 and is parallel to pollution problem source 8 for soil.
6. An increased temperature of aquatic ecosystems is resulting from pollution problem number 7. A temperature change will always influence the organisms and thereby the resulting biological and chemical composition of the ecosystem. By more radical temperature changes, a major part of the organisms inhabiting the ecosystems may be replaced by other organisms.
7. Global warming, that is a result of an increased concentration of carbon dioxide and other greenhouse gases in the atmosphere, will similarly have an effect as point 6, be able to change radically the biological and chemical composition of ecosystems. The pollution problem source is number 1.
8. Effects 2–7 all imply a change of the ecology of the ecosystems. It will be able to change the ecological balance, the diversity, and the

resistance of the ecosystems (see also Section 3.12) and the ecological services utilized by us and offered by of the ecosystems.

In Section 3.11, it was discussed that ecosystems are organized hierarchically. The effects listed above (2–8) mainly focus on the ecosystem level, but all the effects may influence any of the hierarchical levels, which may of course have an impact on all the other levels as all the levels are connected. These considerations may best be illustrated by a toxic substance that in most cases influences the biochemistry of the cells and thereby the entire cell, maybe further on the organs and the organisms, which may again change the populations and therefore also the ecosystems. The influences may even be global, which is for instance the case with the global warming and the effects that change the global cycling of important elements.

Usually, we are interested in our effort to develop a proper integrated environmental management to uncover particularly the effects on:

A. The cell biochemistry—to be able to understand what quantities may cause the adverse effect, which biochemical pathways are involved, and how it would be possible by revealing the biochemical processes to neutralize the adverse effect.
B. The organism, because there are organisms that are more sensitive than others and are key organisms in their ecosystems and therefore of particular interest to be maintained in the ecosystems. The influence on the population dynamics should be determined. This effect may often be extrapolated to the population level.
C. The ecosystem level, because ecosystems are fundamental units and systems in nature, and loss of ecosystems is most often a disaster for the surrounding ecosystems and for human dependent on the ecosystem services.
D. The global level, because it may be a significant threat to the life on Earth. The climate and the global cycles are vital for the entire nature and the conditions of human life.

The effect is in addition to the concentration influenced strongly by the exposure time. Obviously, the longer the exposure, the bigger the effect. Further details are presented in Section 7.3. It is therefore necessary to know the effect as a result of interaction between these two main effect factors.

6.3 Effects of Inorganic Compounds

The present periodic table consists of 118 elements, but more than 20 elements placed at the end of the table are extremely unstable and have therefore no environmental interest with the exception of their radioactivity. Ninety-one of the elements are of natural occurrence, namely, the first 92 elements from hydrogen to uranium except element number 43, technetium. These 91 elements can be classified into four groups (the numbers before the elements indicate the number in the periodic table):

A. Elements that are present in biological material in relatively high concentrations: (1) hydrogen, (6) carbon, (7) nitrogen, (8) oxygen, (11) sodium, (12) magnesium, (14) silica, (15) phosphorus, (16) sulfur, (17) chlorine, (18) potassium, (19) calcium, and (26) iron.
B. Elements that are trace elements. They may often be present in biological material in small concentrations: (5) boron, (9) fluorine, (24) chromium, (25) manganese, (27) cobalt, (28) nickel, (29) copper, (30) zinc, (34) selenium, (42) molybdenum, and (53) iodine. Too high concentrations may, however, cause environmental problems.
C. Elements that have ecotoxicological effect. They are harmful even in very small concentrations and are therefore the elements that we have to consider in environmental management are ecotoxicological elements and compounds. They are the inorganic compounds that may cause pollution problems associated with a toxic effect. They are (13) aluminum, (33) arsenic, (38) strontium, (47) silver, (48) cadmium, (50) tin, (51) antimony, (52) tellurium, (56) barium, (80) mercury, (81) thallium, (82) lead, (83) bismuth, (84) polonium, (86) radon, (88) radium, (90) thorium, (92) uranium, and (94) plutonium. The last six elements are also radioactive in addition to being toxic. They are normally present in small concentrations, and the main concern is their radioactivity. Five elements belonging to group B, namely, chromium, copper, nickel, selenium, and zinc, may be the cause of ecotoxicological/environmental concern due to their frequent occurrence in harmful environmental concentrations. The overview in Section 6.3 includes these five elements and encompasses totally 18 elements.
D. All other elements have no or only minor environmental interest due to their occurrence in low concentrations or low biological effects.

Table 6.2 gives the composition of average freshwater plants on the wet weight basis. The table presents 19 elements out of the 24 elements that are included in groups A and B. The following elements are not included in Table 6.2: fluorine, chromium, nickel, selenium, and iodine, because they are present in very small concentrations in freshwater plants—below 0.000002%.

The elements in groups B and C show bioaccumulation and biomagnification, which are two important processes to be considered in all ecotoxicological evaluations.

TABLE 6.2

Average Freshwater Plant Elemental Composition on a Wet Weight Basis

Element	Plant Content (%)
Oxygen	80.5
Hydrogen	9.7
Carbon	6.5
Silica	1.3
Nitrogen	0.7
Calcium	0.4
Potassium	0.3
Phosphorus	0.08
Magnesium	0.07
Sulfur	0.06
Chlorine	0.06
Sodium	0.04
Iron	0.02
Boron	0.001
Manganese	0.0007
Zinc	0.0003
Copper	0.0001
Molybdenum	0.00005
Cobalt	0.000002

Source: Wetzel, R.G. 1983. *Limnology*. 2nd edition. New York: Saunders College Publishing, 828 pp.

The 24 elements in classes A and B encompass the elements that are needed for the entire biosphere. All 24 elements are not necessarily essential for all organisms, but each organism requires most of these 24 elements for their growth. These elements are threshold agents, see Jørgensen (2000). They are needed in a certain concentration for the growth of the organisms, but if they are present in too high concentrations they may be harmful for the environment. A typical example is the eutrophication problem, that is a result of too high concentrations of nutrients, but the nutrients (with emphasis on nitrogen, carbon, phosphorus, and sulfur) are absolutely necessary for all living organisms. We have included in the following list of elements five class B elements as mentioned above, because they are frequently present in varying concentrations in the environment that may cause ecotoxicological effects. Nonthreshold agents or gradual agents are harmful for the environment in practically any concentration. The harmful effect (group C elements) may be proportional to the concentration or there may be another relationship between the effect and the concentration, for instance, the harmful effect is proportional to the concentration in the exponent 2.

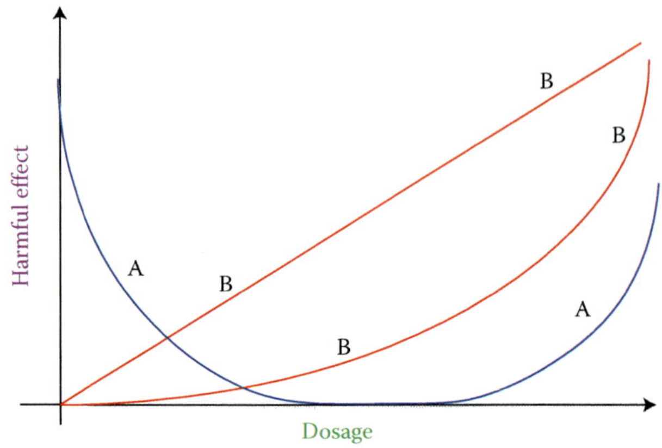

FIGURE 6.1
The harmful effect is plotted versus the dosage or concentration for (A) threshold agents and (B) nonthreshold agents.

Figure 6.1 illustrates the harmful effect versus the concentration for threshold and nonthreshold agents. An overview of group C—the radioactive elements + five of the elements in group B—all with a clear ecotoxicological effect—are given in this chapter with emphasis on the environmental problem caused by the elements or its compounds. Heavy metals are an important soil pollution problem. Three of the group C elements that are heavy metals require a particular environmental attention: mercury, lead, and cadmium, because they have caused very serious environmental and health problems at relatively low concentrations.

A wide variety of terms is used in an inconsistent and confusing manner to describe uptake and retention of pollutants by organisms using different paths and mechanisms. However, three terms are now widely applied and accepted for these processes:

1. *Bioaccumulation* is the uptake and retention of pollutants by organisms via *any* mechanism or pathway. It implies that both direct uptake from air and water and uptake from food are included.
2. *Bioconcentration* is the uptake and retention of pollutants by organisms directly from water through gills or epithelial tissue. This process is often described by means of a concentration factor, indicating the ratio of the concentration in the organism divided by the concentration in the medium.
3. *Biomagnification* is the process whereby pollutants are passed from one trophic level to another and it exhibits increasing concentrations in organisms related to their trophic status. The concentration in top-level carnivorous organisms may often be several magnitudes higher than that in the medium level.

6.4 The Pollution and Ecotoxicological Problems of 18 Elements

As already mentioned, there are 18 elements that require a particular attention in environmental management due to their ecotoxicological effect—see also Jørgensen and Fath (2010). They are (13) aluminum, (24) chromium, (28) nickel, (29) copper, (30) zinc, (33) arsenic, (34) selenium, (38) strontium, (47) silver, (48) cadmium, (50) tin, (51) antimony, (52) tellurium, (56) barium, (80) mercury, (81) thallium, (82) lead, and (83) bismuth. A short overview of the environmental and ecotoxicological problems of these 18 elements is provided below, including their applications and environmental effects.

Table 6.3 gives an overview of the concentration in the earth's crust, in average soil, and in seawater. If environmental concentrations are significantly higher than these concentrations in the earth's crust, in average soil, and in seawater are recorded, there is a high probability for an ecotoxicological effect. The last column of the table gives the most important applications of the elements and their compounds.

Table 6.4 provides the LD50 and LC50 values as expressions for the toxicity. The third column in this table indicates whether there are environmental effects and risks that require particular attention. For instance, arsenic can be found in unacceptably high concentration in some locations particularly in groundwater; see Murphy and Guo (2003). Due to the toxicity including the carcinogenic effect of arsenic, treatment of the water is absolutely essential. An effective treatment of water is possible by precipitation/coagulation and ion exchange.

Table 6.4 is the first coarse overview. The toxicology and ecotoxicology are highly complex and thus it would be impossible to summarize in one table. It is therefore strongly recommended to seek further information, if there is slightest suspicion of an ecotoxicological effect of actual concentrations of the 18 elements listed in Tables 6.3 and 6.4.

6.5 Effects of Organic Compounds

A short overview of the most toxic organic compounds is given below. These compounds are very important from an environment point of view as they pose a threat to the environment and have several harmful effects. The overview applies the following classification:

A. Petroleum hydrocarbons
B. PCBs and dioxins

TABLE 6.3

Typical Concentrations and the Major Applications of the 20 Elements That Have Particular Environmental Concerns

	Concentration			
Element	Earth's Crust, (mg/kg)	Soil (mg/kg)	Seawater (μg/L)	Applications
13.Al	81,000	10,000–300,000	1	Constructions
51.Sb	0.5	1	0.3	Alloys, plastic
33.As	5	1–40	2.6	Chemicals
56.Ba	425	500	30	Chemicals, glass
82.Pb	16	20	0.02	Batteries, soldering, alloys
48.Cd	0.55	0.06	0.1	Surface treatment, dyestuff
29.Cu	100	20	0.9	Cables, wires, construction
24.Cr	300	100	0.05	Surface treatment, alloys
80.Hg	0.065	0.08	0.15	Instruments, dental use
28.Ni	100	40	6.6	Alloys, surface treatment
34.Se	0.09	0.2	0.05	Glass, instruments, dyestuff
47.Ag	0.1	0.1	0.3	Photochemicals, electrical components, ornaments
38.Sr	250	315	8100	Chemicals
81.Tl	1	0.1	0.01	Electronics, alloys
52.Te	0.002	0.001	0.001	Alloys
50.Sn	40	10	3	Alloys, soldering, chemicals
83.Bi	0.02	1	0.03	Alloys, chemicals
30.Zn	40	50	10	Surface treatment, alloys, chemicals

C. Pesticides
D. Polycyclic aromatic hydrocarbons (PAHs)
E. Heavy metals
F. Detergents
G. Synthetic polymers and xenobiotics applied in the plastics industry

The most important environmental effects of these 7 classes of compounds and their chemical characteristics are presented in this overview.

6.5.1 Introduction

A short overview of the toxic contaminants that we most often detect in the environment and consider as the compounds that pose the most threat is given. As we are using about 100,000 different chemicals in the modern society, we can expect to find 100,000 different compounds plus intermediates of these compounds resulting from a wide spectrum of decomposition

TABLE 6.4

Overview of Toxicity Expressed by LD50 (Lethal Doses Showing 50% Mortality) and LC50 (Lethal Concentration in a Medium, Water, or Air, Showing 50% Mortality)

Element	LD50 (mg/kg)	LC50 (mg/L)	Ecotoxicological Attention
13.Al	770 (mice)	3900 (zooplankton)	Low ecotoxicological effects
51.Sb	100 (rats)	9000 (fish)	Industrial exposure
33.As	41 (rats)	74,000 (cyclops)	Groundwater, carcinogenic effect, nerve inflammation
56.Ba	8–23 (rats)	14,500 (cyclops)	Heart problems, diarrhea
82.Pb	130 (rats)	6 ppm (air, rats)	Anemia, brain damage
48.Cd	80 (rats)	65 (cyclops)	Itai-itai, kidney damage
29.Cu	220 (rats)	10 (cyclops)	Highly toxic to plants
24.Cr	3250 (rats)	100,000 (shrimp)	Carcinogenic effect as Cr (IV)
80.Hg	8 (mice)	5 (cyclops)	Central nervous system, mental retardation, teratogenic effects
28.Ni	1620 (rats)	510 (cyclops)	Allergy, carcinogenic, and teratogenic effects
34.Se	7 (rats)	2500 (cyclops)	Liver and kidney damage
47.Ag	129 (mice)	30 (zooplankton)	Skin damage
38.Sr	148 (rats)	125,000 (cyclops)	—
81.Tl	16 (rats)	—	—
52.Te	83 (rats)	—	Teratogenic effects
50.Sn	41 (mice)	55,000 (zooplankton)	Liver and kidney damage
83.Bi	13 (rats)	—	Liver and kidney damage
30.Zn	975 (rats)	100 (zooplankton), 10,000 (fish)	Minor carcinogenic effects

processes which take place in the environment. To be able to overview that many chemical compounds, their fates, effects, and the associated risks, it is necessary to make a classification of the compounds. The properties and the effects of the above listed seven classes of contaminants, that represent the most toxic compounds we use and the most harmful compounds for the environment have been briefly discussed in the following. For a more comprehensive coverage of toxic chemical compounds in the environment, refer to Loganathan and Lam (1994); Newman and Unger (2003); Hoffman et al. (1995); and Schuurmann and Markert (1998).

6.5.2 Petroleum Hydrocarbons

Petroleum hydrocarbons include a variety of organic compounds. Hydrocarbons (compounds composed of carbon and hydrogen) constitute only 50%–90% of petroleum. They are n-alkanes, branched alkanes, cycloalkanes, and aromatics. Cycloalkanes comprise usually the largest portion of hydrocarbons in petroleum, whereas aromatics are usually present to the extent of 20% or less. Characteristic compounds are benzene, alkyl substituted benzenes, and fused ring polycyclic aromatic hydrocarbons (PAHs; see also this group D). Some of these aromatics are carcinogenic and these are probably the group that is of greatest environmental concern. In addition to hydrocarbons, petroleum contains sulfur and nitrogen compounds, such as, thiophene, ethanethiol, and pyridine derivates.

Petroleum compounds are emitted or discharged into all spheres.

Evaporation removes the lower molecular weight, more volatile components of the petroleum mixture. Hydrocarbons with vapor pressure equal to that of n-octane (0.019 atm at room temperature) or greater will be lost quickly via evaporation.

The lower molecular weight hydrocarbons also tend to be the most water soluble; but for the same molecular weight, aromatic compounds are more soluble than cycloalkanes, which are more soluble than branched alkanes with the n-alkanes being the most insoluble in water. Petroleum products are discharged directly into the sea by accidents or by violation of international regulations. In a massive discharge of petroleum products, which has been recorded in several ship accidents, most of the petroleum initially floats on the surface of marine waters as a slick. Eventually, most slicks are dispersed widely and form a 0.1 mm thick layer on the water surface. If drift results in landfall, contact with coral or mangrove communities, it may have a disastrous environmental impact.

Fortunately, chemical transformation and degradation processes act on petroleum compounds in the environment. Microbial transformation and photo oxidation are of particular importance.

The aromatic compounds of petroleum are the most toxic substances present. They are lethal to crustaceans and fish with LC50 in the range of 0.1–10 mg/L.

6.5.3 PCBs and Dioxins

PCBs and dioxins are characterized as aromatic compounds with a high content of chlorine. As the names cover polychlorinated biphenyls and polychlorinated dibenzo (1,4) dioxins, there are many different individual compounds under these labels. For instance, about 209 different PCB compounds are known, although only about 130 are found in commercial mixtures. Figures 6.2 and 6.3 show the molecular structures and names of some of the most common PCBs and dioxins.

The applications of PCBs have been quite diverse (capacitor oil, plasticizers, printer's ink, etc.), but due to investigations in the 1960s and 1970s in which it was found that PCBs occurred widely in the environment and significant bioaccumulation took place, voluntary restrictions were introduced and all "open" applications were banned.

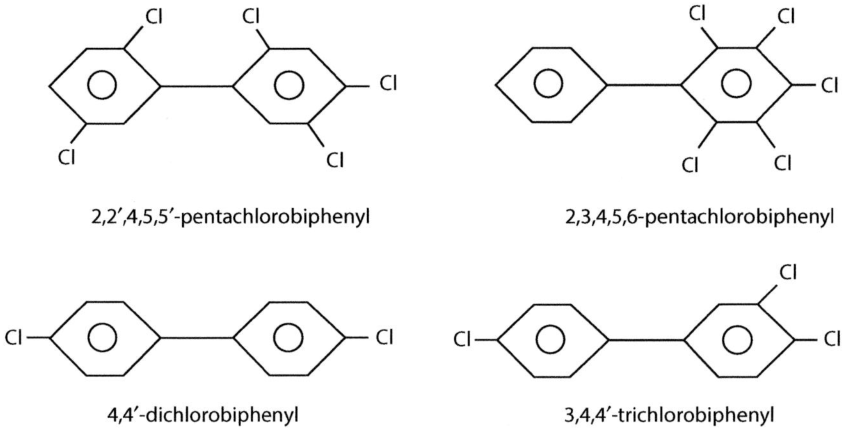

FIGURE 6.2
The molecular structure and names of four common PCBs.

Dioxins are not deliberately produced, but are byproducts of chemical processes involving chlorine, for instance, production of various organochlorine and bleaching of pulp and of combustion processes if chlorine containing compounds are present.

Both PCBs and dioxins are characterized by a low water solubility and a high K_{ow} (the ratio of solubility in octanol and the solubility in water; most components have low $K_{ow} > 5$). Both groups of compounds are very persistent to decomposition processes, which explains why they are strong bioaccumulators, although dioxins have a UV-VIS absorption spectrum that results in significant absorption from solar radiation. Some dioxins have a half-lifetime in the troposphere of a few days.

6.5.4 Pesticides

Pesticides are used to remove pests and they have probably, due to their direct use in nature, been the most criticized environmental contaminants. Usage of DDT and related insecticides accelerated during the 1940s and the subsequent decades until environmental doubt occurred in the mid-1960s. Since 1970, DDT has been banned in most industrialized countries, but is still used in developing countries for instance in India, where it has resulted in very high body concentrations in the Indian population. All the chlorinated hydrocarbon insecticides are banned in most industrialized countries due to their persistence and ability to bioaccumulate (K_{ow} is high; K_{ow} = the ratio solubility in octanol and the solubility in water).

Pesticides can be divided into the following classes depending on their use and their chemical structure:

Herbicides comprise carbamates, phenoxyacetic acids, triazines, and phenylureas.

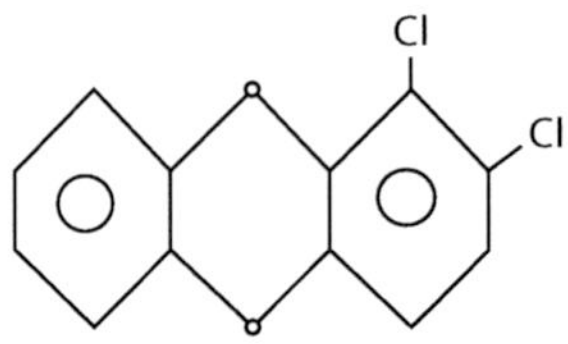

1,2-dichlorodibenzo (1,4) dioxin

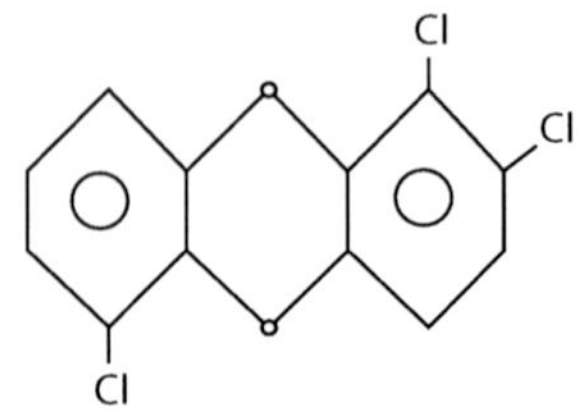

1,2,6-trichlorodibenzo (1,4) dioxin

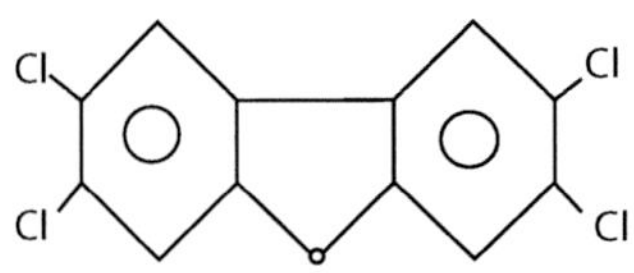

2,3,7,8-tetrachlorodibenzofuran

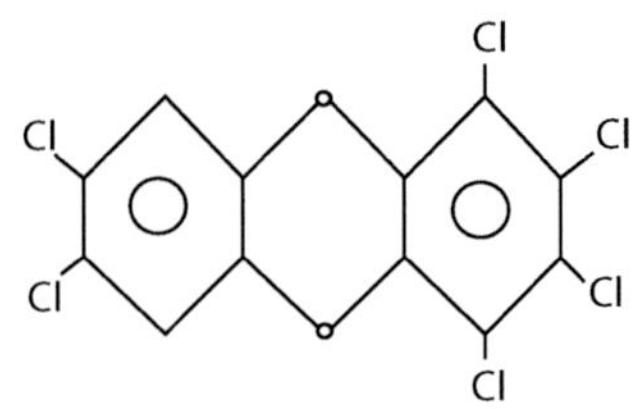

1,2,3,4,7,8-hexachlorodibenzo (1,4) dioxin

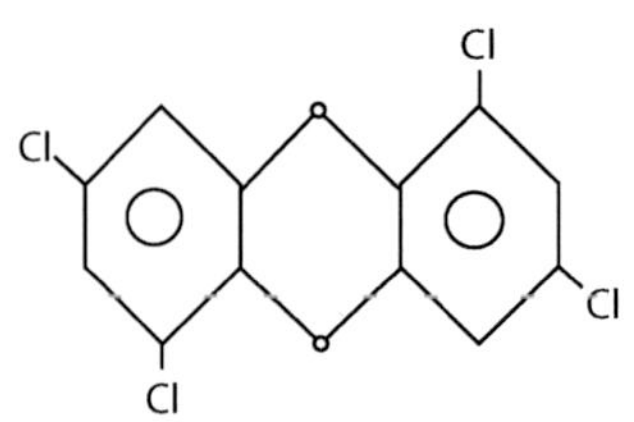

1,3,6,8-tetrachlorodibenzo (1,4) dioxin

FIGURE 6.3
Molecular structure and names of five common dioxins.

Insecticides encompass organophosphates, carbamates, organochlorines, pyrethrins, and pyrethroids.

Fungicides are dithiocarbamates, copper, and mercury compounds. See also these two metals in Table 6.4.

The pesticides are chemically an extremely diverse group of substances, as they only have in common their toxicity to pests. A few of the most important molecules are shown in Figure 6.4. They are mostly produced synthetically, although the natural pesticide pyrethrin has achieved some commercial success.

Chlorohydrocarbons are strongly bioconcentrated as already emphasized. In addition, they are very toxic to a wide range of biota, particularly to aquatic biota.

Organophosphates are almost equally toxic to biota, but due to these compounds' lack of persistence, higher solubility in water, and lower bioaccumulation capacity, they are still in use.

Carbamates are relatively water soluble and have limited persistence. They are however, toxic to a wide range of biota. They act by inhibiting cholinesterase.

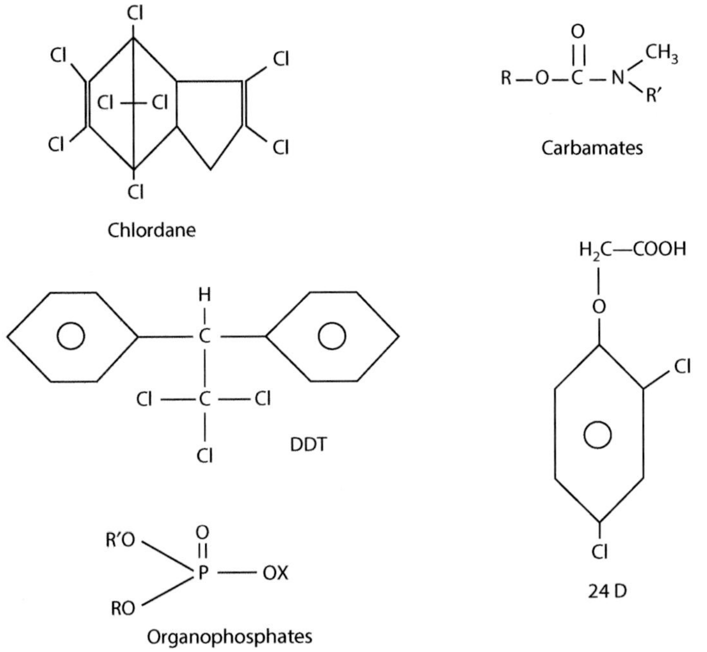

FIGURE 6.4
Molecular structure and names of five common pesticides.

The pyrethins have a complex chemical structure and a high molecular weight. Thus, they are poorly soluble in water and tend to be lipophilic. They are however readily degraded by hydrolysis. They are less harmful to use than most of the other pesticides due to their very low mammalian toxicity.

Phenoxyacetic acid is a very effective herbicide but contains trace amounts of tetrachloro-dibenzo-dioxin.

Pesticides are banned in organic agriculture where they are replaced by other methods, for instance mechanical and biological methods (use of predator insects).

6.5.5 PAHs

PAHs are molecules containing two or more fused 6C-aromatic rings. They are ubiquitous contaminants of the natural environment, but the growing industrialization has increased the environmental concern about these components. Two common members are naphthalene and benzo(a)pyrene; see Figure 6.5. PAHs are usually solids with naphthalene (lowest molecular weight) having a melting point of 81°C.

The natural sources of PAHs in the environment are forest fires and volcanic activity. The anthropogenic sources are coal-fired power plants, incinerators, open burning, and motor vehicle exhaust. As a result of these sources,

PAHs occur commonly in air, soil, and biota. They are lipophilic compounds able to bioaccumulate. The low molecular weight compounds are moderately persistent; whereas for example benzo(a)pyren with a higher molecular weight persists in aquatic systems for up to about 300 weeks.

They are relatively toxic to aquatic organisms and have LC50 values for fish in the range of 0.1–10 mg/L. The major environmental concern regarding PAHs is that many PAHs are carcinogenic. It has been shown (Andersen, 1998) that benzo(a)pyrene is an endocrine disrupter and it cannot be excluded, that many more PAHs have the environmental adverse effect of disturbing the hormone balance of nature.

Human exposure to PAHs occurs through tobacco smoking as well as through compounds in food and the atmosphere.

6.5.6 Organometallic Compounds

Organometallic compounds have metal carbon bonds, where the carbon atoms are part of an organic group. The best known example is probably tetraethyl lead which is used as an additive to gasoline. It has now been phased out of use in many countries—all industrialized countries—due to

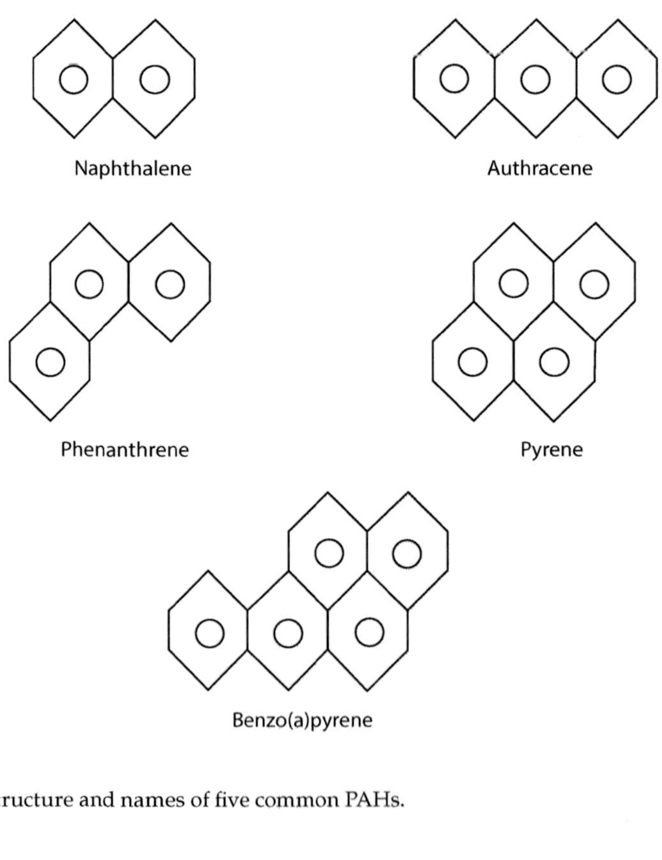

FIGURE 6.5
Molecular structure and names of five common PAHs.

its environmental consequences. Organometallic compounds can be formed in nature from metal or metal ions, for instance dimethylmercury, or are produced for various purposes as catalysts for example, organoaluminum; as pesticides for example, organoarsenic and organotin compounds; as stabilizers in polymers for example, organotin compounds; and as gasoline additive for example, organolead compounds. Organometallic compounds exhibit properties that are different from those of the metal itself and inorganic derivatives of the metal. They have, for instance, a relatively higher toxicity than the metals.

Most organometallic compounds are relatively unstable and undergo hydrolysis and photolysis easily. Most organometallic compounds have weak polar carbon–metal bonds and are often hydrophobic. They therefore only dissolve in water to a small extent and are readily sorbed onto particulates and sediments.

The most harmful organometallic compounds from an environmental point of view are organomercury, organotin, organolead, and organoarsenic which are all very toxic to mammals.

6.5.7 Detergents (and Soaps)

Detergents and soaps contain surface active agents (surfactants) which are classified according to the charged nature of the hydrophilic part of the molecule:

Anionic: Negatively charged

Cationic: Positively charged

Nonionic: Neutral, but polar

Amphoteric: A zwitterion containing positive and negative charges

They are produced and consumed in large quantities and are mostly discharged into the sewage system and end up in the wastewater plant. The early surfactants contained highly branched alkyl hydrophobes that were resistant to biodegradation. These surfactants are largely obsolete today having been replaced by linear alkyl benzene sulphonates (LAS) and other biodegradable surfactants.

The toxicity to mammals is generally low for all surfactants, whereas the toxicity to aquatic organisms is relatively high (LC50 from about 0.1 to about 77 mg/L). The toxicity will generally increase with the carbon chain length (see Figure 6.6).

Many surfactants bind strongly to soils and sediments which implies that, to the extent that they are not biodegraded in a biological treatment plant, they will mainly be found in the sludge phase.

Synthetic polymers and xenobiotics applied in the plastics industry form a very diverse group of compounds from a chemical view point. Synthetic

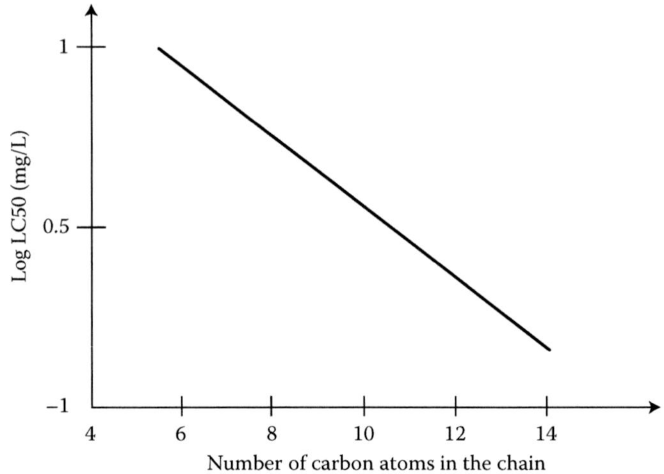

FIGURE 6.6
Log LC50 is plotted versus number of carbon atoms in the chain for LASs. As seen, increased chain length implies increased toxicity. (Reprinted from *Principles of Pollution Abatement*, Jørgensen, S.E., 520, Copyright 2000, from Elsevier.)

polymers are useful (plumbing, textiles, paint, floor, covering, and as the basic material for a wide spectrum of products) because they are resistant to biotic and abiotic processes of transformation and degradation. These properties, however, also cause environmental management problems associated with the use of these components. In addition, several xenobiotic compounds are used as additives, softener, stabilizers, and so on in synthetic polymers to improve their properties. Some of these additives are very toxic and may cause other and additional environmental problems, for instance phthalates are widely used in the plastic industry and it has been demonstrated that phthalates have endocrine disrupter effects.

After use, synthetic polymers are usually incinerated together with industrial and household garbage (solid waste). The presence of PVC implies that hydrochloric acid is formed and also dioxins to a certain extent, but is strongly dependent on the incineration conditions. As it is difficult to separate different types of plastics, it has been discussed to phase out the use of PVC, but due to PVC's unique properties this has not yet been decided.

EXERCISES AND PROBLEMS

1. A lake has a phosphorus concentration of 0.08 mg/L. What is the expected maximum concentration of phytoplankton in the lake? We presume that P is the limiting element.
2. List at least 6 factors that influence the primary productivity.

References

Andersen, H.R. 1998. Examination of endocrine disruptors. Thesis, DFU, Copenhagen University.
Ehrlich, P.R. and P.H. Raven. 1969. Differentiation of population. *Science* 165, 1228–1232.
Hoffman, D.J. et al. 1994. *Handbook of Ecotoxicology*. Boca Raton, London, and New York: Lewis Publishers.
Jørgensen, S.E. 2000. *Principles of Pollution Abatement*. Amsterdam: Elsevier, 520 pp.
Jørgensen, S.E. and B. Fath. 2010. *Ecotoxicology*. Amsterdam, Oxford: Elsevier, 390 pp.
Loganathan B.G. and P.K.S. Lam. 1994. *Global Contamination Trends of Persistent Organic Chemicals*. Boca Raton, London, and New York: CRC Press, 639 pp.
Murphy, T. and J. Guo. 2003. *Aquatic Arsenic Toxicity and Treatment*. Leiden: Beckhuys Publishers, 165 pp.
Newman M.C. and M.A. Unger. 2003. *Fundamentals of Ecotoxicology*. 2nd edition. Boca Raton, London, and New York: CRC Press.
Schuurmann, G. and B. Markert. 1998. *Ecotoxicology*. New York: John Wiley, 902 pp.
Wetzel, R.G. 1983. *Limnology*. 2nd edition. New York: Saunders College Publishing, 828 pp.

7

Environmental Risk Assessment and Surveying Analysis of Environmental Problems

7.1 Environmental Risk Analysis

A brief introduction to the concepts of environmental risk assessment (ERA) is given below to introduce the readers to the concepts and ideas that combine the knowledge about the concentration of a polluting component with its properties to assess an environmental risk. The environmental concentration of a polluting component is usually known by analyses, or it can be estimated or found more or less indirectly by environmental modeling (see, for instance, Jørgensen and Fath, 2011). The topic of the next section is the properties of the polluting components that are usually more difficult to assess than the environmental concentrations. An integration of all sources and their effects are needed before we go to the diagnosis (in Part IV). The last section in this chapter discusses this integration, which may be denoted as an analysis.

ERA is an analysis leading to determination of a risk of adverse effects on the environment or on the human health, caused by a chemical component. Environmental concentrations in different environmental compartments are determined or predicted and compared with no-effect concentrations. ERA reveals if measures are needed to limit the potential environmental consequences of a substance, and it can furthermore be pointed out which testing and knowledge are needed to perform the ERA. ERA is based on the same basic concepts as environmental impact assessment (EIA). EIA includes the use of the diagnosis tools presented in Part IV, and some EIAs have included even the solution tools (Part V) and sometimes even the final integration of all the steps (presented in Part VI). Both ERA and EIA use models and indicators or other diagnostic tools, but EIA is usually, to a higher extent, integrated with the diagnosis, while ERA is often performed independently of the diagnosis.

Treatment of industrial wastewater, solid waste, and smoke is very expensive. Consequently, the industries attempt to change their products and production methods in a more environmentally friendly direction to reduce the treatment costs. They, therefore, need to know how much the different chemicals, components, and processes are polluting our environment. Or expressed differently: What is the environmental risk of using a specific material or chemical compared with alternatives? If industries can reduce their pollution just by switching to another chemical or process, they will consider doing so to reduce their environmental costs or to improve their green image. An assessment of the environmental risk associated with the use of a specific chemical and a specific process gives industries the possibility of making the right selection of materials, chemicals, and processes to the benefit for the economy of the enterprise and the quality of the environment.

Similarly, we need to know the environmental risks of all chemicals used in the society so as to phase out the most environmentally threatening chemicals and set standards for the use of all other chemicals. The standards should ensure that there is no serious risk in using the chemicals, provided that the standards are followed carefully. Modern abatement of pollution therefore includes ERA, which may be defined as the process of assigning magnitudes and probabilities to the adverse effects of human activities. The process involves identification of hazards such as the release of toxic chemicals to the environment by quantifying the relationship between an activity associated with an emission to the environment and its effects. The entire ecological hierarchy is considered in this context, including the effects on the cellular (biochemical) level, organism level, population level, ecosystem level, and for the entire ecosphere.

The application of ERA is rooted in the recognition of the following:

1. The elimination cost of all environmental effects is impossibly high.
2. Practical environmental management decisions must always be made on the basis of incomplete information.

We use about 100,000 chemicals in such amounts that they might threaten the environment, but we only know about 1% of what we need to know to be able to make a proper and complete ERA of these chemicals. Later in this chapter, there is a short introduction to the available estimation methods to apply if we cannot find information about the properties of chemical compounds in the literature. A list of the relevant properties is also given in this context, and what these properties mean for the environmental impact are discussed.

ERA is in the same family as EIA, which attempts to assess the impact of a human activity. EIA is predictive, comparative, and concerned with all possible effects on the environment, including secondary and tertiary (indirect) effects, while ERA attempts to assess the probability of a given (defined) adverse effect as a result of human activity.

Both ERA and EIA use models to find the expected environmental concentration (EEC) which is translated into impacts for EIA and into risks of specific effects for ERA. An overview of ecotoxicological models is given in Jørgensen (1995) and Jørgensen and Fath (2011).

Legislation and regulation of domestic and industrial chemicals with respect to the protection of the environment have been implemented in Europe and North America for decades. Both regions distinguish between existing chemicals and introduction of new substances. For the existing chemicals, the European Union (EU) requires a risk assessment to humans and environment according to a priority setting. An informal priority setting (IPS) is used for selecting chemicals among the 100,000 listed in the European Inventory of Existing Commercial Chemical Substances. The purpose of an IPS is to select chemicals for detailed risk assessment from among the EEC high production volume compounds, that is, >1000 t/year (about 2500 chemicals). Data necessary for the IPS and an initial hazard assessment are called Hedset and cover issues such as environmental exposure, environmental effects, exposure to humans, and human health effects.

At the UNCED (United Nations' Conference on Environmental and Development) meeting in Rio de Janeiro in 1992 on the Environment and Sustainable Development, it was decided to create an Intergovernmental Forum on Chemical Safety (IGFCS, Chapter 19 of Agenda 21). The primary task is to stimulate and coordinate global harmonization in the field of chemical safety, covering the following principal themes: assessment of chemical risks, global harmonization of classification and labeling, information exchange, risk reduction programs, and capacity building in chemicals management.

Uncertainty plays an important role in risk assessment (Suter, 1993). Risk is the probability that a specified harmful effect will occur or in the case of a graded effect, the relationship between the magnitude of the effect and its probability of occurrence.

Risk assessment has emphasized risks to human health and, to a certain extent, has ignored ecological effects. However, some chemicals that have no or only little risk to human health cause severe effects on ecosystems such as aquatic organisms. Examples are chlorine, ammonia, and certain pesticides. An up-to-date risk assessment comprises considerations of the entire ecological hierarchy which is the ecologist's worldview in terms of levels of organization. Organisms interact directly with the environment, and it is organisms that are exposed to toxic chemicals. The species-sensitivity distribution is therefore more ecologically credible (Calow, 1998). A reproducing population is the smallest meaningful level in an ecological sense. However, populations do not exist in vacuum, but require a community of other organisms of which the population is a part. The community occupies a physical environment with which it forms an ecosystem.

Moreover, both the various adverse effects and the ecological hierarchy have different scales in time and space which must be included in a proper ERA; see Figure 7.1, where log spatial (indicated as length) scales of various

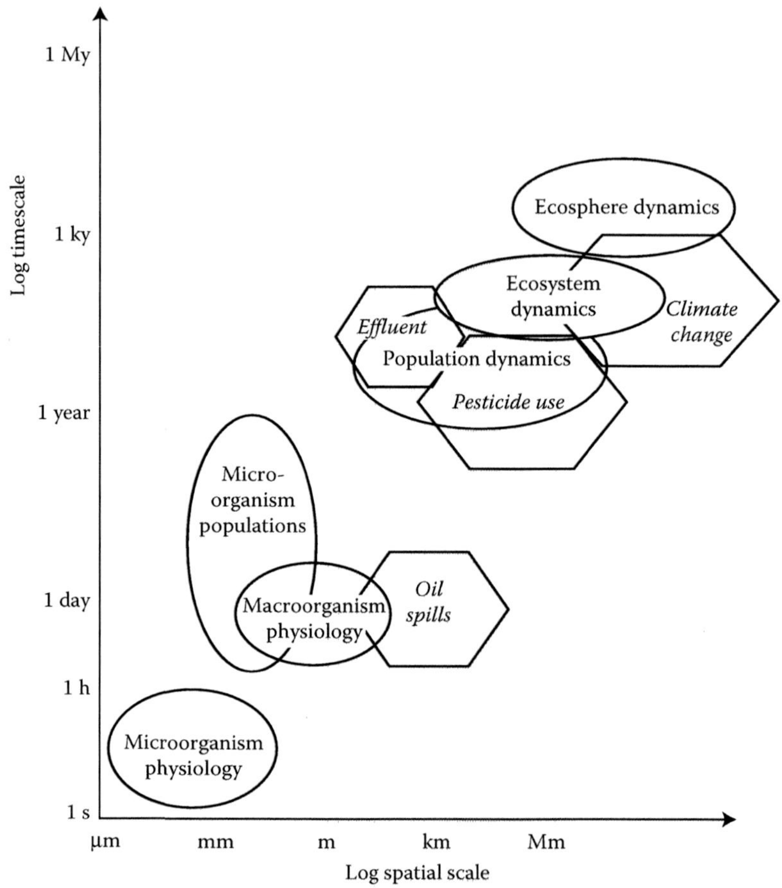

FIGURE 7.1
The spatial (ads length) and timescale for various hazards (hexagons, italic) and for the various levels of the ecological hierarchy (circles, nonitalic).

hazards are plotted versus log of the timescale to recover after the hazard and for ecological processes at different levels of the ecological hierarchy. For example, oil spills occur at a spatial scale similar to those of populations, but they are briefer than population processes. Therefore, a risk assessment of an oil spill requires considering reproduction and recolonization on a longer timescale to determine the magnitude of the population response and its significance to the natural population variance.

EXAMPLE 7.1

The time required to recover after an oil spill with

A. A damage area of 100 km × 100 km in the tropic climate is 5 years.

B. A damage area of 200 km × 200 km in the tropic climate is 10 years.
C. A damage area of 100 km × 100 km in the arctic climate is 40 years.

Explain this difference.

Solution

The recovery time is a function of the circumvent relative to the area and temperature. The circumvent in case B relative to the area is half the value in case A. It means that the recovery processes (ecological processes) in case B are twice the recovery processes in case A, but as four times as much area has to recover, it will take twice as long time in case B as in case A.

The temperature in case C is about 30 degree lower than that in case A, and the rate of ecological processes doubles for every 10 degree higher temperature. This means that the ecological processes will have half the rate at a 10 degree lower temperature, one-quarter the rate at a 20 degree lower temperature, and eight times lower rate at a temperature 30 degrees lower. Therefore, case C will require 5 years times 8 = 40 years.

Uncertainties in risk assessment are taken most commonly into account by the application of safety factors. Uncertainties have three basic causes:

1. The inherent randomness of the world (stochasticity)
2. Errors in execution of assessment
3. Imperfect or incomplete knowledge

The inherent randomness refers to uncertainty that can be described and estimated but cannot be reduced because it is characteristic of the system. Meteorological factors such as rainfall, temperature, and wind are effectively stochastic at levels of interest for risk assessment. Many biological processes such as colonization, reproduction, and mortality also need to be described stochastically.

Human errors inevitably are attributes of all human activities. This type of uncertainty includes incorrect measurements, data recording errors, computational errors, and so on.

Uncertainty is addressed using an assessment (safety) factor from 10 to 1000. The choice of assessment factor depends on the quantity and quality of toxicity data; see Table 7.1. The assessment or safety factor is used in step 3 of the ERA procedure presented below. Other relationships than the uncertainties originating from randomness, errors, and lack of knowledge may be considered when the assessment factors are selected, for instance, cost–benefit. This implies that the assessment factors for drugs and pesticides may be given a lower value due to their possible benefits.

Lack of knowledge results in an undefined uncertainty that cannot be described or quantified. It is a result of practical constraints on our ability

TABLE 7.1

Selection of Assessment Factors to Derive PNEC (See Also Step 3 of the Procedure Presented Below)

Data Quantity and Quality	Assessment Factor
At least one short-term LC_{50} from each of the three trophic levels of the base set (fish, zooplankton, and algae)	1000
One long-term NOEC (nonobserved effect concentration, either for fish or for daphnia)	100
Two long-term NOECs from species representing two trophic levels	50
Long-term NOECs from at least three species (normally fish, daphnia, and algae) representing three trophic levels	10
Field data or model ecosystems	Case by case

to accurately describe, count, measure, or quantify everything that pertains to a risk estimate. Clear examples are the inability to test all toxicological responses of all species exposed to a pollutant and the simplifications needed in the model used to predict the EEC.

The most important feature distinguishing risk assessment from impact assessment is the emphasis in risk assessment on characterizing and quantifying uncertainty. Therefore, it is of particular interest in risk assessment to analyze and estimate the analyzable uncertainties. They are natural stochasticity, parameter errors, and model errors. Statistical methods may provide direct estimates of uncertainties. They are widely used in model development.

The use of statistics to quantify uncertainty is complicated in practice by the needs to consider errors in both the dependent and independent variables and to combine errors when multiple extrapolations should be made. Monte Carlo analysis is often used to overcome these difficulties; see, for instance, Bartell et al. (1992). Model errors include an inappropriate selection or aggregation of variables, incorrect functional forms, and incorrect boundaries. The uncertainty associated with model errors is usually assessed by field measurements utilized for calibration and validation of the model (see Chapter 8). The modeling uncertainty for ecotoxicological models is in principle not different from other models, as stated in Chapter 8.

Chemical risk assessment may be divided into nine steps, which are shown in Figure 7.2. The nine steps correspond to questions which the risk assessment attempts to answer to quantify the risk associated with the use of a chemical. These steps are presented in detail as follows with reference to Figure 7.2.

Step 1: Which hazards are associated with the application of the chemical? This involves gathering data on the types of hazards—possible environmental damage and human health effects. The health effects include congenital, neurological, mutagenic, endocrine disruption

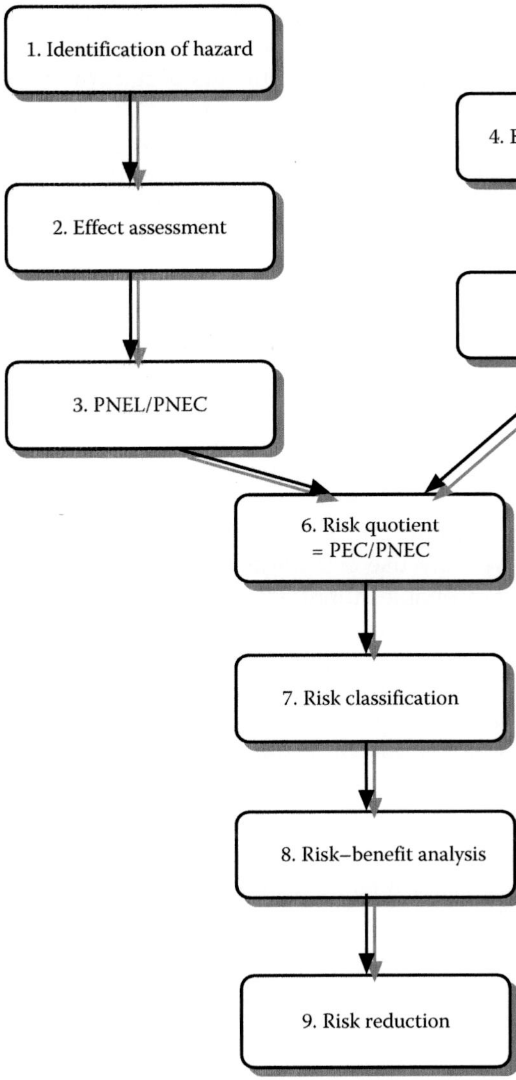

FIGURE 7.2
The presented procedure in nine steps to assess the risk of chemical compounds. Steps 1–3 require extensive use of ecotoxicological handbooks and ecotoxicological estimation methods to assess the toxicological properties of the chemical compounds considered, while step 5 requires the selection of a proper ecotoxicological model.

(so-called estrogen), and carcinogenic effects. It may also include characterization of the behavior of the chemical within the body (interactions with organs, cells, or genetic material). What is the possible environmental damage including lethal effects and sublethal effects on the growth and reproduction of various populations?

As an attempt to quantify the potential danger posed by chemicals, a variety of toxicity tests have been devised. Some of the recommended tests involve experiments with subsets of natural systems, for instance, microcosms or with entire ecosystems. The majority of testing new chemicals for possible effects has, however, been confined to studies in the laboratory on a limited number of test species. Results from these laboratory assays provide useful information for quantification of the relative toxicity of different chemicals. They are used to forecast effects in natural systems, although their justification has been seriously questioned (Cairns et al., 1987).

Step 2: What is the relation between dose and response of the type defined in step 1? It implies knowledge of NEC (noneffect concentration), LD_x- (the dose which is lethal to x% of the organisms considered), LC_y- (the concentration which is lethal to y% of the organisms considered), and EC_z-values (the concentration giving the indicated effect to z% of the considered organisms), where x, y, and z express a probability of harm. The answer can be found by laboratory examination or we may use estimation methods. Based upon these answers, a most probable noneffect level (NEL) is assessed. Data needed for steps 1 and 2 can be obtained directly from scientific libraries, but are increasingly found via online data searches in bibliographic and factual databases. Data gaps should be filled with estimated data. It is very difficult to get complete knowledge about the effect of a chemical on all levels from cells to ecosystem. Some effects are associated with very small concentrations, such as the estrogen effect. It is therefore far from sufficient to know NEC, LD_x-, LC_y-, and EC_z-values.

Step 3: Which uncertainty (safety) factors reflect the amount of uncertainty that must be taken into account when experimental laboratory data or empirical estimation methods are extrapolated to real situations? Usually, safety factors of 10–1000 are used. The choice is discussed above and will usually be in accordance with Table 7.1. If good knowledge about the chemical is available, then a safety factor of 10 may be applied. If, on the other hand, it is estimated that the available information has a very high uncertainty, then a safety factor of 10,000 may be recommended. Most frequently, safety factors of 50–100 are applied. NEL times the safety factor is named the predicted NEL (PNEL). The complexity of ERA is often simplified by deriving the predicted noneffect concentration (PNEC) for different environmental components (water, soil, air, biotas, and sediment).

Step 4: What are the sources and quantities of emissions? The answer requires thorough knowledge of the production and use of the chemical compounds considered, including an assessment of how much of the chemical is wasted in the environment by production and use. The chemical may also be a waste product which makes it very difficult to determine the amounts involved. For instance, the very toxic dioxins are waste products from incineration of organic waste.

Step 5: What is (are) the actual exposure concentration(s)? The answer to this question is named the predicted environmental concentration (PEC). Exposure can be assessed by measuring environmental concentrations. It may also be predicted by a model when the emissions are known. The use of models is necessary in most cases either because we are considering a new chemical or because the assessment of environmental concentrations requires a very large number of measurements to determine the variations in concentrations in time and space. Furthermore, it provides an additional certainty to compare model results with measurements, which implies that it is always recommended to both develop a model and make at least a few measurements of concentrations in the ecosystem components when and where it is expected that the highest concentration will occur. Most models will demand an input of parameters, describing the properties of the chemicals and the organisms, which also will require an extensive application of handbooks and a wide range of estimation methods. The development of an environmental, ecotoxicological model requires extensive knowledge of the physical–chemical–biological properties of the chemical compound(s) considered. The selection of a proper model is discussed in this chapter and in Chapter 2.

Step 6: What is the ratio PEC/PNEC? This ratio is often called the risk quotient. It should not be considered an absolute assessment of risk but rather a relative ranking of risks. The ratio is usually found for a wide range of ecosystems such as aquatic ecosystems, terrestrial ecosystems, and groundwater.

Steps 1–6, shown in Figure 7.3, agree with Figure 7.2 and the information given above.

Step 7: How will you classify the risk? Risk valuation is made to decide on risk reductions (step 9). Two risk levels are defined: (1) the upper limit, that is, the maximum permissible level (MPL), and (2) the lower limit, that is, the negligible level, NL. It may also be defined as a percentage of MPL, for instance, 1% or 10% of MPL.

The two risk limits create three zones: a black, unacceptable, high-risk zone > MPL; a gray, medium-risk level; and a white, low-risk level < NL. The risk of chemicals in the gray and black zones must be reduced. If the risk of the chemicals in the black zone cannot be reduced sufficiently, then it should be considered to phase out the use of these chemicals.

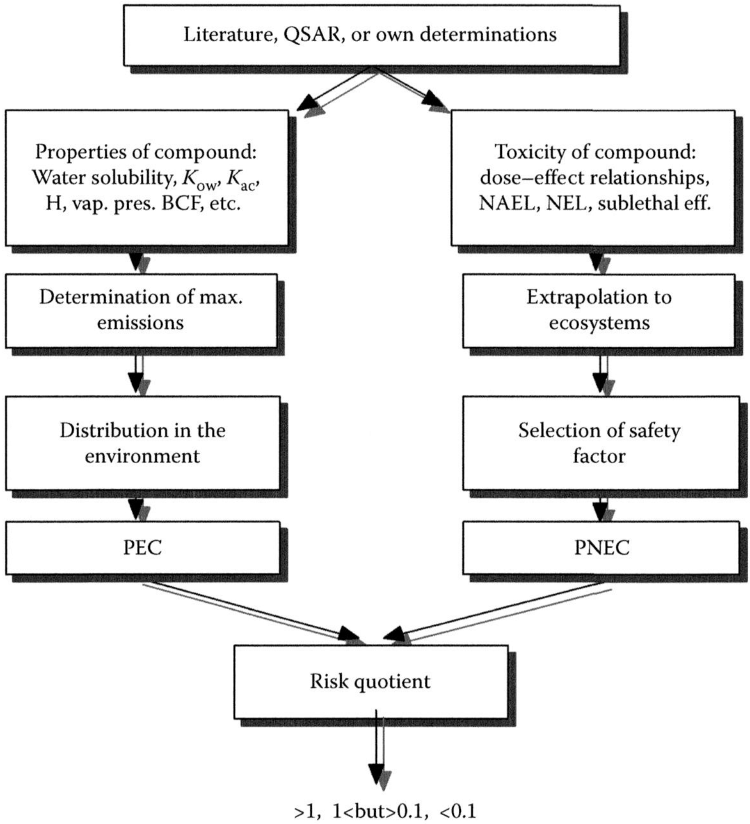

FIGURE 7.3
Steps 1–6 are shown in more detail for practical applications. The result of these steps leads naturally also to assessment of the risk quotient.

Step 8: What is the relation between risk and benefit? This analysis involves an examination of socioeconomic, political, and technical factors, which are beyond the scope of this volume. The cost–benefit analysis is difficult because the costs and benefits are often of a different order and dimension.

Step 9: How can the risk be reduced to an acceptable level? The answer to this question requires deep technical, economic, and legislative investigation. An assessment of the alternatives is often an important aspect in risk reduction.

Steps 1, 2, 3, and 5 require knowledge of the properties of the focal chemical compounds, which again implies an extensive literature search and/or selection of the best feasible estimation procedure. In addition to "Beilstein," it can be recommended to have at hand the following very useful handbooks of environmental properties of chemicals and methods for estimation of these properties in case literature values are not available:

- *Handbook of Ecological Parameters and Ecotoxicology* (Jørgensen et al., 1991). In the year 2000, published as a CD called ECOTOX. It contains three times the amount of parameters in the 1991 book edition. See also Chapter 2 for further details about ECOTOX.
- *Handbook of Environmental Degradation Rates* (Howard et al., 1991).
- *Handbook of Environmental Data on Organic Chemicals* (Verschueren, 2007). Several editions have been published, the latest in 2007.
- *Illustrated Handbook of Physical-Chemical Properties and Environmental Fate for Organic Chemicals, Volume I. Mono-Aromatic Hydrocarbons. Chloro-Benzenes and PCBs* (Mackay et al., 1991). *Volume II. Polynuclear Aromatic Hydrocarbons, Polychlorinated Dioxins, and Dibenzofurans* (Mackay et al., 1992). *Volume III. Volatile Organic Chemicals* (Mackay et al., 1992).
- *Handbook of Estimation Methods in Environmental Chemistry and Ecotoxicology* (Jørgensen et al., 1997).

Steps 1–3 are sometimes denoted as effect assessment or effect analysis and steps 4 and 5 as exposure assessment or effect analysis. Steps 1–6 may be called risk identification, while ERA encompasses all the nine steps presented in Figure 7.2. Particularly, step 9 is very demanding, as several possible steps in reduction of the risk should be considered, including treatment methods, cleaner technology, and substitutes to replace the examined chemical.

In North America, Japan, and the EU, medicinal products are considered similarly to other chemical products as there is, in principle, no difference between a medicinal product and other chemical products. It is also possible to perform an ERA where the human population is in focus. Ten steps corresponding to Figure 7.3 are shown in Figure 7.4, which is not significantly different from Figure 7.3. The principles for the two types of ERA are the same. Figure 7.4 uses the nonadverse effect level (NAEL) and nonobserved adverse effect level (NOAEL) to replace the PNEC, and the PEC is replaced by the tolerable daily intake (TDI).

This type of ERA has particular interest for veterinary medicine which may contaminate food products for human consumption. For instance, the use of antibiotics in pig feed has attracted a lot of attention, as they may be found as residue in pig meat or may contaminate the environment through the application of manure as natural fertilizer.

Selection of a proper ecotoxicological model is a first initial step in the development of an environmental exposure model, as required in step 5. It will be discussed in more detail in the next part, Chapter 8.

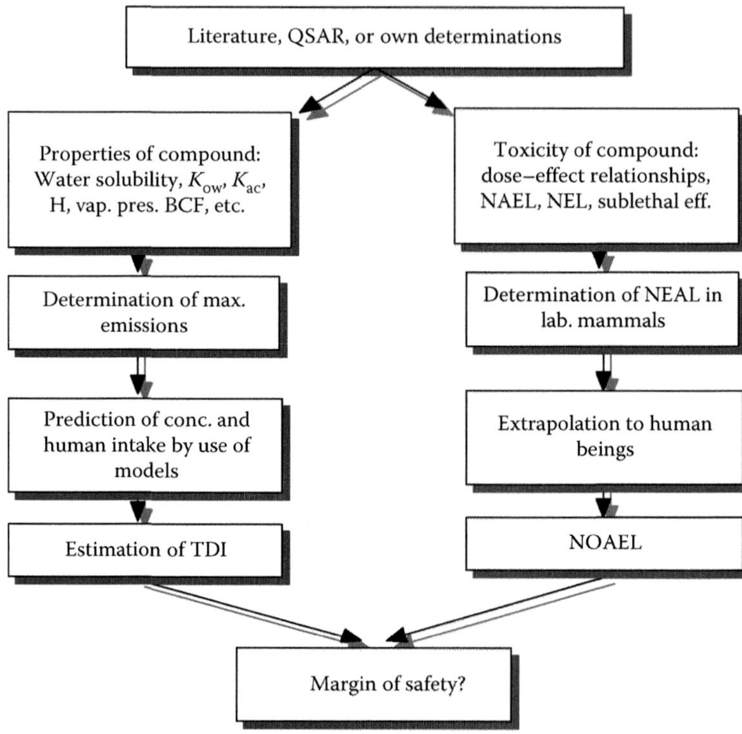

FIGURE 7.4
Environmental risk assessment for human exposure. This leads to a margin of safety which corresponds to the risk quotient in Figures 7.2 and 7.3.

7.2 Ecotoxicological Properties

Slightly more than 100,000 chemicals are produced in such an amount that they threaten or may threaten the environment. They cover a wide range of applications: household chemicals, detergents, cosmetics, medicines, dyestuffs, pesticides, intermediate chemicals, auxiliary chemicals in other industries, additives to a wide range of products, chemicals for water treatment, and so on. They are viewed as mostly indispensable in modern society, resulting in increased production of chemicals, about 40-fold, during the last four decades. A minor or even a major proportion of these chemicals reaches the environment through their production, transport, application, or disposal. In addition, the production or use of chemicals may cause more or less unforeseen waste or byproducts, for instance, chloro compounds from the use of chlorine for disinfection. As we would like to have the benefits of using the chemicals but cannot accept the harm they may cause, this conflict raises several urgent questions which we already have discussed in this chapter. These questions cannot be answered without models, and we cannot develop models without knowing the most important parameters, at least within some ranges. Organization for Economic Cooperation and Development (OECD) has made a review of the properties that we should know for all chemicals. We need to know the boiling point and melting point to determine the chemical form (as solid, liquid, or gas) found in the environment. We must know the distribution of the chemicals in the five spheres: hydrosphere, atmosphere, lithosphere, biosphere, and technosphere. This will require knowing the solubility in water, the partition coefficient water/lipids, Henry's constant, the vapor pressure, the rate of degradation by hydrolysis, photolysis, chemical oxidation and microbiological processes, and the adsorption equilibrium between water and soil—all as a function of the temperature. We need to discover the interactions between living organisms and the chemicals, which implies that we should know the biological concentration factor (BCF), the magnification through the food chain, the uptake rate and the excretion rate by the organisms and where in the organisms that the chemicals will be concentrated, not only for one organism but for a wide range of organisms. We must also know the effects on a wide range of different organisms. This means that we should be able to find the LC_{50}- and LD_{50}-values, the MAC and NEC values (for the abbreviations and the definitions used, see Table 7.2), the relationship between the various possible sublethal effects and concentrations, the influence of the chemical on fecundity, and the carcinogenic and teratogenic properties. We should also know the effect on the ecosystem level. How do the chemicals affect populations and their development and interactions, that is, the entire network of the ecosystem?

Table 7.3 gives an overview of the most relevant physical–chemical properties of organic compounds and their interpretation with respect to the behavior in the environment, which should be reflected in the model.

TABLE 7.2

The Exotoxicoloigcal Properties and Their Definitions

Parameter (Property)	Definition
LC_{50}	Lethal concentration. 50 indicates the % mortality, other mortality may be applied. Often is also the duration of the experiment indicated 48 or 96 hours are usually applied.
LD_{50}	Lethal doses. 50 indicates the % mortality, other mortality may be applied. Often is also the duration of the experiment indicated 48 or 96 hours are usually applied.
MAC	Maximum allowable concentration.
EC	Effect concentration. The effect is indicated, for instance no growth and also the % of organisms affected.
NC	Narcotic concentration effect. The % of test organisms affected is indicated.
HC	Hazardous concentration with indication of the % of test organisms affected.
NEC	Noneffect concentration.

TABLE 7.3

Overview of the Most Relevant Environmental Properties of Organic Compounds and Their Interpretation

Property	Interpretation
Water solubility	High water solubility corresponds to high mobility.
K_{ow} solubility ratio in octanol/water	High K_{ow} means that the compound is lipophilic. This implies that it has a high tendency to bioaccumulate and be sorbed to soil sludge and sediment. BCF and K_{oc} are correlated with K_{ow}.
Biodegradability	This is a measure of how fast the compound is decomposed to simpler molecules. A high biodegradation rate implies that the compound will not accumulate in the environment, while a low biodegradation rate may create environmental problems related to the increasing concentration in the environment and the possibilities of a synergistic effect with other compounds.
Volatilization, vapor	High rate of volatilization (high vapor pressure) implies that the pressure compound will cause an air pollution problem.
Henry's constant, He	H determines the distribution between the atmosphere and the hydrosphere.
pKa	If the compound is an acid or a base, pH determines whether the acid or the corresponding base is present. As the two forms have different properties, pH becomes important for the properties of the compounds. pKa is –log Ka for the process acid → base + H^+
Bioavailability	This concept expresses how available the compound is for biodegradation. Bioavailability depends on a number of factors, included whether the compound is free/labile or bound and the ability of the compound to pass biological membranes.

ERAs also require information about the properties of the chemicals regarding their interactions with living organisms. It might not be necessary to know the properties with the very high accuracy that can be provided by measurements in a laboratory, but it would be beneficial to know the properties with sufficient accuracy to make it possible to utilize the models for management and for risk assessments. Therefore, estimation methods have been developed as an urgently needed alternative to measurements. These are, to a great extent, based on the structure of the chemical compounds, the so-called Quantitative Structure Activity Relationship (QSAR) and Structure Activity Relationship (SAR) methods, but it may also be possible to use allometric principles to transfer rates of interaction processes and concentration factors between a chemical and one or a few organisms to other organisms. This chapter focuses on these methods and attempts to give a brief overview on how these methods can be applied and what approximate accuracy they can offer. A more detailed overview of the methods can be found in Jørgensen et al. (1997).

It may be interesting here to discuss the obvious question: Why is it sufficient to estimate a property of a chemical in an ecotoxicological context with 20%, or sometimes with 50% or higher uncertainty? Ecotoxicological assessment usually gives an uncertainty of the same order of magnitude, which means that the indicated uncertainty may be sufficient from the modeling viewpoint, but can results with such an uncertainty be used at all? The answer is often "yes" because in most cases, we want to ensure that we are (very) far from a harmful or very harmful level. We use (see also Section 7.1 on risk assessment) a safety factor of 10–1000 (most often 50–100). When we are concerned with very harmful effects, such as the complete collapse of an ecosystem or a health risk for a large human population, we will inevitably select a safety factor which is very high. In addition, our lack of knowledge about synergistic effects and the presence of many compounds in the environment at the same time forces us to apply a very high safety factor. In such a context, we will usually go for a concentration in the environment which is magnitudes lower than that corresponding to a slightly harmful effect or considerably lower than the NEC. It is analogous to civil engineers constructing bridges. They make very sophisticated calculations (develop models) that account for wind, snow, temperature changes, and so on, and afterwards they multiply the results by a safety factor of 2–3 to ensure that the bridge will not collapse. They use safety factors because the consequences of a bridge collapse are unacceptable.

The collapse of an ecosystem or a health risk to a large human population is also completely unacceptable. So, we should use safety factors in ecotoxicological modeling to account for the uncertainty. Due to the complexity of the system, the simultaneous presence of many compounds and our present knowledge or rather lack of knowledge, we should, as indicated above, use 10–100 or sometimes even 1000 as a safety factor. If we use safety factors that are too high, the risk is only that the environment will be less contaminated at maybe a higher cost. Besides, there are no alternatives to the use of safety factors. We can step by step increase our ecotoxicological knowledge, but it will take decades before it may be reflected in considerably lower safety factors. A measuring program of all processes and components is impossible due to the high complexity of the ecosystems. This does not imply that we should not use the information of measured properties available today. Measured data will almost always be more accurate than the estimated data. Furthermore, the use of measured data within the network of estimation methods will improve the accuracy of estimation methods. Fortunately, several handbooks on ecotoxicological parameters are available. References to the most important have already been given in Section 7.1. Estimation methods for the physical–chemical properties of chemical compounds were already applied 40–60 years ago, as they were urgently needed in chemical engineering. They are, to a great extent, based on contributions to a focal property by molecular groups and the molecular weight: the boiling point, the melting point, and the vapor pressure as a function of the temperature are examples of properties that were frequently estimated in chemical engineering by these methods. In addition, a number of auxiliary properties result from these estimation methods, such as the critical data and the molecular volume. These properties may not have a direct application as ecotoxicological parameters in ERA, but are used as intermediate parameters which may be used as a basis for estimation of other parameters.

The water solubility, the octanol–water partition coefficient, K_{ow}, and Henry's constant are crucial parameters in our network of estimation methods because many other parameters are well correlated with these two parameters. The three properties can be found for a number of compounds, or estimated with a reasonably high accuracy using knowledge of the chemical structure, that is, the number of various elements, the number of rings, and the number of functional groups. In addition, there is a good relationship between water solubility and K_{ow}; see Figure 7.5. Recently, many good estimation methods for these three core properties have been developed.

During the last couple of decades, several correlation equations have been developed based upon a relationship between the water solubility, K_{ow}, or Henry's constant on the one hand and physical, chemical, biological, and ecotoxicological parameters for chemical compounds on the other. The most important of these parameters are the adsorption isotherms soil–water; the rate of the chemical degradation processes: hydrolysis, photolysis, and chemical oxidation; the BCF; the ecological magnification factor (EMF); the uptake rate; excretion rate; and a number of ecotoxicological parameters. Both the ratio of concentrations in the sorbed phase and in water at equilibrium, K_a, and BCF, defined as the ratio of the concentration in an organism and in the medium (water for aquatic organisms) at a steady state presuming that both the medium and the food are contaminated, may often be estimated with a relatively good accuracy from expressions such as K_a, K_{oc}, or BCF = *a*

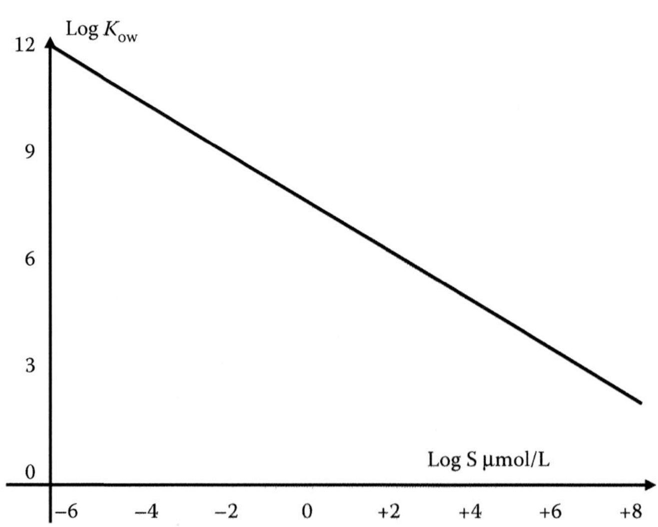

FIGURE 7.5
Log–log relationship between water solubility (unit: μmol/L) and octanol–water distribution coefficient.

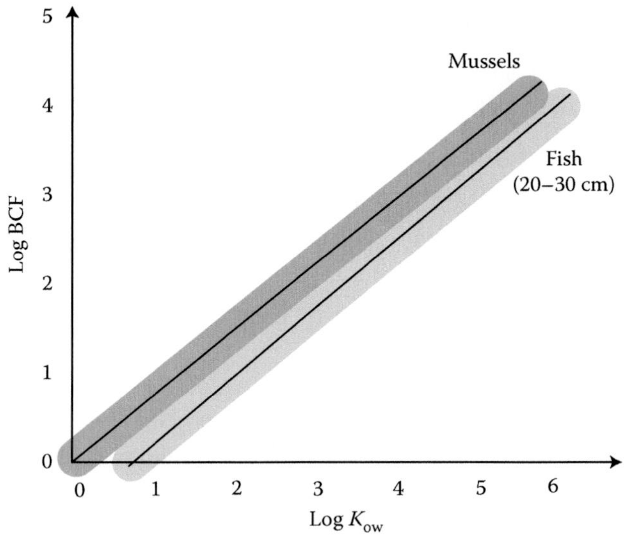

FIGURE 7.6
Two applicable relationships for the octanol–water distribution coefficient and the biological concentration factor for fish and mussels.

log K_{ow} + b. K_{oc} is the ratio between the concentration in soil consisting of 100% organic carbon and in water at equilibrium between the two phases. Numerous expressions with different a and b values have been published (see Jørgensen et al. (1997) and Jørgensen (2000)). Two of these relationships are shown in Figure 7.6.

The biodegradation in waste treatment plants is often of particular interest, in which case the percentage of BOD may be used. It is defined as the 5-day BOD as percentage of the theoretical BOD. It may also be indicated as the BOD_5-fraction. For instance, a BOD_5-fraction of 0.7 will mean that BOD_5 corresponds to 70% of the theoretical BOD. It is also possible to find an indication of BOD_5 percentage removal in an activated sludge plant.

Biodegradation is, in some cases, very dependent on the concentration of microorganisms. Therefore, it may be beneficial to indicate it as a rate coefficient relative to the biomass of the active microorganisms in the units mg/(g dry wt 24 h).

In the microbiological decomposition of xenobiotic compounds, an acclimatization period from a few days to 1–2 months should be foreseen before the optimum biodegradation rate can be achieved. We distinguish between primary and ultimate biodegradation. Primary biodegradation is any biologically induced transformation that changes the molecular integrity. Ultimate biodegradation is the biologically mediated conversion of an organic compound to an inorganic compound and products associated with complete and normal metabolic decomposition.

The biodegradation rate is expressed by a wide range of units:

1. As a first-order rate constant (1/24 h)
2. As half-lifetime (days or hours)
3. Milligram per gram sludge per 24 h (mg/(g 24 h))
4. Milligram per gram bacteria per 24 h (mg/(g 24 h))
5. Milliliter of substrate per bacterial cell per 24 h (mL/(24 h cells))
6. Milligram COD per gram biomass per 24 h (mg/(g 24 h))
7. Milliliter of substrate per gram of volatile solids inclusive microorganisms (mL/(g 24 h))
8. BOD_x/BOD_8, that is, the biological oxygen demand in x days compared with complete degradation (-), named the BOD_x-coefficient
9. BOD_x/COD, that is, the biological oxygen demand in x days compared with complete degradation, expressed by means of COD (-)

The biodegradation rate in water or soil is difficult to estimate because the number of microorganisms varies several orders of magnitude from one type of aquatic ecosystem to the next and from one type of soil to the next.

Models enlisting artificial intelligence have been used as a promising tool to estimate this important parameter. However, a (very) rough first estimation can be made on the basis of the molecular structure and the biodegradability. The following rules can be used to set up these estimations:

1. Polymer compounds are generally less biodegradable than monomer compounds. 1 point for a molecular weight >500 and =1000, 2 points for a molecular weight >1000.
2. Aliphatic compounds are more biodegradable than aromatic compounds. 1 point for each aromatic ring.
3. Substitutions, especially with halogens and nitro groups, will decrease the biodegradability. 0.5 points for each substitution, although 1 point if it is a halogen or a nitro group.
4. Introduction of a double or triple bond will generally mean an increase in the biodegradability (double bonds in aromatic rings are of course not included in this rule). 1 point for each double or triple bond.
5. Oxygen and nitrogen bridges (–O– and –N– (or =)) in a molecule will decrease the biodegradability. 1 point for each oxygen or nitrogen bridge.
6. Branches (secondary or tertiary compounds) are generally less biodegradable than the corresponding primary compounds. 0.5 point for each branch.

Sum the total number of points and use the following classification:

- ≤1.5 points: The compound is readily biodegraded. More than 90% will be biodegraded in a biological treatment plant.
- 2.0–3.0 points: The compound is biodegradable. Probably about 10%–90% will be removed in a biological treatment plant. BOD_5 is 0.1–0.9 of the theoretical oxygen demand.
- 3.5–4.5 points: The compound is slowly biodegradable. Less than 10% will be removed in a biological treatment plant. $BOD_{10} = 0.1$ of the theoretical oxygen demand.
- 5.0–5.5 points: The compound is very slowly biodegradable. It will hardly be removed in a biological treatment plant and a 90% biodegradation in water or soil will take 6 months.
- ≥6.0 points: The compound is refractory. The half-lifetime in soil or water is counted in years.

Several useful methods for estimating biological properties are based upon the similarity of chemical structures. The idea is that if we know the properties of one compound, it may be used to find the properties of similar compounds. If, for instance, we know the properties of phenol, which is named the parent compound, then it may be used to give more accurate estimation of the properties of monochlorophenol, dichlorophenol, trichlorophenol, and so on and for the corresponding cresol compounds. Estimation approaches based on chemical similarity give generally more accurate estimation, but, of course, are also more cumbersome to apply, as they cannot be used generally in the sense that each estimation has a different starting point, namely the parent compound, with known properties.

Allometric estimation methods presume (Peters, 1983) that there is a relationship between the value of a biological parameter and the size of the affected organism. These estimation methods presume that the surface of an organism determines the exchange between the organism and its environment and therefore determine many rates of the organism, particularly those related to the energy balance of organisms. The toxicological parameters LC_{50}, LD_{50}, MAC, EC, and NEC can be estimated from a wide spectrum of physical and chemical parameters, although these estimation equations generally are more inaccurate than the estimation methods for physical, chemical, and biological parameters. Both molecular connectivity and chemical similarity usually offer better accuracy for the estimation of toxicological parameters.

The various estimation methods may be classified into two groups:

A. General estimation methods based on an equation of general validity for all types of compounds, although some of the constants may be dependent on the type of chemical compound or they may be calculated by adding contributions (increments) based on chemical groups and bonds.

B. Estimation methods valid for a specific class of chemical compounds, for instance, aromatic amines, phenols, aliphatic hydrocarbons, and so on. The property of at least one key compound is known. Based upon the structural differences between the key compounds and all other compounds of the considered type—for instance, two chlorine atoms have substituted hydrogen in phenol to get 2,3-dichlorophenol—and the correlation between the structural differences and the differences in the considered property, the properties for all compounds of the considered class can be found. These methods are therefore based on chemical similarity.

Methods of class B are generally more accurate than methods of class A, but they are more cumbersome to use as it is necessary to find the right correlation for each chemical type. Furthermore, the requested properties should be known for at least one key component, which sometimes may be difficult when a series of properties are needed. If estimation of the properties for a series of compounds belonging to the same chemical class is required, then it is tempting to use a suitable collection of class B methods.

Methods of class A form a network which facilitates possibilities of linking the estimation methods together in a computer software system, such as, for instance, EEP (see Figure 7.7), which contains many estimation methods based on relationships between structure and a property or between two properties. The relationship between two properties is most often based on the average result obtained from a number of different equations found in the literature. There is, however, a price for using such "easy to go" software. The accuracy of the estimations is not as good as with the more sophisticated methods based upon similarity in chemical structure, but in many contexts, particularly modeling, the results found by EEP can offer sufficient accuracy. In addition, it is always useful to come up with a first intermediate guess.

The software also makes it possible to start the estimations from the properties of the chemical compound already known. The accuracy of the estimation from use of the software can be improved considerably by having knowledge about a few key parameters such as the boiling point and Henry's constant. As it is possible to get software which is able to estimate Henry's constant and K_{ow} with generally higher accuracy than EEP, a combination of separate estimations of these two parameters prior to using EEP can be recommended. Another possibility would be to estimate a couple of key properties using chemical similarity methods and then use these estimations as known values in EEP. These methods for improving the accuracy will be discussed in the next section. The network of EEP as an example of these estimation networks is illustrated in Figure 7.7. As it is a network of class A methods, it should not be expected that the accuracy of the estimations is as high as it is possible to obtain by the more specific class B methods. By

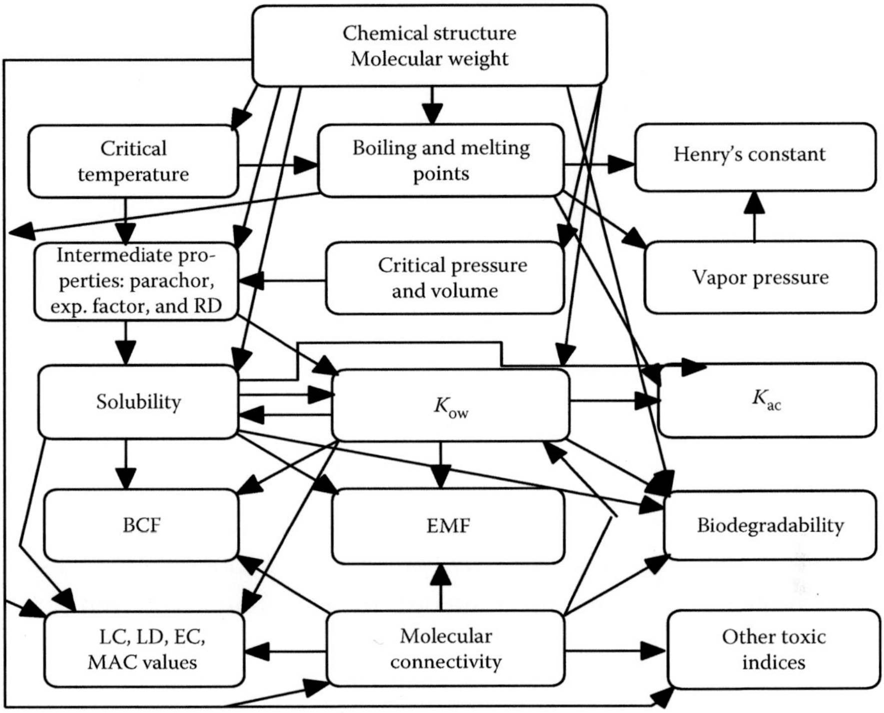

FIGURE 7.7
The network of estimation methods in EEP is shown. An arrow represents a relationship between two or more properties.

EEP it is, however, possible to estimate the most pertinent properties directly and relatively from the structural formula. The last version of EEP contains an estimation of the biodegradation based on a further development of the system presented above.

EEP is based on average values of results obtained by simultaneous use of several estimation methods for most of the parameters. It implies increased accuracy of the estimation, mainly because it gives a reasonable accuracy for a wider range of compounds. If several methods are used in parallel, then a simple average of the parallel results have been used in some cases, while a weighted average is used in other cases where it has been found to be beneficial for the overall accuracy of the program. When parallel estimation methods give the highest accuracy for different classes of compounds, the use of weighting factors seems to offer a clear advantage. It is generally recommended to apply as many estimation methods as possible for a given case study to increase the overall accuracy. If the estimation by EEP can be supported by other recommended estimation methods, then it is strongly recommended to do so.

7.3 Surveying Analysis of Environmental Problems

Part I discussed how to understand and assess the environmental problem and Part II discussed the properties of ecosystems and how to understand the resulting impacts of the environmental problem on nature (the ecosystems). Part III has thus far discussed the sources to the environmental problems and which properties of the pollutants that are decisive for the impacts. ERA has even demonstrated how an integration of a quantification of the sources and the properties of the pollutants can result in a risk assessment. Before we get started on using the diagnostic tools, it is beneficial to present how a survey of the sources quantitatively can be set up with at least a qualitative understanding of the impact associated with the environmental problem.

It is recommended that to get started, to provide the needed survey of the sources by the following points:

1. List all the sources—include at this stage also all suspected sources.
2. Indicate whether the sources are point sources or nonpoint sources.
3. Indicate whether the sources are pulsing or continuous or even better the sources as *f* (time).
4. Quantify all the sources.
5. What is the impact on the ecosystem (nature) or human health as a function of the magnitude of the environmental problem? Is the impact proportional to the magnitude of the environmental problem or how to describe impact as *f* (magnitude of problem)?
6. The impact is often a result, determined by an interaction of various compounds coming from different sources. An example is eutrophication, where the key question is which nutrient is limiting or rather which nutrient can be easily made limiting (see Example 6.1).
7. The duration of the source and thereby the corresponding impact is often decisive for the final effect of the environmental problem.

EXAMPLE 7.2

Point 6 is illustrated with supplementary details to Example 6.1.

A lake is receiving 30 t P and 100 t N per year from mechanical–biological treated wastewater and 3 t P and 150 t N from agricultural drainage water. The ratio N:P of the phytoplankton in the lake is 7:1. The lake is very eutrophic. What to do to solve the environmental problem?

Solution

The point and nonpoint sources give totally a discharge of 33 t P and 250 t N. P is the limiting nutrient. It is difficult to reduce the nutrients of nonpoint sources, particularly phosphorus, while nitrogen can be removed relatively effectively by wetlands. It is, however, very easy and cost moderate (see also Example 6) to remove phosphorus from wastewater even with 95% efficiency. Therefore, the phosphorus can be reduced from 33 t per year to $3 + 30 \times 0.05$ t = 4.5 t per year—a factor 7.33, by removal of phosphorus from wastewater.

The duration of the source (point 7) is an important factor. For toxic substances, the LC_{10}-value, which means the lethal concentration causing 10% mortality among the test animals, is for instance determined for 48 h and 96 h duration of the experiments and the value is of course different for the two different durations of the experiment.

The toxicity of a compound varies with the exposure time. Time/effect relationships are, of course, important in understanding the toxic effect. This has already been mentioned above for air pollutants where it is reflected in the air quality standards. Figure 7.8 shows a toxicity curve of log exposure time versus log LC_{50} threshold. LC_{50} indicates when the curve becomes asymptotic to the time axis. The *threshold* LC_{50} is usually (should be) magnitudes greater than the concentration found in nature or the threshold values reflected in environmental standards. Figure 7.9 gives another illustration of the importance of the exposure time. The exposure duration is considered by the standards used for the air quality. The standards are different for annual geometric means, maximum 24 h concentration, maximum 1 h concentration, and so on.

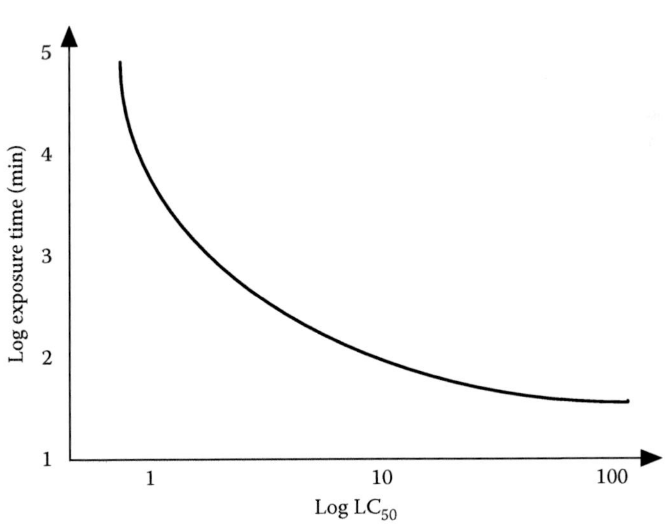

FIGURE 7.8
Log exposure time versus log LC_{50} (mg copper ions/L). The plot indicates LC_{50} for green algae in hard water at pH = 7.0.

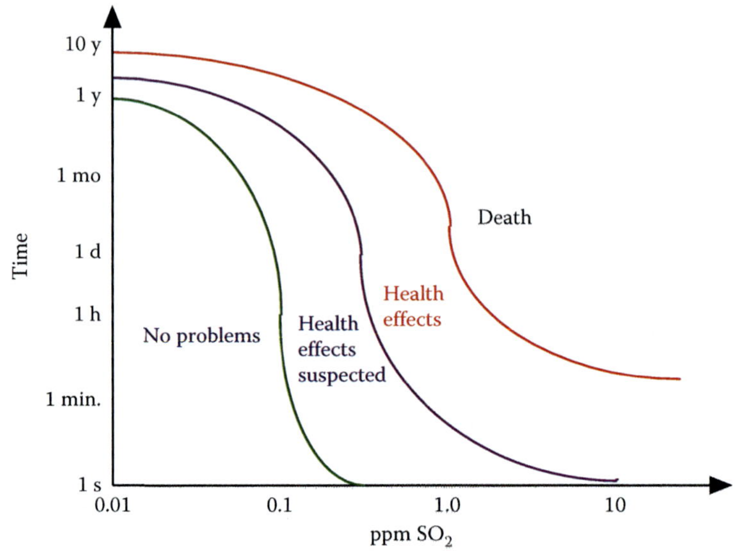

FIGURE 7.9
Effects of sulfur pollution on human health.

EXERCISES AND PROBLEMS

1. Explain why the concentration of most micropollutants in an organism increases with the time (age) and weight of the organism.
2. What would be the difference in biodegradability of a branched alkyl sulfonate with 12 carbon atoms and 7 branches and a completely linear alkyl sulfonate with the same number of carbon atoms?
3. The following contaminants have been found in the soil of an industrial site: benzene, phenol, and pyrene. Evaluate the potential for groundwater contamination by the three compounds.

References

Bartell, S.M., R.H. Gardner, and R.V. O'Neill. 1992. *Ecological Risk Estimation*. Boca Raton, FL: Lewis Publishers.

Cairns Jr., J., K.L. Dickson, and A.W. Maki. 1987. *Estimating Hazards of Chemicals to Aquatic Life*. Philadelphia, PA: American Society for Testing and Materials.

Calow, P. 1998. Ecological risk assessment: Risk for what? How do we decide? *Ecotoxicol. Environ. Saf.* 40, 15–18.

Howard, P.H. et al. 1991. *Handbook of Environmental Degradation Rates*. Boca Raton, FL: Lewis Publishers.

Jørgensen, S.E. 1995. State of the art of ecological modelling in limnology. *Ecol. Model.* 78, 101–116.
Jørgensen, S.E. 2000. *Pollution Abatement*. Amsterdam: Elsevier, 488 pp.
Jørgensen, S.E. and B. Fath. 2011. *Fundamentals of Ecological Modelling*. 4th edition. Amsterdam: Elsevier, 400 pp.
Jørgensen, S.E., B. Halling-Sørensen, and H. Mahler. 1997. *Handbook of Estimation Methods in Ecotoxicology and Environmental Chemistry*. Boca Raton, Boston, London, New York, and Washington, DC: Lewis Publishers, 230 pp.
Jørgensen, S.E., S.N. Nielsen, and L.A. Jørgensen. 1991. *Handbook of Ecological Parameters and Ecotoxicology*. Amsterdam, Oxford: Elsevier.
Mackay, D., W.Y. Shiu, and K.C. Ma. *Illustrated Handbook of Physical-Chemical Properties and Environmental Fate for Organic Chemicals*. Boca Raton and New York: Lewis Publishers.
Suter, G.W. 1993. *Ecological Risk Assessment*. Chelsea, MI: Lewis Publishers, 362 pp.
Peters, R.H. 1983. *The Ecological Implications of Body Size*. Cambridge: Cambridge University Press, 329 pp.
Verschueren, K. 2007. *Handbook of Environmental Data on Organic Chemicals*. Amsterdam: Van Nostrand Reinhold.

Part IV

Developing a Diagnosis

8

Ecological/Environmental Models as Diagnostic Tools

8.1 Physical and Mathematical Models

Living organisms are able to make models. The more complex the organism is, the more complex models it is able to make. Mankind has always made more or less descriptive models of the environment, which often have been used to solve problems or to obtain knowledge about how nature functions. When you use mathematics as a useful tool in our description of nature, we talk about mathematical models in contrast to physical models, which try to imitate the physical appearance of the modeled object, for instance a model of a ship.

A model gives a simplified picture of reality and will never be able to contain all the features of the real system, because then it would be the real system itself. It is, however, important that a model contains the characteristics of the system, which are essential in the context of the management or scientific problem to be solved or comprehended.

Geographical maps, that are also models, clearly illustrate the properties of models. We have different types of maps, which are used for different purposes, for instance, maps for airplanes, cars, railways, ships, archaeological sites, site-specific geological characteristics, and so on. It is similar for ecological and environmental models. There are also lake models, which are used to manage the eutrophication or the acidification of lakes, or lake models that provide information about sustainable fishery, or lake models that focus on the oxygen profile, and so on. Geographical maps have different scales, depending on their use and the available data. If we want to find our way in a town, we would like to have maps that show all the streets, while it is sufficient to have a map of the motorways if we just want to go by car from Copenhagen to Paris. It is the same for ecological models. If we want to manage the average eutrophication of a large lake, it is probably not necessary to distinguish between the levels of eutrophication in different parts of the lake, which, however, would be necessary if we want to ensure a minimum eutrophication level in all parts of the lake.

Let us summarize the comparison of geographical maps with models—particularly ecological models in the following points:

1. Different scales for different purposes.
2. Different types for different purposes.
3. A map or a model cannot contain all detailed information.
4. They are useful simplifications of reality, which is useful in a given context.

The philosophy behind the use of a model might be best illustrated by an example. We have for many years used physical models of ships to determine the profile which gives a ship the smallest resistance in water. Such a model will have the shape and the relative main dimensions of the real ship, but will not contain all the details, such as the instrumentation, the layout of the cabins, and so on. Such details are of course irrelevant to the objectives of that model. Other models of the ship serve other aims: blue prints of the electrical wiring, layout of the various cabins, drawings of pipes, and so on.

Correspondingly, an ecological model must contain the features that are of interest to the management or to the scientific problem that we wish to solve by the model. An ecosystem is a much more complex system than a ship. It implies that it is a far more complicated matter to capture the main features of importance for an ecological problem. However, intense research during past decades has made it possible today to set up many workable and applicable ecological models.

An ecological model focuses on the objects of interest for a considered well-defined problem. It would disturb the main objectives of a model to include too many irrelevant details. There are many different ecological models of the same ecosystem, as the model edition is selected according to the objectives of the model.

The model might be physical, such as the ship model used for the resistance measurements, which may be called micro cosmos or it might be a mathematical model, which describes the main characteristics of the ecosystem and the related problems in mathematical terms.

Physical models will only be touched upon very briefly in this chapter, which will focus entirely on the construction of mathematical models, particularly applications in environmental management. The field of ecological modeling has developed rapidly during the last three decades essentially due to three factors:

1. The development of computer technology, which has enabled us to handle very complex mathematical systems.
2. A general understanding of environmental problems, including that a complete elimination of pollution is not feasible (denoted zero discharge), but that proper pollution control with the limited

economical resources available requires serious consideration of the influence of pollution impacts on ecosystems.

3. Our knowledge of environmental and ecological problems has increased significantly. We have particularly gained more knowledge as to quantitative relations in ecosystems and between ecological properties and environmental factors.

Models may be considered a synthesis of what we know about the ecosystem with reference to the considered problem, in contrast to a statistical analysis, which only will reveal the relationships between the data. A model is able to include our entire knowledge about the system:

- Which components interact with which other components, for instance that zooplankton grazes on phytoplankton?
- Our knowledge about the processes are often formulated as mathematical equations which have been generally proved valid.
- The importance of the processes with reference to the problem.

This is to mention a few examples of knowledge which may often be incorporated in an ecological model. This implies that a model can offer a deeper understanding of the system than a statistical analysis. It is therefore a stronger tool in research and can result in a better management plan on how to solve an environmental problem. This does not of course mean that statistical analytical results are not applied in the development of models. On the contrary, models are built on all available knowledge, including knowledge gained by statistical analyses of data, physical–chemical–ecological knowledge, the laws of nature, common sense, and so on. That is the advantage of modeling.

8.2 Models as a Management Tool

The idea behind the use of ecological management models is demonstrated in Figure 1.1 (Chapter 1). Urbanization and technological development have had an increasing impact on the environment. Energy and pollutants are released into ecosystems, where they may cause more rapid growth of algae or bacteria, damage species, or alter the entire ecological structure. An ecosystem is extremely complex, and it is therefore an overwhelming task to predict the environmental effects that such emissions may have. It is here that a model comes into the picture. With sound ecological knowledge, it is possible to extract the components and processes of the ecosystem which are particularly involved in a specific pollution problem to form the basis of the ecological model (see also the discussion in Chapter 1). As indicated in

Figure 1.1, the resultant model can be used to select the environmental technology eliminating the emission most effectively.

Figure 1.1 represents the idea behind the introduction of ecological modeling as a management tool, which occurred around the year 1970. The environmental management of today is more complex and applies therefore a wider spectrum of tools. Today, we have as an alternative and supplement to environmental technology, cleaner technology, ecotechnology, environmental legislation, international agreements and sustainable management plans. Ecotechnology is mainly applied to solve the problems of nonpoint or diffuse pollution, often originating from agriculture. The significance of nonpoint pollution was hardly acknowledged before around 1980. Global environmental problems furthermore play a more important role today than 20 or 30 years ago, for instance the reduction of the ozone layer and climatic changes due to the greenhouse effect. Global problems can hardly be solved without international agreements and plans. Figure 1.2 attempts to illustrate the more complex picture of environmental management today.

As mentioned above, ecological systems are usually characterized by a much higher complexity than physical systems. Fortunately, we now have very powerful computers that can help us develop complex models. Still, however, it is a crucial question in the development of ecological models: Which are the important processes and components in the model context? To answer this question, we need to know the problem and the ecosystem. Simplifications in calculations and in the development of ecological models require that we can distinguish the essential processes and components from the minor important ones, which should not be included in the model, because they only make the model unwieldy and difficult to overview.

Our problems are getting more and more complex and more and more comprehensive due to globalization, among other factors. It is therefore becoming more and more difficult to survey all the factors which may influence the problems. Science is using an analysis of the problem and often separates a problem and a few factors having influence on the problem in a laboratory; but many of the real-life problems are too complex to be solved by this scientific method. A synthesis of many factors and their interaction is needed and it implies that we have to develop a model to be able to survey the many possibilities. Some of the interactions and processes may be known from science; but for real-life problems including environmental problems there are many interactions operating simultaneously. Fortunately, models have been developed to deal with the high complexity. Today, we have a wide experience in the development of ecological models. Many thousands of models have been published during the last three decades. The journal, *Ecological Modelling*, today publishes more than 4000 pages per year. Ecological and environmental models are able to strongly support our decisions in ecological and environmental management.

So, building a model helps us perceive the interconnections and to connect previous experience with present observations and simulate what would happen in the future. Model development often requires a multidisciplinary approach, because processes and interconnections from many disciplines may be an integrated part of the problem. The experience of the last several decades with the development of ecological models has shown us the clear advantages that models are able to offer ecology and environmental management today:

- Models are a synthesis of all of what we know—observations, theoretical knowledge, knowledge about rates and sizes, knowledge about food items, etc.
- Models are tools to overview complex systems.
- Models make quantification possible by the use of mathematical formulations and computers.

It should be emphasized, that all models may be considered a synthesis of all what we know about the problem and system. Models give a much more profound picture of the problem and the system than for instance a statistical treatment of observations, because a model can include *all* our theoretical knowledge.

8.3 Modeling Components

Models have five types of components.

1. Forcing functions, or external variables, which are functions or variables of an external nature that influence the state of the system. The problem to be solved can often be formulated as follows: if certain forcing functions vary, how will this variation influence the state of the ecosystem? The model is used to predict what will change in the system, when forcing functions are varied with time. The forcing functions under our control are often called control functions. The control functions in ecological models could for instance be inputs of toxic substances into ecosystems and in eutrophication models the control functions are inputs of nutrients. Natural forcing functions in contrast to control functions are for instance the climatic variables, which influence biotic and abiotic components and process rates.
2. State variables describe, as the name indicates, the state of the system. The selection of state variables is crucial to the model structure, but often the choice is obvious. If, for instance, we want to model the bioaccumulation of a toxic substance in an ecosystem, the state variables should be the organisms in the most important food chains

and concentrations of the toxic substance in the organisms. In eutrophication models the state variables will be the concentrations of nutrients and phytoplankton. When the model is used in a management context, the values of state variables predicted by changing the forcing functions can be considered the results of the model, because the model contains relations between the forcing functions and the state variables.

3. Mathematical equations are used to represent biological, chemical and physical processes. The total set of equations describes the relationship between the forcing functions and state variables. The same type of processes may be found in many different environmental contexts, which implies of course that the same equations can be used in different models. This does not imply, however, that the same process is always formulated by use of the same equation.

   First, the considered process may be better described by another equation because of the influence of other factors. Second, the number of details needed or desired to be included in the model may be different from case to case due to a difference in the complexity of the system and the problem. Some modelers refer to the description and mathematical formulation of processes as submodels. A comprehensive overview of submodels can be found in the ecological literature. See, for instance, the *Handbook of Ecological Processes* (Palmeri et al. 2013).

4. Parameters are coefficients in the mathematical representation of processes. They may be considered constant for a specific ecosystem or part of an ecosystem. In causal models the parameter will have a scientific definition, for instance, the excretion rate of a toxic substance from a fish. Many parameters are not indicated in the literature as constants but as ranges, but even that is of great value in the parameter estimation, as will be discussed further in the following text. In Jørgensen et al. (2000), a comprehensive collection of parameters in environmental sciences and ecology can be found. Our limited knowledge of parameters is one of the weakest points in ecological modeling, a point that will be touched on often throughout the book. Furthermore, the application of parameters as constants in our models are unrealistic due to the amount of feedback in many real ecosystems. The flexibility of ecosystems is inconsistent with the application of constant parameters in the models. A new generation of models that attempts to use varying parameters according to some ecological principles seems a possible solution to the problem, but further development in this direction is absolutely necessary before we can achieve an improved modeling procedure reflecting the processes in real ecosystems.

5. Universal constants, such as the gas constant and atomic weights, are also used in most models.

Mathematical models can be defined as formal expressions of the essential elements of a problem in mathematical terms. The first recognition of the problem is often verbal. This may be recognized as an essential preliminary step in the modeling procedure, which will be treated in more detail in the next section. The processes that are linking the state variables and the forcing functions are, however, easily expressed as mentioned above by the use of mathematical equations. In the most applied type of ecological models—dynamic biogeochemical models—differential equations are applied to express the changes of the state variables: accumulation = input − output according to the mass conservation principle. These equations express simply, that accumulation = inputs − outputs, where inputs and outputs are processes which are expressed by algebraic equations.

Many ecological researchers have discovered how powerful ecological models are to test scientific hypotheses about ecosystem reactions. Ecosystems are complex systems and it is therefore not easy to make direct experiments with entire ecosystems in most cases. It is, however, possible to make changes for instance in the forcing functions and observe the reaction of ecosystems as changes of the state variables. In such cases, it will be necessary to develop a model of the focal ecosystem with the involved forcing functions and state variables to be able to assess whether the observations are in accordance with a proposed theory or hypothesis. These possibilities have opened up for a wider use of ecological models in ecology, particularly system ecology, of course.

8.4 Modeling Procedure

Figure 8.1 presents a procedure which is often applied in model development. The procedure can be considered the experience gained by the development of several different ecological models during the 1970s. The open question at that time was: how can we ensure that the model represents the integrated knowledge about the system and the problem as correctly as possible?

As seen, the first step in all model making is to define the problem. Even that would often require a multidisciplinary team because it is important to quantify all the sources of the problem. The system is defined together with the problem because the problem affects either a man-made or a natural system. The problem can of course not be solved unless the processes, reactions and the interconnections in a real system are considered. It does not imply that all details associated with the problem and the system should be considered by the model development. It is important on an early stage of the model development that it is made clear for all members of the model team that the focus is the system—not the details, remembering that the system

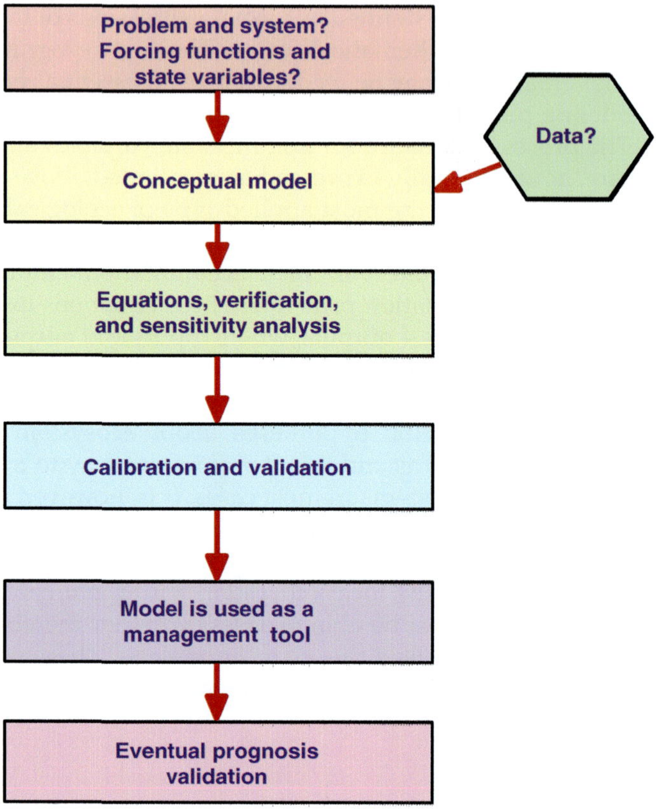

FIGURE 8.1
A tentative modeling procedure is shown.

is more than sum of its parts. On the other hand, it is important to include all the important sources to the problem for the ecosystem in the model. The selection of model complexity will be discussed in slightly more detail later in this chapter.

EXAMPLE 8.1

Find the connectivity of the model in Figure 3.4 (Chapter 3).

Solution

It is advantageous to calculate the connectivity of a model, which is defined as the ratio of realized connections to the number of possible connections. The conceptual diagram shows 14 connections + 5 forcing functions. The number of possible connections for 7 state variables would be $7 * 7 - 7 = 42$. It means that the connectivity is 14/42 or 0.333. It is generally accepted that the connectivity should be between 0.25 – 0.5. If it is lower than 0.25 the connections are not sufficient to ensure a good recycling of the elements and if it is more than 0.5 the food net becomes too rigid.

It is beneficial to use a mass or energy balance to choose the most important components to be included in the model. Let us illustrate the considerations by use of an example. Let us anticipate that it is an open question, whether birds should be included in an eutrophication model. Birds contribute to the inputs of nutrients by their droppings. If the nutrients—nitrogen and phosphorus—coming from the bird droppings are insignificant compared with the amounts of nutrients coming from drainage water, precipitation and wastewater, inclusion of birds as a contributing component in the model is an unnecessary complication that only would contribute to the uncertainty. There are, however, a few cases where birds may contribute as much as 25% or at least more than 5% of the total inputs of nutrients. In such cases, it is of course important to include birds as a model component. A mass balance is needed to uncover the main sources of the problem—in the example of the inputs of nutrients.

In this first step the problem and the system—for ecological models the system is an ecosystem, several interacting ecosystems or a part of an ecosystem—are described by the use of words. The verbal model is, however, difficult to visualize and it is, therefore, conveniently translated in the second step of the modeling procedure into a conceptual diagram, which contains the state variables, the forcing functions and how these components are interrelated by mathematical formulations of processes.

As shown in Figure 8.1, the construction of the conceptual diagram considers the available data. Due to the calibration of the model (step 4), it is generally recommended to include only state variables that have observations. In a few cases, it may be possible to include one or two state variables that are not observed, if it is difficult to get the observations and it is obvious how the nonobserved state variables are related to the other observed state variables.

Figure 3.4 illustrates a conceptual diagram of the phosphorus cycle in a lake. Compare this with the nitrogen cycle in Figure 3.3. The state variables are nitrate, ammonium (which is toxic to fish in the unionized form of ammonia), nitrogen in phytoplankton, nitrogen in zooplankton, nitrogen in fish, nitrogen in sediment and nitrogen in detritus.

The next step comprises of the mathematical formulation. A differential equation is formulated. As mentioned above for each state variable: accumulation = inputs – outputs. The accumulation for the state variable S is dS/dt, while the inputs and outputs are processes, expressed by algebraic equations. For the conceptual diagram of the phosphorus cycle in Figure 3.4, for instance, all the shown processes should and could be expressed by an equation.

The conceptual diagram in Figure 3.4 indicates the state variables as boxes and the processes as arrows between the boxes, for instance, process number 1. The forcing functions are symbolized by arrows to or from a state variable, for instance 15 and 16. It is of course possible to use other symbols for the modeling components. The software STELLA which will be used to illustrate the development of models uses boxes for state variables, thick arrows with a symbol of a valve for the processes, thick arrows coming or going to

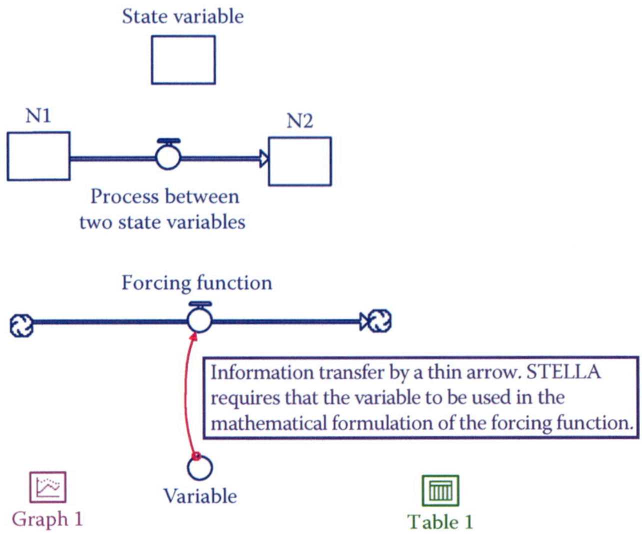

FIGURE 8.2
The symbols applied to erect a conceptual diagram using STELLA. State variables are boxes for which differential equations are erected as accumulation = inputs − outputs. Processes are thick arrows with the symbol of a valve. Forcing functions are thick arrows starting or ending as a cloud. Circles are variables in general. Graph 1 and Table 1 indicate that the results can be presented as graphs or as tables.

a cloud for the forcing functions (requires a constant, an equation, a table, or a graph), a thin arrow to indicate the transfer of information, and a variable (forcing function, parameter, and/or a state variable calculated by an algebraic expression from another state variable, and so on); see Figure 8.2.

There are other symbolic languages for development of conceptual diagrams, for instance, Odum's energy circuit language. It has many more symbols than STELLA and is therefore more informative but of course also more time-consuming to develop. For an overview of the most applied symbolic languages including Odum's energy circuit language, see Jørgensen and Fath (2011). In this book, we will either use a simple conceptual diagram as shown in Figure 3.4 or the STELLA diagram (see Figure 8.2).

Figure 8.3 illustrates the idea behind the use of differential equations for state variables. In mathematics, the differential equations are solved analytically, while the equations are solved numerically by computers. A time step is selected for the model calculations by computers. The shorter the time step the closer will the computer calculations come to the real time variations of inputs and outputs. It is recommended to test different time steps and use the longest time step which would not give any significant change of the model results by decreasing the time step. Significant changes are of course evaluated relative to the accuracy of the observations that are used as the basis for the development of the model.

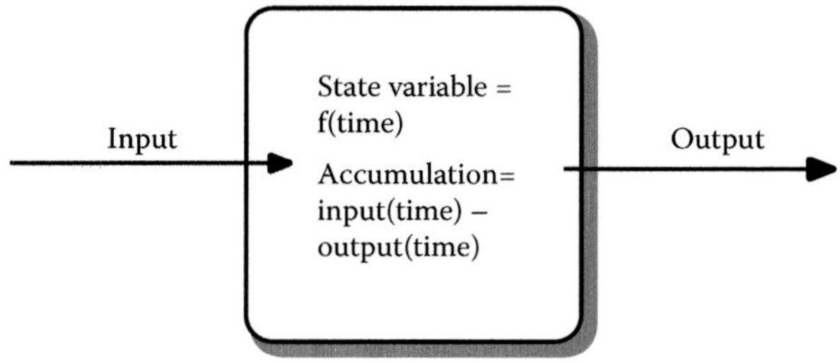

FIGURE 8.3
The idea behind the use of differential equations is illustrated. The differential equation accounts for the increase of a state variable due to what is denoted in the figure as accumulation for a selected time step. The accumulation is input–output. In theoretical mathematics, the time step is infinitely small, but when the equation is solved numerically with the use of computers, we are selecting a time step. The shorter the time step, the closer is the solution to the mathematical solution.

The software STELLA erects the differential equations from the diagram. The time derivate of the state variables will be equal to all the inputs = all process arrows going into the state variables minus all outputs = all process arrows going out from the state variable. The processes must, however, be formulated as an algebraic equation. Table 8.1 gives an overview of the most applied process equations.

There are several modifications of these expressions. For instance tr is often used for a threshold concentration, in the Michaelis–Menten expression. The concentration is replaced by the concentration – tr. The concentration has therefore to exceed tr to generate any rate. Grazing and predation are often used to multiply the Michaelis–Menten expression by (1 – concentration/carrying capacity) similar to what is used in the logistic growth expression, see process 5 in Table 8.1. This implies that when the food is abundant (concentration is high), another factor determines the growth for instance of the space or the nesting area.

Table 8.1 provides information about the applications of the six expressions for which processes these equations are most often used. Most processes of ecological relevance are covered by these six expressions, although there of course are a few processes that will require another mathematical formulation.

The third step of the modeling procedure includes verification and sensitivity analysis. Verification is a test of the internal logic of the model. Typical questions in the verification phase are: Does the model react as expected? Is the model stable in the long term? Does the model follow the law of mass conservation? Is the use of units consistent? It implies that all equations should be checked for a consistent use of units. Verification is to some extent a subjective assessment of the behavior of the model. To a large extent verification will occur during the use of the model before the calibration phase.

Notice that during verification it is possible to perform "Gedanken Experiments" similar to the way in which Einstein tested his scientific work. We can for instance test a eutrophication model by its reactions to the

TABLE 8.1

Process Equations

1. A constant flow rate—also denoted by zero-order expression.
2. A first-order rate expression, where the rate is proportional to a variable for instance a concentration of a state variables: rate = $dC/dt = k*C$. This expression corresponds to exponential growth. The following expression is obtained by integration $C(t) = C_o * e^{kt}$. First-order decay has the rate = $dC/dt = -k * t$ and $C(t) = C_o * e^{-kt}$.
3. A second-order rate expression, where the rate is proportional to two variables simultaneously.
4. A Michaelis–Menten expression or Monod kinetics known from kinetics of enzymatic processes. At small concentrations of the substrate the process rate is proportional to the substrate concentration, while the process rate is at maximum and constant at high substrate concentrations, where the enzymes are fully utilized. The same expression is used when the growth rate of plants are determined by a limiting nutrient according to Liebig's minimum law. A graph of this expression is shown in Figure 3.2. The so-called Michaelis–Menten constant or the half-saturation constant corresponds to the concentration that gives half the maximum rate. At small concentrations of substrate or nutrients, the rate is very close to a first-order rate expression, while it is close to a zero-order rate expression at high concentrations. Notice furthermore that the rate is regulated from a first-order to a zero-order expression more and more as the concentration increases.
5. A first-order rate expression with a regulation due to limitation by another factor, for instance the space or the nesting areas. It is expressed by introduction of a carrying capacity. The general first-order expression is applied regulated by the following factor: (1 – concentration/carrying capacity). When the concentration reaches the carrying capacity the factor becomes zero and the growth stops. This process rate expression is denoted as logistic growth and in Figure 8.4 it is illustrated including the graphic differences between exponential growth by a first-order reaction expression and logistic growth. These two growth expressions are often applied in population dynamic models.
6. Rates govern by diffusion uses often a concentration gradient dC/dx to determine the rate, as expressed in Fick's laws: rate = $k * (dC/dx)$ (Fick's First Law).

following ideas: we rent a helicopter and buy 10,000 kg of phosphorus fertilizer and drop it instantly to the lake. The experiment could by use of the model be made at no cost, while it would be very expensive to rent a helicopter and buy 10,000 kg of fertilizer. It is another advantage of models: it is easy to perform "Gedanken Experiments."

Sensitivity analysis follows verification. Through this analysis the modeler gets a good overview of the most sensitive components of the model. Thus, sensitivity analysis attempts to provide a measure of the sensitivity of either parameters, or forcing functions, or submodels to the state variables of greatest interest in the model. If a modeler wants to develop a model which is able to predict the toxic substance concentration in carnivorous insects as a result of the use of insecticides, he will obviously choose this state variable as the most important one, maybe besides the concentration of the toxic substance concentration in plants and herbivorous insects.

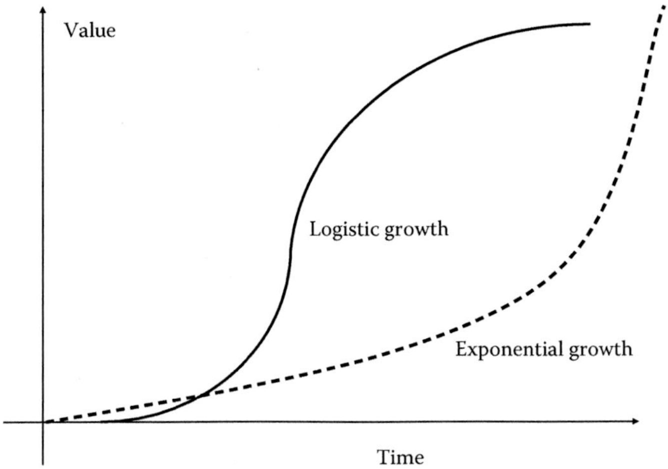

FIGURE 8.4
The exponential growth and the logistic growth are shown. The growth rate is increasing linearly for the exponential growth, while the logistic growth has an increasing growth rate, followed by a declining growth rate. The growth rate for the logistic growth is at maximum where the slope of the logistic curve has maximum.

In practical modeling the sensitivity analysis is carried out by changing the parameters, the forcing functions or the submodels. The corresponding response on the selected state variable is observed. Thus, the sensitivity, S, of a parameter, P, is defined as follows:

$$S = \frac{[\partial x/x]}{[\partial P/P]}, \tag{8.1}$$

where x is the state variable under consideration.

The relative change in the parameter value is chosen based on our knowledge of the certainty of the parameters. If, for instance, the modeler estimates the uncertainty to be about 50%, he will probably choose a change in the parameters at ±10% and ±50% and record the corresponding change in the state variable(s). It is often necessary to find the sensitivity at two or more levels of parameter changes as the relationship between a parameter and a state variable is rarely linear.

A sensitivity analysis makes it possible to distinguish between high-leverage variables, whose values have a significant impact on the system behavior and low-leverage variables, whose values have minimal impact on the system. Obviously, the modeler must concentrate his effort on improvements of the parameters and the submodels associated with the high-leverage variables.

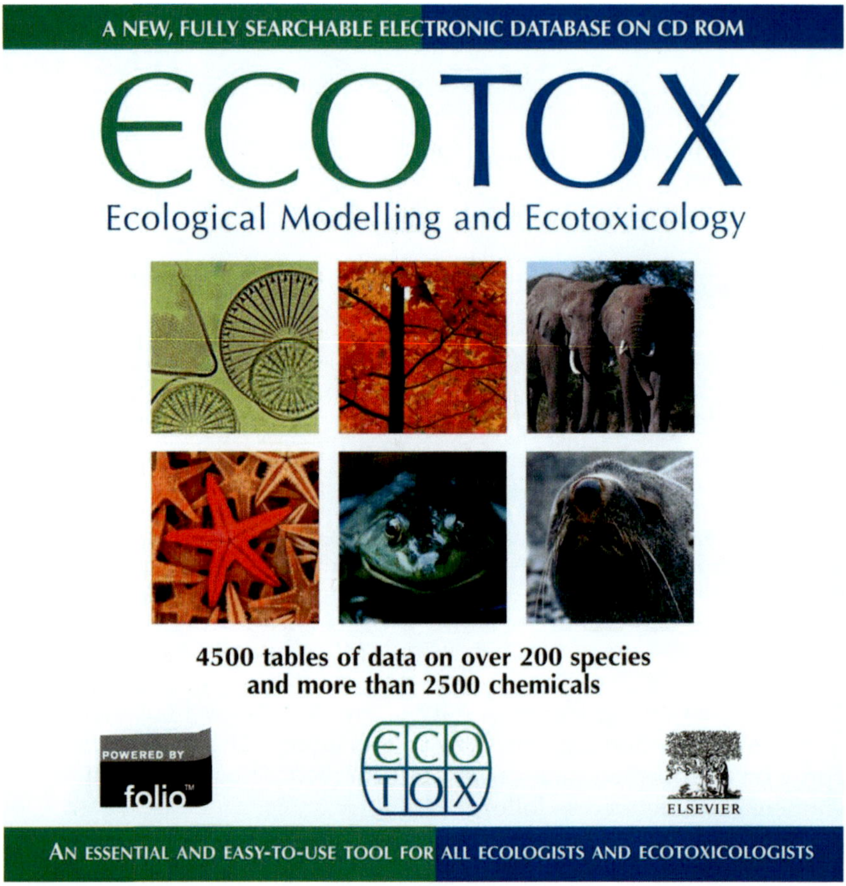

FIGURE 8.5
One of the major literature sources for parameters. (Adapted from Jørgensen, L.A., S.E. Jørgensen, and S.N. Nielsen. 2000. ECOTOX CD.)

A sensitivity analysis of forcing functions is not in principle different from the sensitivity analysis of parameters, except that forcing functions are often indicated as a table function of the time or by an equation. The table values or the equation are multiplied by factors that consider the possible uncertainty. If the uncertainty is estimated to be from for instance 10%–50% the factors 0.5, 0.9, 1.1, and 1.5 are used and it is recorded how much these factors would change the focal state variable. Based upon these results it is possible to determine how accurately we have to know the forcing functions as a function of the time for a required certainty for the focal state variable.

A sensitivity analysis of submodels (different process equations) can also be carried out. Then the change in a state variable is recorded when the equation of a submodel is deleted from the model or changed to an alternative expression, for instance, with more details built into the submodel. Such results may be used to make structural changes in the model. If the sensitivity, for instance, shows that it is crucial for the model results to use a more detailed given submodel, this result should be used to change the model correspondingly. The selection of the complexity and the structure of the model should therefore work hand in hand with the sensitivity analysis.

The fourth modeling step encompasses calibration and validation. The scope of the calibration is to improve the parameter estimation. Some parameters in causal ecological models can be found in the literature, not necessarily as constants but as approximate values or intervals. To cover all possible parameters for all possible ecological models including ecotoxicological models, we need, however, to know more than 10 billion parameters. It is therefore obvious that in modeling there is a particular need for parameter estimation methods. A description of these methods can be found in Jørgensen and Fath (2011). Chapter 7 mentions applicable methods for the development of ecotoxicological parameters. Under all circumstances it is a great advantage to give even approximate values of the parameters before calibration is started, because it is much easier to search for instance for a value between 1 and 4 than to search between 0 and 8.

Even where all parameters are known within intervals either from the literature or from estimation methods, it is in most cases necessary to calibrate the model. Several sets of parameters are tested by the calibration and the various model results of state variables are compared with the observed values of the same state variables. The parameter set that gives the best agreement between model results and measured values is chosen.

The need for the calibration can be explained by use of the following characteristics of ecological models and their parameters:

1. Most parameters in environmental science and ecology are not known as exact values. Therefore, all literature values for parameters (Jørgensen et al., 1991, 2000; see Figure 8.5) have a certain uncertainty. Parameter estimation methods must be used, when no literature value can be found. It is important particularly for ecotoxicological models.
2. All models in ecology and environmental sciences are simplifications of nature. The most important components and processes may be included, but the model structure does not account for all possible details. To a certain extent the influence of some unimportant components and processes can be taken into account by the calibration. This will give values for the parameters slightly different from the real, but still unknown, values in nature. The difference may partly account for the influence from the omitted details.
3. By far most models in environmental sciences and ecology are "lumped models," which implies that one parameter may represent

> the average values of several species. As each species has its own characteristic parameter value, the variation in the species composition with time will inevitably give a corresponding variation in the average parameter used in the model. Adaptation and shifts in species composition will require other approaches, the so-called structurally dynamic models (SDMs).

Calibration cannot be carried out randomly if more than a couple of parameters have been selected for calibration. If, for instance, 10 parameters have to be calibrated and the uncertainties justify the testing of 10 values for each parameter, the model has to be run 10^{10} times, which of course is an impossible task. Therefore, the modeler must learn the behavior of the model by varying one or two parameters at a time and observing the response of the most crucial state variables. In some (few) cases, it is possible to separate the model into several submodels, which can be calibrated approximately independently. Although the calibration described is based to some extent on a systematic approach, it is still a trial-and-error procedure. However, procedures for automatic calibration are available. This does not mean that the trial-and-error calibration described above is redundant. If the automatic calibration should give satisfactory results within a certain frame of time, it is necessary to calibrate only 4–8 parameters simultaneously. In any circumstances it will become easier to find the optimum parameter set, the smaller the uncertainties of the parameters are, before the calibration gets started.

In the trial-and-error calibration, the modeler has to set up, somewhat intuitively, calibration criteria. For instance, you may want to simulate accurately the minimum oxygen concentration for a stream model and/or the time at which the minimum occurs. When you are satisfied with these model results, you may then want to simulate the shape of the oxygen concentration versus time curve, and so on. You calibrate the model step by step to achieve these objectives step by step.

If an automatic calibration procedure is applied, it is necessary to formulate objective criteria for the calibration. A possible criterion could be to find the parameter set which yields the smallest standard deviation between modeled and observed values of all the state variables or a number of selected state variables. Often, the modeler is most interested in a good agreement between model output and observations for one or two state variables, which are considered the main model results, while he is less interested in a good agreement with the other state variables. He may furthermore choose weights for the various state variables to account for the emphasis he puts on each state variable in the model. For a model of the effect of an insecticide he may put emphasis on the toxic substance concentration in carnivorous insects and he may consider the toxic substance concentrations in plants, herbivorous insects and soil to be of less importance. He may, therefore, choose a weight of 10 for the first state variable and only one or two for the subsequent three.

If it is impossible to calibrate a model properly, it is not necessarily due to an incorrect model, but may be due to poor quality of the data. The quality of the data is crucial for calibration. It is, furthermore, of great importance that the observations reflect the real dynamics of the system. If the objective of the model is to give a good description of one or a few state variables, it is of course particularly essential that the data can show the dynamics of just these internal variables. The frequency of the data collection should therefore reflect the dynamics of the focal state variables. This rule has unfortunately often been violated in modeling.

It is strongly recommended that the dynamics of all state variables are considered before the data collection program is determined in detail. Frequently, some state variables have particularly pronounced dynamics in specific periods—often in the spring—and it may be of great advantage to have a dense data collection in this period in particular. Jørgensen et al. (1981) and Jørgensen and Fath (2011) show how a dense data collection program in a certain period can be applied to provide additional certainty for the determination of some important parameters.

From these considerations, recommendations can now be drawn up about the feasibility of carrying out a calibration of a model in ecology:

1. Find as many parameters as possible from the literature; see Jørgensen et al. (1991 and 2000); see Figure 8.5. Even a wide range for the parameters should be considered very valuable, as approximate initial guesses for all parameters are urgently needed.
2. If some parameters cannot be found in the literature, which is often the case, estimation methods should be used. For crucial parameters it may be better to determine them by experiments *in situ* or in the laboratory.
3. A sensitivity analysis should be carried out to determine which parameters are most important to be known with high certainty.
4. The use of an intensive data collection program for the most important state variables should be considered to provide a better estimation for the most crucial parameters. It is generally not possible to uncover the dynamics of state variables by observations with too little frequency that do not reflect the dynamics of the system. For instance if a model considers the diurnal variation of photosynthesis and respiration for plants, or the diurnal variations are of utmost important for the core problem of the model, it is absolutely necessary to build the model on basis of observations several times per 24 hours. Further details about the method see Jørgensen and Fath (2011).

8.5 Overview of Available Ecological Models

It is feasible in the modeling literature to find models of the following ecosystems:

1. Lakes
2. Estuaries
3. Rivers
4. Coastal zones
5. Coastal lagoons
6. Open sea ecosystems
7. Grasslands
8. Savannas
9. Forests
10. Polar ecosystems
11. Mountain ecosystems
12. Coral reefs
13. Wetlands (various types of wetlands, wet meadows, bogs, swamps, forested wetlands, marshes, and floodplains)
14. Deserts
15. Agricultural systems
16. Aquacultures
17. Wastewater systems

It is furthermore possible to find models focusing on the following environmental problems:

1. Oxygen depletion
2. Eutrophication
3. Acidification
4. Pollution by toxic organic compounds, including pharmaceuticals and endocrine disruptors
5. Pollution by heavy metals
6. Control of fishery
7. Pollution of groundwater
8. Planning of landscapes
9. Global warming and climate changes
10. Decomposition of the ozone layer and its effects

11. Groundwater pollution
12. Spreading of fire
13. Air pollution

It is possible in the modeling literature to find papers dealing with almost all combinations of these $13 \times 17 = 221$ types of models, although it will be hard to find combinations of pollution problems and polar ecosystems or mountain ecosystems, at least for mountain ecosystems above the timberline. Furthermore, models of the decomposition of the ozone layer, mainly the chain of chemical processes in the ozone layer, have been developed. It implies that if we classify models in accordance to the ecosystem *and* the pollution problem they attempt to solve, we will have approximately 200 classes of models—slightly less than the above mentioned 221 possibilities.

In addition, a number of ecological models have been developed to answer ecological scientific questions as for instance how important is such and such process for the resistance (buffer capacity) of the considered ecosystem or what is the results of such and such chain of processes? By a statistical examination of the journal *Ecological Modelling* (2012–2013, it was found that about 15% of the models published in the journal were papers focusing on a better understanding of the importance of components and processes and forcing functions for the reaction of ecosystems.

We can divide the ecosystem models, 1–17, into five groups according to how many different models are possible to find in the literature and not counting minor modifications as a change of model:

I. Ecosystems that have been modeled heavily and where it is possible to find several hundreds of different models in the model literature: rivers, lakes, forests, and agricultural systems.

II. Ecosystems for which it is possible to find in the order of a hundred of different models: estuaries, wetlands, and grasslands.

III. Ecosystems that have been modeled many times but still less than a hundred different models have been developed: wastewater systems and aquaculture systems.

IV. Ecosystems that have been modeled more than 10 times but less than 25–30 times: coral reefs.

V. Ecosystems which have only been modeled a couple of times up to a handful of times: polar ecosystems, savannas, and mountain ecosystems above the timberline.

Similarly, we can divide the models of pollution problems into five groups:

I. Heavily modeled pollution problems: oxygen depletion, eutrophication, organic toxic substances including pharmaceuticals in the environment, air pollution problems, and global warming including impacts of climate changes.

II. Environmental problems that have been modeled approximately a hundred times: heavy metal problems, acidification, and groundwater pollution.
III. Environmental problems that have been modeled many times but less than a hundred different models have been developed: fire spreading and overfishing.
IV. Environmental problems that have been modeled more than 10 times but less than 25–30 times: endocrine disruptors.
V. Environmental problems for which it is not possible to find more than at the most a couple of models or a handful of models in the model literature: application of GMOs.

8.6 Model Types

A number of new model types have been developed mainly during the last two decades to answer a number of relevant modeling problems or questions which arose as a result of the increasing use of ecological models in the 1970s. Seven relevant modeling questions already formulated around 1980 as a result of the model experience gained in the 1970s were attempted to be answered with the new model types, mainly developed from 1990 to 2015:

1. How can we describe the spatial distribution which is often crucial to understanding ecosystem reactions and to selecting the best environmental strategy?
2. Ecosystems are middle number systems (Jørgensen, 2002) in the sense that the number of components are magnitudes smaller than the number of atoms in a system. All the components are different and that is often important for a proper description of the ecosystem reactions to consider the differences in properties among individuals.
3. The species are adaptable and may change their properties to meet the changes in the prevailing conditions—which means the forcing functions. Furthermore, the species may by more drastic changes be replaced by other species better fitted to the combinations of forcing functions. How then to account for these changes? The networks may also be changed if more biological components with very different properties are replaced by other species. How do we then account for these structural changes?
4. Can we model a system that only has a poor database—a few data of only low quality?

5. The forcing functions and several ecological processes are in reality stochastic. How do we account for the stochasticity?
6. Can we develop a model, when our knowledge is mainly based on a number of rules, properties, and propositions?
7. Can we develop a model based on data from a wide spectrum of different ecosystems—this means that we have only a very heterogeneous database.

These problems could not be solved by the "old" model types: biogeochemical models, population dynamic models, and steady-state models.

Spatial models often based on the use of GIS—Geographical Information System—have been developed to come up with an answer to question 1.

Individual-based models—abbreviated IBMs—are able to give an answer to question 2. Software, which can be used to develop IBMs is even available to facilitate IBM development. The software can also be utilized to cover spatial distribution (see question 1).

Structurally dynamic models—SDMs—have been developed to solve the problem expressed in question 3.

Fuzzy models can be used to make models based on a poor and maybe only a semiquantitative database.

Stochastic models were applied in the 1970s in a very few cases, but they are further developed today, although the application is still not wide, probably because it is not frequently that there is an urgent need to include stochastic processes in ecological models.

IBMs can in many cases meet the demands expressed in question 6.

Artificial neural networks—ANNs—give a good solution to the problem formulated in question 7.

Sometimes, ecotoxicological models are considered a special type of model, although they are developed similar to other biogeochemical models. They have been widely used, particularly in the last 10–15 years, because the need for these models is very obvious: for environmental risk assessment of chemicals. The two chapters on heavy metal pollution and medicine (pharmaceuticals) in the environment represent these types of models. All the other types of models can be found in the various chapters as examples presented in more detail or mentioned by the model overview given in most chapters.

Table 8.2 provides a summary of model statistics, based on the number of publications in the journal *Ecological Modeling*. The percentage application of the most general model types is compared for the period 1975–1980 with the period 2001–2009; see also Jørgensen et al. (2009). Ecotoxicological models are

TABLE 8.2

Application of the Most General Model Types for the Period 1975–1980 with the Period 2001–2009

	% Application 1975–1980	% Application 2001–2009
Dynamic biogeochemical models	62.5	30.8
Steady-state biogeochemical models	0	1.8
Population dynamic models	31.0	24.2
Spatial models	0	19.9
Structurally dynamic models	1.5	8.2
Individual-based models and cellular automata	0	5.9
ANN and use of artificial intelligence	0	5.4
Fuzzy models	0.5	1.8
Ecotoxicological models	0	2.0

included as a model type. Although they are constructed similarly to biogeochemical models, they are, as indicated above, often considered as a separate model type because they are an important subset of biogeochemical models and have some characteristic properties. In the table, we have distinguished between nine model types:

1. Dynamic biogeochemical models
2. Steady-state biogeochemical models
3. Population dynamic models
4. Spatial models
5. Structurally dynamic models, SDM
6. Individual-based models, IBM
7. Ecotoxicological models
8. Fuzzy models
9. ANN models

The number of papers published during the period 2001–2010 is about nine times the number of papers published per year during the period 1975–1980. This means that the number of dynamic biogeochemical model papers published recently is still more than 4.5 times the number published during the late 1970s, and that the number of papers on structurally dynamic modeling has increased by a factor of almost 50 during approximately the last 40 years. It is of course not surprising as the new types of models have been developed because there was an urgent need to answer the seven modeling problems listed earlier.

EXAMPLE 8.2

Which types of models would you apply to solve the following problems:

a. Protection of a lion population in a national park?
b. Construction of a wetland for removal of nitrate from agricultural drainage water?

Solution

a. A population dynamic model
b. A biogeochemical model

With the present spectrum of model types it is possible to solve all the major modeling problems that started to be raised as mentioned above already in the 1970s. This development has of course increased the application of ecological models in general and particularly the use of the new model types. However, it is also clear that all the problems cannot be solved completely satisfactory. We still have a few problems that may not be possible to solve by use of a single model type. A very complicated problem probably requires the use of hybrid models, that is, by a combination of the model types presented here.

New model types will inevitably be developed during the coming decade to solve the most complicated ecological and environmental problems which we have still not properly solved. It is, however, agreed among ecological modelers, that we currently have sufficient model types to solve most of ecological modeling problems being faced by us today. A few modeling case studies need, however, a proper model development in real situations.

8.7 Institutionalized or Mediated Modeling

There are thousands of ecological–environmental models of ecosystems available in the literature. It is, however, often better to develop a tailored model for a given environmental problem for a considered ecosystem. Every ecosystem is different and the environmental problems are different in the sense that they have different implications for different ecosystems and for the local population dependent on the natural resources of the ecosystems. The tailored model can be developed by an expert team, the model is denoted an expert model, or by a brainstorming meeting of *all* local stakeholders together with a model team. This latter type of model is denoted as a mediated/institutionalized model. The experience shows, however, that both standard (nontailored) models and expert models, are often not applied, or at least not sufficiently to set up management plans, because they are not accepted by local organization as an integral part of their environmental management, and they do not consider the relevant socioeconomic aspects for most of the models. The best is therefore to develop a mediated or institutionalized model, that is developed step by step and accepted by local stakeholders together with a very few nonlocal modeling experts. The following procedure can be applied:

1. Arrange a 1-week brainstorming session where the modelers, the NGOs, the local scientists, and managers formulate the problem and the characteristics of the ecosystem and how the problems associated with the ecosystem interfere with other local environmental–economic–social problems. A multidisciplinary and interdisciplinary team is needed. An overview of the knowledge about the ecosystem is also provided.
2. The result of the workshop is a conceptual diagram of the model which includes the other interfering local problems. It cannot be excluded that two or more models are needed to solve complex problems. It cannot be excluded that supplementary knowledge and observation are needed.
3. The modeling team develops an easy-to-go software for the model according to the conceptual diagram.
4. The model and its possibility as a supporting tool in management are demonstrated by the modelers for the entire brainstorming team. The entire team is working with the models together for at least 3 days.
5. The managers and other personnel that should be able to run the models are trained (probably 4–10 days depending on the model complexity) in the use of the model.
6. The entire team has an annual or biannual follow-up meeting to discuss the application of the model and possible minor modifications of the model.

The following questions arise generally as result of model development:

- How is it possible to consider all the different aspects of the problem and its solution, when several aspects of the problem are to be considered?—There may be different natural science aspects, for instance geology, zoology, botany, and chemistry, or it may even be a combination of the environmental, economic, and social aspects of the problem.
- The answer is that by implementing a very wide spectrum of expertise in the modeling team, but this gives rise to the next obvious question: how to ensure a good cooperation of the team members,

when they represent many different disciplines and therefore many different opinions and "languages?"

- How is it possible to consider *all* relevant ecosystem properties at the same time?
- How is it possible to integrate insights from many different disciplines, for instance from ecology, economics, and social science, or even different disciplines of natural sciences?
- How can we ensure that all important stakeholders are included in the modeling process?
- How is it possible to integrate impacts and knowledge at different scales?
- How is it possible to understand the very root of the problems and their sources and have this understanding been reflected in the modeling and the final model result?
- How is it possible to build the best of a consensus among the different opinions and disciplines?

Institutionalized or mediated modeling (abbreviated IMM) can answer these questions. The main idea is to represent without exception *all* stakeholders, policy makers, managers, and scientists that have knowledge and ideas about the problem, the system and the possible solutions in the modeling procedure. The model is developed as a result of an integrated brainstorming meeting, where all ideas, opinions, disciplines, and knowledge are represented. For the development of most mediated models, though of course dependent on the complexity of the problem and the system, several days are required to reach a satisfactory model which can be used as a tool to solve the problem. The advantages of IMM are that (partly taken from van den Belt, 2004):

1. The level of shared understanding increases.
2. A consensus is built around the structure of a complex problem for a complex system, because all interests are represented in the stepwise model development.
3. The result of the modeling process, the model, serves as a tool to disseminate the insights gained by the modeling procedure.
4. The effectiveness of the decision making is increased, because the mediated model makes it possible for the policy makers and the stakeholders to see the consequences of the action plans over longer timescales.
5. Team building is developed parallel to the model development.
6. The process is emphasized over the product.
7. State-of-the-art knowledge is captured, organized, and synthesized.

It is characteristic for a team developing a mediated model that what could be denoted as "groupiness" is increased because

I. The individual members perceive clearly that they are a part of the group
II. The members become oriented toward a common goal
III. Interaction between group members takes place
IV. The interdependence is realized and acknowledged
V. A structure of roles/status and norms is built

8.8 Model Selection

Standard models are available for some basic ecosystem problems as for instance eutrophication, water level problems, acidification, oxygen depletion, distribution of air pollutants, bioaccumulation of toxic substances, water treatment by use of wetlands, and so on. However, it is better to tailor the model according to the combination of observations, data, problem, and ecosystem characteristics. Expert models are more expensive to apply than standard models because they are more time-consuming. Development of mediated models based on brainstorming meetings will inevitably cost more than the application of standard models or even more than expert models due to the cost of the brainstorming meetings. In most cases, however, institutionalized/mediated models are a good investment because the models *are* applied and *accepted* by the local organization, the links to other local environmental and socioeconomic problems are considered, and more holistic and sustainable solutions to the environmental problems are proposed.

Which model to apply among the standard models is to a high extent a question about the database, the problems and the characteristics of the lake. This question is discussed below on basis of a brief overview of the available models. A comprehensive overview of models can be found in Jørgensen et al. (1997, 2009), Jørgensen and Fath (2011), and Jørgensen (2011). Table 8.3 gives an overview and synthesizing results of different environmental strategies and different model complexity.

The cost indicated in the table is a range, where the higher values should be valid for large ecosystems with complex hydrodynamics and the lower values for relatively small ecosystems with simple hydrodynamics. The cost indicated is only for the model development, but includes the cost of the meetings leading to the model for the tailored models. If the database is insufficient the cost of updating the database must of course be added to the overall costs. The costs for the standard models are mainly manpower for the calibration and validation of the model and organization of the database to

TABLE 8.3

Possible Model Strategies

Model	Database Required	Cost (10^3 dollars)
Tailored expert model, high complexity	Comprehensive	400–1000
Tailored mediated model, high complexity	Comprehensive	500–1200
Tailored expert model, medium complexity	At least for the st. var.	300–600
Tailored mediated model, medium complexity	At least for the st. var.	350–700
Tailored mediated simple model	Small	120–300
Standard model, complex	Comprehensive	100–250
Standard model, medium complexity	At least for the st. var.	50–150
Standard model, simple	Small	25–75

be accessible for the model. For the expert models, the cost includes the cost of the experts meeting and the experts' time and for the mediated models, the cost also includes the brainstorming meetings.

As an example of standard models, UNEP-IETC offers downloadable standard models of different complexity for lakes and reservoirs:

I. A *one-layer model* that is based on four variables only (total nitrogen in the water phase, total phosphorus in the water phase, phosphorus in the sediment, and nitrogen in the sediment). The model is not able to offer a high accuracy of the prognoses; but can be developed based on a limited database. The model is not considering two or more layers, but it is possible to account for the formation of a thermocline by a correction factor. Chlorophyll a concentration, zooplankton concentration, productivity, fish concentration, and transparency can all be calculated based upon general correlations. It means that the calculations of course have a relatively high standard deviation.

II. A *two-layer model* that gives better prognoses but also requires a much more comprehensive database.

III. A *structurally dynamic model for deep lakes* that considers adaptations and shifts in species composition. It is easier to use than II because it contains facilities to automatically calibrate some of the model parameters. It is recommended to use this standard model instead of II. It can capture the structural changes as a result of biomanipulation and other ecotechnological restoration methods.

IV. A *structurally dynamic model for shallow lakes* that is also easier to use than II. It is able to cover the structural changes that may result from competition between submerged vegetation and phytoplankton.

If the environmental problem is important and complex, and it is not possible to provide sufficient finance for the development of a mediated or institutionalized model, it is recommended to use a standard or expert model; but it is *very important* that the selection is made by a meeting between local scientists, local managers and model experts. It is, however, the best long-term solution as indicated above to develop a mediated/institutionalized model.

Table 8.3 gives an overview of the model possibilities and gives some first guidance on which model to apply in which case. Follow, however, the discussion of the selection problem presented below to make the final decisions together with the information summarized in Table 8.3. Compare the cost of the model with the economic importance of the ecosystems, including the values of the ecosystem services. If the ecosystem is large and of great local importance, it would most probably be the best long term solution to invest more money in a proper management model including the erection of a suitable database that is able to support the model. The cost of developing a good model for a given case is often 1% or less of the economic value of the ecosystem. In other words, if we avoid a more than 1% mistake in our management decisions by using a good model, we have more than paid for the model. If the freshwater ecosystem is only important for a limited population, and it is a smaller ecosystem, it may be sufficient to use a standard model and in some cases where the needed observations are not available for a complex standard model, a simple model seems to be the best selection at least in the first hand. For instance when eutrophication is the problem for a lake, the above mentioned one-layer model, will most probably be able to do the job in the first hand. It is always beneficial to apply a model which is easily accessible at least in the first stage. The results can often be used to make a better model selection afterward because the unanswered questions are revealed.

Models are being used increasingly in environmental management, primarily because they are the only tools which have the ability to quantitatively relate the impact on an ecosystem with the consequences for the state of the ecosystem. Ecosystem models have been developed particularly during the last few decades, and it is not surprising they have found a wide application in ecosystem management.

The field of ecological and environmental modeling has developed rapidly during the last two to three decades essentially due to the following factors:

- The development of computer technology enabling us to handle very complex mathematical systems.
- A general understanding of pollution problems, including the knowledge that a complete elimination of all pollution sources ("zero discharge") is not feasible, but rather that proper pollution control with often-limited economic resources requires serious consideration of the influence of various pollution impacts on ecosystems.
- Our knowledge of environmental and ecological problems has increased significantly. We have particularly gained more knowledge about quantitative relations in ecosystems, and between ecological properties and environmental factors (see also the discussion in Chapter 1).

8.9 Models as a Strong Management Tool: Problems and Possibilities

A management problem to be solved often can be formulated as follows: If certain forcing functions (management actions) are varied, what will be their influence on the state of an ecosystem? The model is used to answer this question or, in other words, to predict what will change a system when the control functions that are managed by humans are varied over time and space.

Typical *control functions* are the consumption of fossil fuel, the regulation of the water level in a river by a dam, the discharge of pollutants, or a fisheries policy.

It is important that, to a certain extent, the manager should take part in the entire development of a management model, since he will ultimately define the modeling objectives and select the modeling scenarios. The success of the application of a management model, to a large degree, is dependent on an open dialog between the modeler and the manager.

A further complexity is the construction of ecological-economic models. As we gain more experience in constructing ecological and economic models, more and more of them will be developed. It often is feasible to find a relation between a control function and economic parameters. If a lake for instance is a major water resource, an improvement in its water quality will inevitably result in a reduction in the treatment costs of drinking water if the same water quality is to be provided. It is also possible sometimes to relate the value of a recreational area to the number of visitors, and to how much money they spend on average in the area. In many cases, however, it is difficult to assess a relationship between the economy and the state of an ecosystem. For example, how can we assess the economic advantages of increased transparency in a water body or increased visibility due to reduced air pollution? Ecological-economic models are useful in some cases, but should be used with caution, and the relations between the economy and environmental conditions critically evaluated, before the model results are applied.

Data collection is the most expensive step of the model construction. For many lake models, it has been found that needed data collection comprises 80%–90% of the total model costs. Because complex models require much more data than simple ones, the selection of the complexity of environmental management models should be closely related to the costs involved in the environmental problem to be solved. Thus, it is not surprising that development of the most complex environmental management models has generally been limited to large ecosystems, where the economic involvement is great.

The predictive capability of environmental models can always be improved in a specific case by expansion of the data collection program, and by a correspondingly increased model complexity, provided that the modelers are sufficiently skilled to know in which direction further expansion of

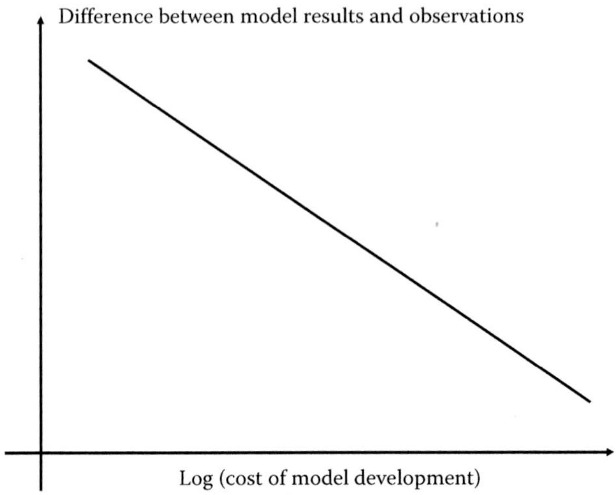

FIGURE 8.6
The more a modeler invests in a model and in data collection, the closer he/she will come to realistic predictions. However, the modeler will always gain less for the next dollar invested, and will never be able to give completely accurate predictions. With good approximations, log (costs) versus the difference between model results and observations is a straight line, corresponding to an exponentially decreasing difference with an increasing investment in the model development.

the entire program must develop in order to improve the model's predictive capabilities.

The relation between the economy of the project and the accuracy of the model is presented in the form shown in Figure 8.6. The reduction in the discrepancy between model predictions and reality is lower for the next dollar invested in the project, because the log (cost) versus difference between model results and observations gives a straight line. But it is also clear from the shape of the curve that the associated errors can hardly be completely eliminated. All model predictions have a standard deviation associated with them. This fact is not surprising to scientists, but it often is not understood or appreciated by decision makers, to whom the modeler typically presents his or her results.

Engineers use safety factors to assure that a building or a bridge will last for a certain period of time, with a very low probability of breakdown, even under extreme conditions. No reputable engineer would propose using a smaller, or no safety factor to save some concrete and reduce the costs. The reason is obvious: Nobody would want to take the responsibility for even the smallest probability of a building or bridge collapse.

When decision makers are going to make decisions on environmental issues, the situation is strangely different. Decision makers in this situation want to use the standard deviation to save money, rather than assuring a high environmental quality under all circumstances. It is the modeler's duty, therefore, to carefully explain to the decision maker all the consequences of the various decision possibilities. A standard deviation of a prognosis for an environmental management model cannot, however, always be translated into a probability, because we do not know the probability distribution. It might be none of the common distribution functions, but it is possible to use the standard deviation qualitatively or semiquantitatively, translating the meaning of the model results by the use of words. Civil engineers are more or less in the same situation, and have been successful in the past convincing decision makers of appropriate steps to be taken in various situations. There is no reason that environmental modelers cannot do the same.

It is often advantageous to attack an environmental problem in the first place with the use of simple models. They require much fewer data, and can give the modeler and decision maker some preliminary results. If the modeling project is stopped at this stage for one or another reason, a simple model is still better than no model at all, because it will at least give a survey of the problem.

Simple models, therefore, are good starting points for the construction of more complex models. In many cases, the construction of a model is carried out as an iterative process, and a step-wise development of a complex structurally dynamic model may be the result. As previously mentioned, the first step is the development of a conceptual model. It is used to get a survey of the processes and state variables in the concerned ecosystem. The next step is the development of a simple calibrated and validated model. It is used to establish a data collection program for a more comprehensive effort closer to the final selected version. However, the third model will often reveal specific model weaknesses, the elimination of which is the goal of the fourth version of the model. At first glance, this seems to be a very cumbersome procedure. However, because data collection is the most expensive part of modeling, constructing a preliminary model for optimization of the data collection program will ultimately require less financial resources.

At first, a simple mass-balance scheme is recommended for biogeochemical models. The mass balance will indicate what possibilities exist for decreasing or increasing the concentration of a chemical or pollutant, which is a crucial issue for environmental management.

Point sources of pollution are usually easier to control than anthropogenic nonpoint sources which, in turn, are more easily controlled than natural pollutant sources. A distinction can be made between local, regional, and global pollutant sources. Because the mass balance indicates the relative quantities from each source, it is possible to identify which sources should receive the initial attention (e.g., if a nonpoint regional source of pollutants is dominant, it would be pointless to concentrate first on eliminating small, local point sources, unless the latter also might have some political influence on regional decisions).

It has already been recognized that the modeler and the decision maker should communicate with each other. It is recommended, in fact, that the decision maker is invited to follow the model construction process from its very first phases, in order to become acquainted with the model strength and limitations. It is also important that the modeler and the decision maker together formulate the model objectives and interpret the model results. Moreover, they should work together in all phases of the modeling exercise. Having the modeler first build a model, and then transfer it to a decision-maker accompanied by a small report on the model, is not recommended.

Communication between the decision maker and the modeler can be facilitated in many ways, and it is often the primary responsibility of the modeler to do so. If a model is built as a menu system, it might be possible to teach the decision maker how to use the model in only a few hours, thereby also increasing his or her understanding of the model and its results. If an interactive approach is applied, it is possible for the decision maker to visualize a wide range of possible decisions. The effect of this approach is increased by the use of various graphic methods to illustrate the best possible decision in regard to what happens with the use of various management strategies. Under all circumstances it is recommended that time be invested in developing a good graphic presentation of the model results to a decision maker. Even if he or she has been currently informed about a model project through all its phases, the decision maker will not necessarily understand the background and assumptions of all the model components. Thus, it is important that the model results, including the main assumptions, shortcomings, and standard deviations underlying them, are carefully presented with the use of an illustrative method.

It is clear that we are not yet sufficiently advanced in environmental modeling efforts to solely use model results to define management programs, even utilizing expert systems and decision-support systems. A model should never be used in this way by a decision maker, but rather should be considered one useful tool in the management decision-making process. This implies that modeling results should be clearly and illustratively presented, and be considered a significant component in discussions about selecting specific courses of action. Other elements to be considered in such discussions would include potential side effects, interpretation of model predictions, and the implications of the accuracy of the prognosis.

A good environmental model can be a powerful tool in the decision-making process for management actions. A wide range of environmental problems has been modeled to varying degrees over the last 10–15 years. They have generally been of importance in assistance to decision makers. With the continuing rapid growth in the use of environmental models, the situation will only improve in the future. However, we have not achieved the same level of experience for all environmental problems.

The use of models in environmental management is definitely growing. They have been widely used in several European countries, in North America and in Japan. Further, environmental agencies in more and more countries are making use of model applications. Through such journals as *Ecological Modelling* and the *International Society for Ecological Modelling* (ISEM), it is possible to follow the progress in the field. This "infrastructure" of the modeling field facilitates communication and accelerates the exchange of experiences, thereby enhancing the growth of the entire field of ecological modeling.

It should not be forgotten in this context that models have always been applied in science. The difference between present and previous models is that today, with modern computer technology, we are able to work with very complex models. At the same time, however, it is a continuing temptation to construct models that are too complex. It is easy to add more and more equations and state variables to a computer program, but much harder to get the data needed for calibration and validation of the model. Even if very detailed knowledge about a problem is available, it is not possible to develop a model capable of accounting for the complete input–output behavior of a real ecosystem and being valid for all frames (Ziegler, 1976). This type of model is called "the base model" by Ziegler, and would be very complex and require such a large number of computational resources that it would be almost impossible to simulate. The base model of a problem in ecology will never be fully known because of the complexity of natural ecosystems, and our inability to observe all its states. Thus, a model may be made more realistic, up to a point, by adding more connections. However, addition of new parameters after that point is reached does not contribute further to an improved simulation capacity. Indeed, more parameters imply more uncertainty because of the possible lack of information about the flows the parameters quantify.

For a given quantity of data, the addition of new state variables or parameters beyond a certain model complexity does not increase our ability to model an ecosystem, but only adds to the unaccountable uncertainty. The question that can be formulated with relation to this problem is: *How can we select the complexity and the structure of the model to assure the optimum knowledge gained or the best answer to the question posed by the model?*

Costanza and Sklar (1985) examined 88 different models, and were able to show that the best model efficiency is obtained by a medium complexity or as they expressed it: the best model is a trade-off between knowing nothing about everything—the model is too complex—and knowing everything about nothing—the model is very accurate but too simple.

Which components and processes are most important for the concerned problem? Know the ecosystem and the problem being addressed before a model is selected, including the model complexity. Thus, the conclusion is: a proper overview and a holistic picture of a given ecosystem are crucial for the right selection of model complexity. Some details are of course needed to understand how the ecosystem works on the systems level. Thus, the additional conclusion is therefore that, although we can never know everything about an ecosystem needed to make a complete model (i.e., with inclusion of all details), good, workable models that expand our knowledge of ecosystems can be produced, particularly of their properties as systems. Ulanowicz (1979) points out that a correct, very precise predictive model cannot be constructed. Thus, it would often be most fruitful to build a model that illustrates general trends, taking into account the probabilistic nature of the environment.

Further, models are, and can be applied, as management tools (e.g., see Jørgensen, 1986; Jørgensen and Fath, 2011). All in all, models should be considered as tools—tools to provide an overview of complex ecosystems. For a complex system, a few interactive state variables already make it impossible to get an intuitive or logical overview on how the system reacts to perturbations or other changes. In such cases, models are the good tools to help to obtain an overview or picture of the properties of an ecosystem on the systems level.

There are only two possibilities for getting around the dilemma rooted in the model complexity issue: either limit the number of state variables in the model or describe the system by the use of holistic methods and models, preferably by using higher level scientific laws. *The trade-off for the modeler is between knowing much about little, or little about much!*

More complex models require more data and more knowledge about the ecosystem. They imply higher costs, but they should be justified by the importance of the pollution study. Thus, it is recommended to proceed in a step-wise manner toward a more complex learning from simpler models before the complex model is constructed. A procedure for this step-wise development is presented and discussed by Jørgensen and Fath (2011).

Another crucial problem is associated with a model generality. Can a model used in, for instance, for one lake be used unchanged for other case studies? The answer is not a simple yes or no. However, experience shows that simpler models can be used more generally than more complex models. They contain a description of the basic processes characterizing aquatic ecosystems (e.g., nutrient uptake by phytoplankton, which is dependent on the nutrient concentration and the concentration of phytoplankton; mineralization of detritus). The more complex models inevitably will contain more site-specific process descriptions, which may not be important for all lake and reservoir case studies. Thus, more complex models will generally have to be modified from case to case. This is illustrated in Table 8.4, in which the experience gained with the general use of an eutrophication model containing 17–20 state variables is presented.

EXERCISES AND PROBLEMS

1. Find the connectivity of the model conceptualized in Figure 5.1.
2. Develop a conceptual model for the biomagnification of a toxic substance through a food chain with primary producers, primary consumers, and secondary consumers.

TABLE 8.4

Survey of Eutrophication Studies Based upon the Application of a Modified Glumsø Model

Ecosystem	Modification	Level[a]
Glumsø, version A	Basis version	7
Glumsø, version B	Nonexchangeable nitrogen	7
Ringkøbing Firth	Boxes, nitrogen fixation	5
Lake Victoria	Boxes, thermocline, other food chain	4
Lake Kyoga	Other food chain	4
Lake Mobuto Sese Seko	Boxes, thermocline, other food chain	4
Lake Fure	Boxes, nitrogen fixation, thermocline	7
Lake Esrom	Boxes, Si-cycle, thermocline	4
Lake Gyrstinge	Level fluctuations sediment exposed to air	4–5
Lake Lyngby	Basis version	6
Lake Bergunda	Nitrogen fixation	2
Broia Reservoir	Macrophytes, 2 boxes	2
Lake Great Kattinge	Resuspension	5
Lake Svogerslev	Resuspension	5
Lake Bue	Resuspension	5
Lake Kornerup	Resuspension	5
Lake Søbygaard	SDM	7
Lake Balaton	Adsorption to suspended matter	2
Roskilde Fjord	Complex hydrodynamics	4
Lagoon of Venice	Ulva/Zostera competition	6
Lake Annone	SDM	6
Lake Balaton	SDM	6
Lake Mogan, Ankara	Only P cycle, competition submerged vegetation /phytoplankton + SDM	6
Stadsgraven, Copenhagen	4–6 interconnected basins	5 (Level 6: 1993)
Internal lakes of Copenhagen	5–6 interconnected basins	5

[a] Level 1: Conceptual diagram selected.
Level 2: Verification carried out.
Level 3: Calibration using intensive measurements.
Level 4: Calibration of an entire model.
Level 5: Validation. Object function and regression coefficient are found.
Level 6: Validation of a prognosis for significant changed loading or development of SDM (structurally dynamic model; see Sections 8.6 and 8.9).
Level 7: Validation of a prognosis and development of SDM.

3. Give examples of problems where the development of IMM would be a good solution. List the stakeholders interested in the problem.
4. Mention at least four factors that could limit the carrying capacity of a population in an ecosystem. With factors determining the carrying capacity, you should obviously use a logistic growth model, but

how could you consider the influence of four factors on the carrying capacity in your equations?

5. Develop a conceptual model for the oxygen balance in an aquatic ecosystem. The aquatic ecosystem receives discharge of organic matter and ammonium.

References

Costanza, R. and F.H. Sklar. 1985. Articulation, accuracy and effectiveness of mathematical models: A review of freshwater wetland applications. *Ecol. Model.* 27, 45–69.

Jørgensen, L.A., S.E. Jørgensen, and S.N. Nielsen. 2000. ECOTOX CD.

Jørgensen, S.E. 1986. Structural dynamic model. *Ecol. Model.* 31, 1–9.

Jørgensen, S.E. 2002. *Integration of Ecosystem Theories: A Pattern*. 3rd edition. Dordrecht, Boston, and London: Kluwer Academic Publisher, 428 pp.

Jørgensen, S.E. 2009. *Ecological Modelling: An Introduction*. Southampton: WIT, 190 pp.

Jørgensen, S.E., T. Chon, and F. Recknagel. 2009. *Handbook of Ecological Modelling and Informatics*. Wessex Institute of Technology (WIT), Southampton, UK, 432 pp.

Jørgensen, S.E. and B. Fath. 2011. *Fundamentals of Ecological Modelling*. 4th edition. Amsterdam: Elsevier, 400 pp.

Jørgensen, S.E., B. Halling-Sørensen, and H. Mahler. 1997. *Handbook of Estimation Methods in Ecotoxicology and Environmental Chemistry*. Boca Raton, Boston, London, New York, and Washington, DC: Lewis Publishers, 230 pp.

Jørgensen, S.E., L.A. Jørgensen, L. Kamp Nielsen, and H.F. Mejer. 1981. Parameter estimation in eutrophication modelling. *Ecol. Model.* 13, 111–129.

Jørgensen, S.E., S. Nors Nielsen, and L.A. Jørgensen. 1991. *Handbook of Ecological Parameters and Ecotoxicology*. Amsterdam: Elsevier. (Published as CD under the name ECOTOX, with L.A. Jørgensen as first editor in 2000.)

Palmeri, L., A. Barausse, and S.E. Jørgensen. 2013. *Ecological Processes Handbook*. Boca Raton, FL: CRC Press, 386 pp.

Ulanowicz, R.E. 1979. Prediction chaos and ecological perspective. In Halfon, E.A. (Ed.), *Theoretical Systems Ecology*. New York: Academic Press, pp. 107–117.

Van den Belt, M. 2004. *Mediated Modeling*. Washington, DC: Island Press, 340 pp.

Ziegler, B.P. 1976. *Theory of Modelling and Simulation*. New York: Wiley, p. 435.

9

Ecological Indicators as Diagnostic Tools

9.1 The Role of Ecosystem Health Assessment (EHA) in Environmental Management

The idea of applying an assessment of ecosystem health in environmental management emerged in the late 1980s. The parallel to the assessment of human health is very obvious. We go to our doctor to get a diagnosis (What is wrong? Why do I not feel completely healthy?) and hopefully initiate a cure to bring us back to normal (= healthy conditions). Your doctor will apply several indicators/examinations (pulse, blood pressure, sugar in the blood, and urine, etc.) before he will come up with a diagnosis and a proper cure. The idea behind the assessment of ecosystem health is similar; see Figure 9.1. We observed that an ecosystem is not healthy and want a diagnosis: what is wrong? What caused this unhealthy condition? And what can we do to bring the ecosystem back to normal? To answer these questions, and also to follow the results of the "cure," ecological indicators are applied, to provide synoptic information about the state of ecosystems, and their main attribute is the combination of several environmental factors in single values, which may be useful in terms of management and in the development of ecological concepts compliant to the general public's understanding (Marques et al., 2009).

Ecological indicators, in general, address the ecosystem's structure and/or functioning, accounting for a certain aspect or component, for instance nutrient concentration, water flow, macroinvertebrate and/or vertebrate diversity, plant diversity and productivity, erosion symptoms and, on occasion, ecological integrity at a system's level. They seek to provide quantitative representations of either the forces that steer an ecosystem, of responses to forcing functions, or of previous, current, or future states of an ecosystem, and are expected to reveal conditions and trends that will help in development planning and decision-making processes.

The application of ecological indicators in general has not been exempt of criticisms, the first of which is precisely what intends to be their main attribute, that is, that aggregation results in an oversimplification of the ecosystem under observation. Besides, problems arise from the fact that indicators account not only for numerous specific system characteristics, but also other

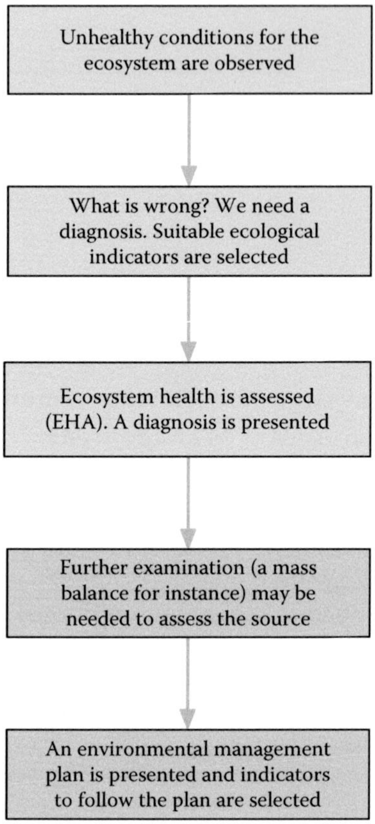

FIGURE 9.1
This figure illustrates how ecological indicators are used for EHA and how to follow the effects of the environmental management plan.

kinds of factors, such as physical, biological, ecological, socioeconomic, and so on. Therefore, indicators should forcibly be handled following the right criteria and in situations that are consistent with its intended use and scope; otherwise they may drive to confusing interpretations of data. Numerous attempts have therefore been carried out to operationalize the idea, and again and again environmental managers and ecologists have asked the question: which ecological indicators should we apply?

Obviously, the level of quality of any given indicator will always remain a matter of perspective. For instance, from a relatively holistic viewpoint, O'Connor and Dewling (1986) proposed, more than two decades ago, five criteria to define a suitable indicator for the assessment of ecosystem degradation, which in our opinion can still be considered up-to-date. A good indicator should be: (1) relevant; (2) simple and easily understood in layman's terms; (3) scientifically justifiable; (4) quantitative; and (5) cost-effective.

Also, Hellawell (1986) detailed the following characteristics as being, from a toxicological perspective, ideal as an indicator species: (1) easy to identify and to sample; (2) have universal distribution; (3) be a resource of economic importance; (4) easy to cultivate and maintain in laboratory conditions; and (5) exhibit bioaccumulative ability and low genetic variability. Such features are obviously strictly related to the concept of a bioaccumulator.

For a field ecologist, a good ecological indicator could be defined by the following characteristics (Marques et al., 2009): (1) ease in handling; (2) sensitivity to small variations of environmental stress; (3) independence of reference states; (4) applicability in extensive geographical areas and in the greatest possible number of communities or ecological environments; and (5) relevance to policy and management needs.

UNESCO (2003) also listed the characteristics that environmental indicators should present: (1) have an agreed scientifically sound meaning; (2) be representative of an important environmental aspect for the society; (3) provide valuable information with a readily understandable meaning; (4) be meaningful to external audiences; (5) help in focusing on necessary information to answer important questions; and (6) assist the decision-making process by being efficient and cost-effective in terms of use.

Dale and Beyeler (2001) contributed by considering the following as the most suitable qualities of a good ecological indicator: (1) be easily measured; (2) be sensitive to stresses on the system; (3) respond to stress in a predictable manner; (4) predict changes that can be averted by management actions; (5) be anticipatory; (6) be integrative; (7) have a known response to natural disturbances, anthropogenic stresses, and changes over time; and (8) have low variability in its response.

Despite the evident convergence of ideas between different authors, it is nevertheless clear that a good indicator should not gather unanimity of opinions. Moreover, it is obviously not easy to fulfill all these requirements, and in fact, despite the panoply of bio-indicators and ecological indicators that can be found in the literature, these are more often than not only more or less specific for a given kind of stress, or applicable to a particular type of community and/or scale of observation, and rarely has their validity been, in fact, utterly proven.

It is clear today that it is not possible to find one indicator or even a few indicators that can be used generally, as some naively thought when EHA was introduced. Of course there are general ecological indicators that are almost used every time we have to assess ecosystem health; but they are never sufficient to present a complete diagnosis—the general indicators always have to be supplemented by other indicators. Our doctor also has general indicators. He will always take your pulse, temperature, and blood pressure—these are very good general indicators, but he also always has to supplement these general indicators with an other indicator, that he selects according to the description of the diseases given by the patient. It is the same with the ecological doctor. If he observes dead fish but clear water, he will suspect the presence of a toxic substance in the ecosystem, while he will associate dead fish and very muddy water with oxygen depletion. In these two cases he will use two different sets of indicators, although some general indicators may be used in both cases.

The first international conference on the application of ecological indicators for the assessment of ecosystem health was held in Fort Lauderdale, Florida, in October 1990. Since then, there have been several international and national conferences on ecological indicators and on EHA. In 1992, a book with the title *Ecosystem Health* edited by Costanza, Norton, and Haskell was published by Island Press. Blackwell published a book with the same title in 1998, now edited by Rapport, Costanza, Epstein, Gaudewt, and Levins. Blackwell launched a journal with the title *Ecosystem Health* in the mid-1990s with Rapport as the editor-in-chief. Elsevier launched a journal with the title *Ecological Indicators* in the year 2000, now with Felix Mueller as the editor-in-chief. So, as can be seen from this short overview of the development of the use of EHA and ecological indicators to perform the EHA, there has been a significant interest in EHA and ecological indicators.

Some may have expected that EHA to a certain extent would replace ecological modeling as it was a new method to quantify the "disease" of an ecosystem. It is also possible to assess ecosystem health based upon indicators derived from observations only. Nevertheless, although ecological indicators may help in establishing a useful connection between empirical research and modeling—some of them can actually be used in orientors in ecological models (Jørgensen and Bendoricchio, 2001; Marques et al., 2009), indicators can hardly be used to make prognoses and do not give the overview of the ecological components and their interactions as a model does. EHA and ecological modeling are rather two different and complementary tools that together give a better image of the environmental management possibilities than if EHA or ecological modeling were used independently. Today, models are used increasingly, as a tool to perform an EHA. The models are, furthermore, used to give prognosis of the development of selected ecological indicators when a well-defined environmental management plan is followed.

A number of ecological indicators have been applied over the last 20 years to assess ecosystem health. As already stressed, general ecological indicators that can be applied in all cases do not exist or have at least not been found yet. A review of the literature published in the last 20 years about EHA and a selection of ecological indicators reveals that there are holistic indicators that can be applied more generally, and there are problem and ecosystem specific indicators, that will be used repeatedly for the same problems or the same type of ecosystems. All ecosystems are different, even ecosystems of the same type are different, and there are therefore always some very case-specific indicators that are selected on basis of sound theoretical considerations of the combination of the problem and the ecosystem. Our knowledge about human health is much more developed than our knowledge about ecosystem health, and still, there is no general procedure on how to assess a diagnosis for each of the several hundred different possible cases a doctor will meet in his practice. We will, however, attempt in the next section to give an overview of the most applied ecological indicators for different ecosystems and their classification. It is possible to give such an overview—but not to give a general applicable procedure with a generally valid list of indicators. This does not of course mean that we have nothing to learn from case studies. Because the selection of indicators is difficult and varies from case to case, it is of course possible to expand one's experience by learning about as many case studies as possible.

9.2 Criteria for the Selection of Ecological Indicators for EHA

Evolution of complex ecosystems can be described in terms of four major attributes (Von Bertalanffy, 1952):

1. Progressive integration (entails the development of integrative linkages between different species of biota and between biota, habitat, and climate).
2. Progressive differentiation (progressive specialization as systems evolve biotic diversity to take advantage of abilities to partition resources more finely, and so forth).
3. Progressive mechanization (covers the growing number of feedbacks and regulation mechanisms).
4. Progressive centralization (it probably does not refer to a centralization in the political meaning, as ecosystems are characterized by short and fast feedbacks and decentralized control, but to the more and more developed cooperation among the organisms [the Gaia effect] and the growing adaptation to all other components in the ecosystem).

Costanza (1992) summarizes the concept definition of ecosystem health as follows: health as (1) homeostasis, (2) absence of disease, (3) diversity or complexity, (4) stability or resilience, (5) vigor or scope for growth, and (6) balance between system components. He emphasizes that it is necessary to consider all or least most of the definitions simultaneously. Consequently, he proposes an overall system health index, HI = V*O*R, where V is system vigor, O is the system organization index, and R is the resilience index. Costanza probably touches on the most crucial ecosystem properties to cover ecosystem health with this proposal.

Kay and Schneider (1992) use the term "ecosystem integrity" to refer to an ecosystem's ability to main its organization. Measures of integrity should therefore reflect the two aspects of the organizational state of an ecosystem: functional and structural. Function refers to the overall activities of the ecosystem. Structure refers to the interconnection between the components of the system. Measures of function would indicate the amount of energy being captured by the system. It could be covered by measuring the exergy captured by the system. Measures of structure would indicate the way in which energy is moving through the system. The exergy stored in the ecosystem could be a reasonable indicator of the structure.

For the use of ecological indicators for EHA from a practical environmental management point of view, integrating the different and complementary perspectives from the several authors referred to above could lead to seven criteria being proposed:

1. Simple to apply and easily understood by laymen
2. Relevant in the context
3. Scientifically justifiable
4. Quantitative
5. Acceptable in terms of costs
6. Covering all relevant and actual problems
7. Sensitive to possible changes

From a more scientific point of view, on the other hand, the following criteria should be taken into account:

a. Handling easiness
b. Sensibility to small variations of environmental stress
c. Independence of reference states
d. Applicability in extensive geographical areas and in the greatest possible number of communities or ecological environments
e. Possible and reliable quantification

As it is not easy to fulfill all these $7 + 5 = 12$ requirements, the generality of the ecological indicators often applied is only limited. Therefore, the selection should at least in the first hand encompass too many rather than too few indicators, because it is easier to evaluate the redundant indicators after their use. All stakeholders (interested in the sustainable development of a ecosystem) should usually be involved in the selection of indicators. After the selection of a handful or more indicators, each of them should be tested for the above-mentioned seven criteria, which will probably imply that the number of selected indicators is reduced.

Let us take a few of the most common environmental problems as an example:

1. *Eutrophication of aquatic ecosystems*: The transparency, which is easily understood by laymen, should be supplemented by indication of phytoplankton concentration as biomass mg/L or as chlorophyll a in mg/m^3, which is a direct measure of the eutrophication. Also the maximum primary production as mg C/((m^3 or m^2) 24 h) and/or as g C/((m^3 or m^2) y) could be very informative, although it is probably not the most understandable of indicators for laymen. Furthermore, the nutrient concentrations (particularly of course nitrogen and phosphorus) express the eutrophication potential. The ratio between zooplankton and phytoplankton concentration is also in some cases relevant because a relatively high zooplankton concentration may indicate that that the eutrophication is under control.
2. *Toxic substances*: The concentration of the toxic substances in the water or the air in sediment or soil and in organisms late in the food chain. Sediment and soil often have a much higher concentration than the water or the air, but the relevance of these concentrations is dependent on the water solubility and the vapor pressure of the toxic substance. Due to biomagnification, it is recommended to quantify the concentration in organisms at different trophic levels, namely macroinvertebrates, carnivorous fish, and birds, when the concentration in the water is low and therefore very uncertain. Besides measures of concentration, there is an advantage in an ecotoxicological line of evidence within the ecological approach (Triebskorn et al., 2001, 2003; Brack et al., 2005; Sanchez and Porcher, 2009), involving the use of biological effects methods, namely biomarkers and bioassays to help in integrating the chemical and biological information, providing for instance an overall insight into the quality of a water body (Allan et al., 2006; Hagger et al., 2006). Biomarkers and bioassays have indeed been recognized as potentially important lines of evidence to establish cause–effect relationships in ecological quality assessment (Martinez-Haro et al., 2015). More specifically, they improve the capability to ascertain the causes of a failing ecological status, namely in aquatic ecosystems, and whether pollutants are the cause of it, thus closing the gap between ecology and chemistry (Maas et al., 2004; ICES, 2007).
3. *Acidification of ecosystems*: pH, alkalinity, and pH-buffer capacity of the water in aquatic ecosystems and also of rainwater.
4. *Introduction and invasion of exotic species*: The concentration of the introduced species and all the species that are influenced directly by the introduced species. For instance, when the Nile Perch was

introduced in Lake Victoria it would have been relevant to also follow the concentration of Tilapia and Haplochromis. When the Iberian snail invaded North Europe it would have been relevant to follow the Iberian snail as well as the competing snails.

For all environmental management problems, it may be relevant and recommendable to follow the biodiversity. Odum's attributes, see Table 4.1, including the thermodynamic indicators, are very informative, and although they are not fully understandable by laymen, they may be of importance for environmental managers. At least some of this group of indicators should therefore be included in the selection of indicators.

9.3 Classification of Ecosystem Health Indicators

The ecological indicators applied today in different contexts, for different ecosystems and for different problems, can be classified in nine levels from the most reductionist to the most holistic indicators. Ecological indicators for EHA do not include indicators of the climatic conditions, which in this context are considered entirely natural conditions.

Level 1 considers the use of ecotoxicological tools able to provide early-warning signals that allow for preventive measures before the ecological damage occurs, focusing on two types of biological assessment tools, biomarkers and bioassays (Martinez-Haro et al., 2015).

Level 2 covers the presence or absence of specific species. The best known application of this type of indicator is the saprobien system (Hynes, 1971) which classifies streams into four classes according to their pollution by organic matter causing oxygen depletion: oligosaprobic water (unpolluted or almost unpolluted), beta-mesosaprobic (slightly polluted), alpha-mesoprobic (polluted), and polysaprobic (very polluted). This classification was originally based on observations of species that were either present or absent. The species that were applied to assess the class of pollution were divided into four groups: organisms characteristic of unpolluted water, species dominating in polluted water, pollution indicators, and indifferent species. Records of fish in European rivers have been used to find a relationship between water quality and presence (and absence) of fish species by ANN (artificial neural network). The result of this examination has shown that the presence or absence of fish species can be used as a strong ecological indicator for the quality of river water.

Level 3 uses the ratio between classes of organisms. A characteristic example is the Nygård Algae index, though hundreds of indices have been proposed over the last decade.

Level 4 is based on the concentration of chemical compounds. Examples are the assessment of the level of eutrophication on the basis of the total phosphorus concentration assuming that phosphorus is the limiting factor for eutrophication.

Level 5 applies concentration of entire trophic levels as indicators, for instance the concentration of phytoplankton (as chlorophyll a or as biomass per m^3) is used as an indicator for the eutrophication of lakes. A high fish concentration has also been applied as an indicator for good water quality, or birds as an indicator for a healthy forest ecosystem.

Level 6 uses process rates as indication, for instance primary production determinations are used as indicators for eutrophication either as maximum gC/m^2 day or gC/m^3 day or gC/m^2 y or gC/m^3 y. A high annual growth of trees in a forest is used as an indicator for a healthy forest ecosystem and a high annual growth of a selected population may be used as an indicator for healthy environment. A high mortality in a population can, on the other hand, be used as an indication of an unhealthy environment. High respiration may indicate that an aquatic ecosystem has a tendency for oxygen depletion.

Level 7 covers composite indicators as for instance represented by many of Odum's attributes; see Table 9.2. Examples are biomass, respiration/biomass, respiration/production, production/biomass, and ratio primary producer/consumers. Odum uses these composite indicators to assess whether an ecosystem is at an early stage of development or is a mature ecosystem. It is presumed that mature ecosystems have more resistance toward changes due to impacts. These indicators are in accordance with the ecosystem properties presented in Part II.

Level 8 encompasses holistic indicators such as resistance, resilience, buffer capacity, biodiversity, all forms of diversity, size, and connectivity of the ecological network, turnover rate of carbon, nitrogen, and so on, and energy. As will be discussed in the next section, high resistance, high resilience, high buffer capacity, high diversity, big ecological network with a medium connectivity, normal turnover rates are all indications of a healthy ecosystem. These indicators are also rooted in the ecosystem properties and there are already presented in Part II. Additional information about species diversity is provided in Section 9.6.

Level 9 indicators are thermodynamic variables, which we may call super-holistic indicators as they try to see the forest through the trees and capture the total image of the ecosystem without inclusion of details. Such indicators are exergy (work energy content), emergy, exergy destruction (the amount of work energy lost as heat energy), entropy production, power, mass, and/or energy system retention time. The economic indicator cost/benefit (which includes all ecological benefits—not only the economic benefits to society) also belongs to this level. They are already presented together with the ecosystem properties in Part II.

9.4 Indicators of Level 1—Biomarkers and Bioassays

A good review on this topic can be found in Martinez-Haro et al. (2015), carried out under the supervision and coauthored by J.C. Marques. As detailed by these authors, we must first define the terms *biomarker* and *bioassay*. In the search for definitions we found that these are frequently broad, and sometimes used as equivalent terms (Martinez-Haro et al., 2015). For instance, the definition of biomarker by Depledge (1994) "a biochemical, cellular, physiological, or behavioral variation that can be measured in tissue or body fluid samples, or at the level of whole organisms, that provides evidence of exposure to and/or effects of, one or more chemical pollutants (and/or radiations)"; that can also refer to a bioassay. To be able to distinguish between the two terms we use the term biomarker (which was originally developed to monitor human health), to refer to measurements at the cellular or molecular level, whereas bioassays are procedures that usually work at a higher level than the cell (from tissues to communities). Thus, hereinafter, within the context of pollution monitoring and control, we understand a biomarker as

> Quantitative measurements of changes occurring at cellular, biochemical, molecular, or physiological levels, which can be measured in cells, body fluids, tissues, or organs within an organism and which may be indicative of xenobiotic exposure and/or effect.
>
> **(Vidal-Liñán and Bellas, 2013)**

While, the term "bioassay" is here understood as *a procedure that uses living material to establish the relationship between the levels of chemicals and their adverse effects on populations, communities, and ecosystems, and to identify biological resources at risk* (e.g., Cairns and Pratt, 1989). Biomarkers and bioassays are intended to be sensitive (early warning), rapid and cost-effective, compared to the monitoring of community level responses, but it must be kept in mind that both biomarkers and bioassays are conceived as monitoring tools to detect the potential risk of damage to the ecosystem by contaminants present in a given environmental compartment.

Response to environmental stress within a biological system may initially involve changes at the molecular level that may eventually lead to ecosystem scale impacts (Depledge and Fossi, 1994). By definition, metrics recorded using population abundances are sensitive to environmental stress only when deleterious effects altering the community structure have already taken place (Martinez-Haro et al., 2015). Carson (1962) warned us with a paradigmatic example of how the effects of persistent chemical pollutants impairing reproductive traits manifest only after long exposure periods, at a stage where the destructive process may have gone beyond the point where it can be easily reversed (see also Sumpter, 2005; Soares et al., 2009). But the toxicity of those pollutants can be easily identified and quantified by measuring the appropriate biological responses (e.g., eggshell thickness, calcium metabolism) in environmentally or laboratory exposed organisms. Moreover, although responses at the community or population level are directly relevant in terms of ecological effects, it cannot always be proven that differences among sites are due to contaminants or to natural factors; that is, as we ascend in the biological organization level we obtain ecological relevance but we lose specificity, speed, and reproducibility as a routine technique for environmental monitoring. In this context, it is clear that biomarkers and bioassays offer, contrary to the ecological approach, early warning signals reflecting the adverse biological responses toward anthropogenic environmental pollutants. In this case, responses measured at the lower levels of biological organization are usually sensitive and specific responses, indicative of effect and/or exposure to toxicants. For example, disruption of feeding activity at the individual level can be directly related to ecosystem impairment by having an immediate effect at the functional level (e.g., organic matter decomposition); and also indirectly as effects on life history traits at the individual level (e.g., growth, reproduction, survival) that may be propagated to successively higher levels of biological organization (Forrow and Maltby, 2000; Baird et al., 2007; Amiard-Triquet, 2009).

9.4.1 Biomarkers

The development of biomarkers as ecotoxicological tools was motivated by the need for more sensitive early warning indicators of sublethal ecological effects. Nowadays, a wide range of biomarkers is available which can reveal if a studied organism/population has been exposed to or is affected by environmental pollutants and/or environmental stress. Overall, biomarkers used in environmental monitoring are classified into two main categories: biomarker of exposure and effect. A definition of these categories was provided by the WHO (1993): a biomarker of exposure is "an exogenous substance or its metabolite or the product of an interaction between a xenobiotic agent and some target molecule or cell that is measured in a compartment within an organism"; and a biomarker of effect is "a measurable biochemical, physiological, behavioral or other alteration within an organism that, depending upon the magnitude, can be recognized as associated with an established or possible health impairment or disease."

Briefly, examples of well-established biomarkers of exposure are the induction of proteins as metallothioneins, which respond following exposure to certain metal species, or cytochrome P450 monooxygenase system induced following exposure to organic pollutants such as aromatic hydrocarbons, polychlorinated biphenyls, or dioxins. Biomarkers of effect are for example: the enzyme delta-aminolevulinic acid dehydratase (ALAD), inhibited even at small levels of lead; the cholinesterase enzymes (ChE), inhibited following exposure to organophosphates and carbamates pesticides and also to some nonessential metals; the comet assay or micronucleus formation to evaluate DNA or chromosomal damage, respectively, due to genotoxins; or imposex phenomenon (imposition of male secondary sexual characteristics on gastropods females) due to organotin compounds. Several reviews can be found regarding different categories of biomarkers from different perspectives focusing, for example, on the target species (e.g., Depledge and Fossi, 1994; van der Oost, 2003; Valavanidis et al., 2006; Montserrat et al., 2007), the relationships at population and community levels (e.g., Lagadic et al., 1994; Boudou and Ribeyre, 1997; Cajaraville et al., 2000; Clements, 2000; Hyne and Maher, 2003), or the utility in environmental impact assessment (e.g., Galloway et al., 2004a; Depledge and Galloway 2005; Hagger et al., 2006; Schettino et al., 2012).

9.4.2 Bioassays

A bioassay is a biological method, alternative or complementary to a chemical analysis, intended to detect and measure the presence of a substance. In general toxicological studies it is aimed at quantifying the toxicity of individual chemicals or mixtures of known composition by exposing whole living organisms under standardized conditions over a certain period of time (Cairns and Pratt, 1989; Rand et al., 1995), the so called dose–response experiments. In environmental studies, ecotoxicological bioassays consist of the exposure of test or field organisms in controlled conditions to environmental matrices (water, sediment) whose toxicity we wish to assess, and the measurement of ecologically relevant quantitative responses. In order to obtain results relevant at the ecosystem level, the measured responses should have implications on the biological fitness of the individual (e.g., mortality, growth, reproduction, feeding rates; see, e.g., Rand et al., 1995). The observed biological effect in a bioassay is generally the result of the bioavailability of a complex mixture of pollutants that may be present in a sample of water, but is also dependent on physicochemical parameters of the environment (Keddy et al., 1995). Hence, because the environmental conditions of an ecosystem are difficult to replicate in the laboratory, *in situ* bioassays, provide a more realistic exposure scenario than those conducted under laboratory-controlled conditions, by integrating major natural fluctuating environmental variables (Crane et al., 2007), which is particularly relevant for ecosystems where environmental conditions are highly variable (Allan et al., 2006; Wharfe et al., 2007).

The use of liquid and solid-phase bioassays in environmental risk assessment and management has a long international history particularly in North America (e.g., Environmental Protection Agency—USEPA), and also in marine environmental monitoring in many other countries (Bougis et al., 1979; Kobayashi, 1991; Vashchenko and Zhadan, 1993), and over the past two decades has also been established within some European programs (e.g., OSPAR Coordinated Environmental Monitoring Programme, UK National Marine Monitoring Program).

Adverse effects include both lethal and sublethal effects, the latter currently being the most common effects measured in aquatic organisms. Chronic toxic effects may occur when the chemical produces deleterious sublethal effects as a result of a single exposure, but more often they are a consequence of repeated or long-term exposures to low levels of persistent chemicals, alone or in combination. In this case, the most common sublethal effects in aquatic organisms are behavioral (e.g., swimming and feeding) and physiological (e.g., growth, embryo and larval development, reproduction). Some sublethal effects may have little or no effect on the organism because they are rapidly reversible or diminish or cease with time, or in contrast, they may indirectly result in mortality, for example, changes in swimming behavior may diminish the ability of aquatic organisms to find food or to escape from predators.

9.5 Levels 2–4 Indicators

This section gives information about level 2–4 indicators. The next section presents level 5 and 6 indicators. Level 7 indicators are presented in Table 9.1 and levels 8–9 are already covered in Part II, although some additional information about species diversity is given in the next section.

When talking about indicator species, we must distinguish two cases: presence/dominance of species or completely absence/presence in a very small number (concentration). In both cases it is possible to translate into health assessment. The AMBI index, for instance, which accounts for the presence of species indicating a type of pollution and the presence of species indicating an unpolluted situation, has been considered useful in terms of the application of the European Water Framework Directive in coastal ecosystems and estuaries. The index has been shown to be useful for the assessment of other anthropogenic impacts, such as physical alterations in the habitat, heavy metal inputs, and so on. And what is more, it has been successfully applied in Atlantic (North Sea; Bay of Biscay; and South of Spain) and Mediterranean (Spain and Greece) European coasts (see Borja et al., 2000).

TABLE 9.1

The Translation of Physiochemical Parameters to a Saprobic System

Parameters	**Saprobic Classes**			
	4. Degradation	**2. Recovery**	**3. Active Decomp.**	**Clear Water**
Dissolved Oxygen	**0–3 mg/L**	**BOD5 Increasing**	**>60%**	**>90%**
(mg/L or saturation%)	**<50%**	**>70%**		
		BOD5 Decreasing <30% Saturation		
BOD5	High	Increasing <5 mg/L	<5mg/L	<3 mg/L
NH^{4+}	0.5–2 mgN/L	0.3–1.2. mgN/L	<0.2 mgN/L	<0.1 mgN/L
NO^{2-}	0–0.2 mgN/L	0–0.2 mgN/L	~ 0.2 mgN/L	<0.05 mgN/L
NO^{3-}	Very low	1–2 mgN/L	2–6 mgN/L	High
Turbidity	High	Low	Very low	Very clear

Regarding submarine vegetation, there is a number of genera that universally appear when pollution situations occur. Among them, there are the green algae *Ulva, Enteromorpha, Cladophora,* and *Chaetomorpha*; and the red algae *Gracilaria, Porphyra,* and *Corallina.*

High structural complexity species, such *Fucus* and *Laminaria,* belonging to Phaeophyta, are seen worldwide as the most sensitive to any kind of pollution, with the exception of a species of Fucus genus that cope with moderate pollution (Niell and Pazo, 1978). On the other hand, marine Spermatophytae can be considered indicator species of good water quality. In the Mediterranean Sea, for instance, the presence of Phaeophyta *Cystoseira* and *Sargassum* or meadows of *Posidonia oceanica* indicate good water quality. The latter is the most used indicator of water quality in the Mediterranean Sea.

For the development of the AMBI, the soft bottom macrofauna is divided into five groups according to their sensitivity to increasing stress:

I. Species very sensitive to organic enrichment and present under unpolluted conditions.
II. Species indifferent to enrichment, always in low densities with nonsignificant variations with time.
III. Species tolerant to excess of organic matter enrichment. These species may occur under normal conditions, but their populations are stimulated by organic enrichment.
IV. Second-order opportunist species, mainly small-sized polychaetes.
V. First-order opportunist species, essentially deposit feeders.

The following formula is used:

$$\text{AMBI} = \frac{(0 \times \%\text{GI} + 1.5 \times \%\text{GII} + 3 \times \%\text{GIII} + 4.5 \times \%\text{GIV} + 6 \times \%\text{GV})}{100}$$

which leads to the following classification:

Classification	AMBI
Normal	0–1, 2
Slightly polluted	1, 2–3, 2
Moderately polluted	3, 2–5
Highly polluted	5–6
Very highly polluted	6–7

For the application of this index, thousands of taxa have been classified, which are representative of the most important soft bottom communities present in European estuarine and coastal systems. The Marine Biotic Index can be applied using the AMBI© software (Borja et al., 2000, and www.azti.es, where the software is freely available). Similar ideas have been applied for estuaries, benthic ecosystems, and other marine ecosystems (see Marques et al., 2009; Pinto et al., 2009).

As already mentioned, there are many indices and for a more comprehensive coverage of the many indices, Jørgensen et al. (2010) and Marques et al. (2009) can be referred to.

The classification of rivers by the so-called saprobic system (Hynes, 1971) is based on the presence of species. Four degrees of pollution can be identified:

I. Clear water zone
II. Recovery zone
III. Active decomposition zone
IV. Degradation zone

The classification presented here has only four classes, but more complicated and detailed systems are in use. The classification is based on an examination of plant and animal species present in the water at a number of locations. The species are divided into four groups:

1. Organisms characteristic of unpolluted water
2. Species predominating in polluted water
3. Pollution indicators
4. Indifferent species

Normally, the assessment of the ecological conditions of running water is a complicated and time-consuming. An index is found by the following method: each species found is given an average saprobic value, S, by distributing 10 points over the four pollution classes (degrees). Species with a broad distribution have their 10 points distributed evenly over all the degrees. The average value of S is found by multiplying the number of points from each degree by 1, 2, 3, and 4, respectively, and dividing the sum by 10. Furthermore, a higher weight is given to species with a narrow ecological distribution by ranking indicator value from 1 to 5. Finally, the relative abundance of each species is found and the saprobic value S for the sampling site is estimated by use of the following equation:

$$S = \Sigma(H^*s^*G)/\Sigma(H^*G), \tag{9.1}$$

where H is the number of individuals, S is the saprobic value, and G is the indicator value for the species.

However, simple versions of the saprobic system are often used, based on only smell and appearance of macro-invertebrates as pollution indicators. A direct translation to measured chemical concentration is not possible, but it is most often advantageous to measure the chemical concentrations because this is done much faster than the biological classification. A tentative translation of the chemical analysis to the saprobic system is shown in Table 9.1. As seen in the table, it is not sufficient to measure only the BOD5 concentration, defined as the mg/L of oxygen consumed to decompose the organic matter over a period of 5 days, but the ammonium and nitrate concentrations must be included in the examination to be able to obtain a full picture of the conditions. Notice that BOD5 measure the concentration of organic matter by its consumption of oxygen. COD, the chemical oxygen demand is often used supplementary to the BOD5. COD is generally 1.5–3 times higher than the BOD5 because the oxidation by use of chromate, which is used for the determination of COD, oxidizes all organic matter and even some inorganic compounds. Furthermore, assessment of the diurnal variations is important, too, as it is the minimum oxygen concentration which is crucial. It can usually be recorded shortly before sunrise. The chemical examination is generally applied more than the biological classification today, because it is a faster assessment of ecosystem health.

Simpler systems than the saprobic system have been developed, for instance, the Trent Index and the Chandler Index. They use genera or families instead of identification of species, which make them faster to use.

When the ecosystem is unhealthy due to too high a concentration of specific toxic substances, the concentration of one or more focal toxic compounds is of course a very relevant indicator. PCB contamination of the

Great North American Lakes has been applied as an ecological indicator by recording the concentrations of PCB in birds and in water. It is often important to find a concentration in a medium or in organisms where the concentration can be easily determined, is relatively stable (does not fluctuate on a day to day basis) and has a sufficiently high value that is magnitudes higher than the detection limit, which facilitate a clear indication. PCB in birds is high due to biomagnification through the food chain. Similarly, the concentrations of heavy metals in sediment have been applied as an indicator for contamination by heavy metals. The concentrations of heavy metals in sediment are relatively stable and are magnitudes higher than in the water, where the concentration often fluctuates, sometimes from hour to hour.

9.6 Levels 5–6 Indicators

The use of levels 5 and 6 indicators can be illustrated by focusing on the eutrophication problem. Eutrophic means nutrient rich. A too high concentration of nutrients (mainly phosphorus and nitrogen) implies a high concentration of phytoplankton measured for instance by the concentration of chlorophyll a. The abatement of eutrophication requires that the environmental management determines the limiting nutrient(s) (see Chapters 3 and 6) and furthermore which nutrient it is most easy (most cost-moderate) to make limiting for the eutrophication. Indicators are used to assess the eutrophication needs and also to distinguish between different zones of aquatic ecosystems. The zone that is close to the shoreline is named the littoral zone and often has other chemical characteristics from open water. Furthermore, a thermocline separating warm surface water and cold bottom water, or a halocline separating bottom water with a higher salinity from surface water with a lower salinity, may be able to give the surface water, named epilimnion, another chemical characteristic from the bottom water, named hypolimnion.

The eutrophication is often assessed by the use of levels 5 (the concentration of level 1, the trophic level) and 6 indicators (productivity), often together with the concentrations of nutrients (level 4 indicators) and even by identification of the most dominant species (level 3 indicators). The eutrophication is classified in 8 levels. The 8 classes and the corresponding indicators are shown in Table 9.2. Levels 1–7 represent increasing eutrophication, while dystrophic aquatic ecosystems often have brownish water with a high concentration of humic acid, originated from allochthonous material within the catchment area and brought to the ecosystem in various forms of organic matter.

TABLE 9.2

Classification of Eutrophication of Aquatic Ecosystems (for instance, lakes)

Trophic Type	Mean Primary Production mg Cm^{-2} 24 h^{-1}	Phytoplankton		Organic C mg/L	Total P µg/L	Total N µg/L
		µgC/L	µg chlorophyll/L			
Ultra-oligotrophic	<50	<50	0.01–0.5	–	<5	<250
Oligotrophic	50–300	20–100	0.3–3	<1–3	3–10	25–300
Oligo-mesotrophic	–	–	1–5	–	5–10	250–500
Mesotrophic	250–1000	100–300	2–15	1–5	8–25	300–600
Meso-eutrophic	–	–	3–25	3–15	10–30	500–1100
Eutrophic	>1000	>300	10–50	5–30	>100	>1200
Hypereutrophic			>100	30–100	>150	>2000
Dystrophic	<50–500	>50–200	0.1–10	3–30	<1–10	5–200

9.7 Species Diversity

A species is a basic unit of biological classification and a taxonomic rank. A species is often defined as a group of organisms capable of interbreeding and producing fertile offspring. While in many cases this definition is adequate, more precise or differing measures are often used, such as similarity of DNA, morphology or ecological niche, which is defined as the ecological conditions under which species can live and reproduce. In the context of differentiation, the similarity of DNA explains why individuals of the same species can survive the same combination of constraints. We know about 2 million species, but as we still have unexplored areas in nature, particularly in the tropical rainforest, it is estimated that the number of species is much higher than 2 million. As already mentioned, most estimates point to about 5 million species, ±3 million.

The differentiation of species is expressed by several species diversity measures:

1. Number of species S eventually expressed as log S.
2. Shannon's index of general diversity: $H = -\Sigma\, n_i/N \log_2 n_i/N$,

   where n_i is the number of individuals of the *i*th species and N is the total number of individuals. The natural logarithm can also be applied.
3. Evenness index $e = H/\log S$.
4. The Margalef index quantifies the diversity relating specific richness to the total number of individuals. $D = (S - 1)/\log_e N$.

For S = number of species and N = total number of individuals.

The author did not establish reference values.

Species diversity is used to express the possibilities according to the above presented discussion to maintain the functioning of ecosystems by changed forcing functions (constraints). The idea is that the species diversity represents the diversity of characters or properties of the organisms to meet the changed conditions (constraints). Many ecological investigations have shown that (see Gaston, 1996) the two diversities are related.

EXAMPLE 9.1

An ecosystem is dominated by five species and has totally 405 species. The total number of individuals in the ecosystem (microorganisms are not accounted in this context) is 100,000. The numbers of individuals in the five dominant species are 8000, 20,000, 20,400, 22,600, and 25,000. The 400 nondominating species have, with good approximation, 10 individuals each. Find the values of the three diversity indices proposed in the text.

Solution

The number of species is S = 405 or log S = log 405 = 2.603

$$\begin{aligned}H &= -(0.08 * \log 0.08 + 0.2 * \log 0.2 + 0.204 * \log 0.204 + 0.226 * \log 0.226 + 0.25 * \\ &\quad \log 0.25 + 400 * 0.0001 * \log 0.0001) \\ &= -(0.08 * (-1.1) + 0.2 * (-0.7) + 0.204 * (-0.690) + 0.226 * (-0.656) + 0.25 * \\ &\quad (-0.6) + 400 * 0.0001 * (-4)) \\ &= 0.654\end{aligned}$$

$$E = 0.654/2.603 = 0.251$$

Both the diversity and the evenness are small. The diversity for many ecosystems is between 2 and 4.

EXAMPLE 9.2

Species diversity is low toward the poles and high toward the tropics. Give at least a partial explanation to this gradient in species diversity.

Solution

A detailed and full explanation would require inclusion of many factors. Two factors seem to be particularly important:

1. The productivity gradient associated with the latitude is very similar to the species gradient as a function of latitude.
2. The development of species diversity increases as all biological processes generally increases with an increasing temperature.

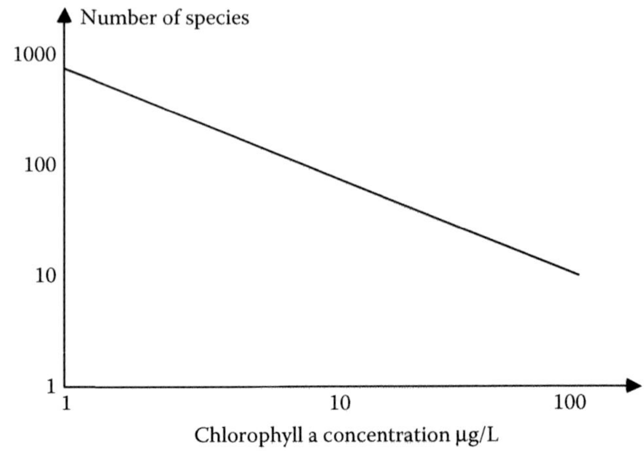

FIGURE 9.2
Weiderholm (1980) obtained the relationship shown for a number of Swedish lakes between the number of species and eutrophication, expressed as chlorophyll a in µg/L.

Species diversity is often a very sensitive indicator for pollution. Figure 9.2 shows how the number of species decrease with eutrophication.

9.8 Comparing Methodologies

Martinez-Haro et al. (2015) analyzed 15 publications in which the ecological condition of ecosystems was concomitantly assessed by applying community based ecological indices and ecotoxicological approaches, namely through the use of biomarker and/or bioassay techniques (Table 9.3). Additionally, these authors analyzed the work developed by Hagger et al. (2008) in which, in order to characterize the ecological health of aquatic ecosystems, biological-effect endpoints (at different biological organization levels) were assessed considering point-source pollution, alien species, and hydro-morphological factors.

Among the different biological elements, those based on benthic invertebrate fauna were the most frequently used, being represented in 14 of the revised works. In fact, benthic invertebrates are one of the key biological components considered for the assessment of benthic compartment health, being one of the most prevalent functional groups used in aquatic assessment (Birk et al., 2012). Regarding the use of biomarkers and bioassays, none of the studies included both types. Furthermore, a balanced use of both types of tools was found in the revised studies: seven studies addressed the use of biomarkers and eight that of bioassays. In the studies addressing

TABLE 9.3

Studies Assessing the Ecological Quality Status of Water Bodies Based Simultaneously on Ecological Indices and Ecotoxicological Assays (Namely Biomarkers and/or Bioassays)

System	Country	Community	Biological Elements	Bioassay	Biomarker	Spp	Comparison	Ref.
R	Germany	IBI	BMI, F		Set	Fish	Multivariate statistics	Böhmer et al., 2001; Dietze et al., 2001; Triebskorn et al., 2001, 2003—VALIMAR
R	UK	Shannon index—H′, ASPT, BMWP, EQI N-taxa, EQI ASPT	BMI	*In situ* feeding rate		Amphipods	Pearson, least-squares regression	Maltby et al., 2002
CW	Spain	AMBI, BENTIX, RBI, ABC	BMI	Embryo toxicity test		Sea urchins	Pearson correlation	Marín-Guirao et al., 2005
R	Belgium	IBI	F		Set	Fish	Simple comparison	Mayon et al., 2006
R	Spain	IPS, IBMWP	BMI		Set	Fish	Simple comparison	Damásio et al., 2007
CW	Spain	Previous works (M-AMBI)	All	Microtox 10-d mortality		Bacterium, Amphipods	Simple comparison	Borja et al., 2008
CW	Spain	H′, Margalef′s d, RBI, EBI	BMI	Embryo toxicity test		Sea urchins	Simple comparison	Cesar et al., 2009
R	Germany, Belgium, Spain	AQEM assessment system	BMI	LC50		Cladoceran, Algae, Fish	Correlation	Von del Ohe et al., 2009 MODELKEY
R	Germany	Multimetric index	BMI	Embryo toxicity test		Fish	Simple comparison	Bartzke et al., 2010

(*Continued*)

TABLE 9.3 (*Continued*)

Studies Assessing the Ecological Quality Status of Water Bodies Based Simultaneously on Ecological Indices and Ecotoxicological Assays (Namely Biomarkers and/or Bioassays)

System	Country	Community	Biological Elements	Bioassay	Biomarker	Spp	Comparison	Ref
R	Spain	QBR, IHF, IBMWP, IASPT	BMI		Set	Trichoptera	PCA	Damásio et al., 2011
TW	Italy	H′, Evenness index, Simpson index, Margalef index—d, R-MaQI	A, BMI, F	*Vibrio fischeri* *Corophium orientale* *Parancentrotus lividus*			Weight-of-evidence	Micheletti et al., 2011
TW	Portugal	AMBI, M-AMBI	BMI		Set	Decapod	PCA	Pereira et al., 2012
TW	France	IFREMER* (AMBI, BENTIX, BOPA, BQI)	BMI		Set	Bivalve	Simple comparison	Tankoua et al., 2012; Dauvin et al., 2007
TW	Portugal	M-AMBI-EDI	BMI, F	ToxScreen		Bacterium	PCA, scaled value index	Azevedo et al., 2013
R	Spain	IBMWP, IMMiT	BMI		Set	Trichoptera	Simple comparison	Prat et al., 2013
Other Related Study								
TW	UK	Environment Agency of England and Wales	O		Set	Bivalve	Simple comparison	Hagger et al., 2008

biomarkers, invertebrates (molluscs and arthropods) represented the target species in four of the seven studies. Biomarkers in vertebrates (namely fish) were applied in three of the seven studies. In all studies addressing biomarkers, the set of responses evaluated intended to capture and establish cause–effect relationships. Among the most used biomarkers are those related to oxidative stress (e.g., enzymes related to GSH; Mayon et al., 2006; Prat et al., 2013) and tissue damage (e.g., both at lipid and DNA levels; Damásio et al., 2011). Additionally, some studies included a wide range of biomarkers covering different levels of organization or biological response, below cell level (Dietze et al., 2001; Triebskorn et al., 2001, 2003). Regarding bioassays, embryo toxicity tests (sea urchin and fish) were mostly used (e.g., Marín-Guirao et al., 2005, Bartzke et al., 2010), being represented in four of the eight studies, followed by bacterium tests represented in three of the studies (e.g., Borja et al., 2008; Azevedo et al., 2013).

Overall, most of the studies were carried out in rivers and transitional waters (8 out of 15 and 4 out of 15, respectively) and only three in coastal waters (Table 9.3). Studies on lakes, however, were not found. The number of studies assessing the status of transitional waters was high despite the fact that estuarine science has been considered the "poor relation" of the aquatic research for many years (Elliott and Whitfield, 2011). This could be due to the fact that transitional waters are natural environments that are difficult to classify and assess due to their inherent variability (Dauvin, 2007; Elliot and Quintino, 2007). In fact, we found the first studies addressing rivers in 2001 (Böhmer et al., 2001), and it was not until 2008 when the first works for transitional waters appeared (Borja et al., 2008). Furthermore, Borja (2005) pointed out that little attention that has been put into coastal and transitional waters: since 1999, the number of scientific papers published in peer-reviewed journals on WFD in marine ecosystems (including transitional waters), only represented 10% of the total papers published over the period 1998–2004. Currently, this percentage increased to 17% (SCOPUS—March 2014—key words: "Water Framework Directive" and "transitional waters" or "estuaries" or "coastal" or "marine," over the period 1998–2014).

To address both approaches together, there are methodologies from simple comparisons based on positive/negative results (e.g., Mayon et al., 2006; Damásio et al., 2007; Bartzke et al., 2010), to complex methodologies based on multivariate statistics (Dietze et al., 2001), principal component analyses (PCA; Dámasio et al., 2011; Pereira et al., 2012; Azevedo et al., 2013), or the integrated weight-of-evidence methodology (Micheletti et al., 2011). The different methodologies used to integrate the assessment from ecological and ecotoxicological approaches, make it difficult to compare studies and to recommend standardizing analytical protocols to carry out the integration. In the studies in which more complex statistical analyses were performed, the use of ecotoxicological tools was in many cases useful to distinguish more clearly the most environmentally disturbed sites (Pereira et al., 2012). In the studies developed by Maltby et al. (2002) and/or

Marín-Guirao et al. (2005), a formal comparison of results obtained under both approaches was performed using Pearson correlations and/or least squares regression analysis. The analysis of raw results, in different ways, is the most common procedure used in the revised studies to treat data from ecotoxicological tools. For example, in the study developed by Cesar et al. (2009), data from both approaches were integrated through PCA to characterize the ecological quality of the study area. Despite this, a formal comparison (that is, namely using statistic methodologies) between community level responses and ecotoxicological results was not performed. Similarly, a formal comparison was not provided by Mayon et al. (2006), Dámasio et al. (2007), or Borja et al. (2008).

Other studies classify the ecotoxicological values in order to assess ecosystem health, as is applied for ecological indices. For example, for the sea urchin embryo toxicity test, Marín-Guirao et al. (2005) associated the results obtained from the toxicity bioassays to the five EcoQS dividing the percentage of normally developed embryos by five (from 0% to 100%). After this, the established threshold values were tested against several sediment quality attributes. Then, the results from ecological indices and ecotoxicological analyses were analyzed through Pearson correlations. Similarly, Barzke et al. (2010) developed an index based on fish embryo toxicity test, the fish teratogenicity index. Observed effects were weighted according to the effect intensity and mathematically combined in an index. Finally, after checking for statistical differences between the sample and reference values, different quality thresholds for the values of the index were determined. Under a similar approach, Hagger et al. (2008) constructed a biomarker response index to simplify the complex biological alterations measured by multiple biomarkers into a single predefined quality class. Their index was composed by biomarkers at different levels of biological organization, weighted accordingly, and ranked based on the biological responses determined from previous laboratory and field studies. The values obtained were then classified into four categories of biological impact to assess the degree of alteration from normal/reference responses.

Community and ecotoxicological approaches were congruent in 60% of the studies revised (Martinez-Haro et al., 2015), which was illustrated by integrated multilevel approaches (Triebskorn et al., 2001, 2003). Active and passive monitoring experiments along with the study of indicators at different levels of organization, that is, biochemical, cellular, individual, and population-level, were addressed, showing that ecological differences, for instance between streams, can be detected by most of the biomarker responses in fish (Böhmer et al., 2001; Dietze et al., 2001; Triebskorn et al., 2001, 2003). Furthermore, although macrozoobenthos communities indicated much clearer long-term differences between the sites than biomarkers, which showed much more temporal variation (Böhmer et al., 2001), the studied biomarker responses were related to effects in fish populations as well as in macrozoobenthic communities (Triebskorn et al., 2001, 2003).

A clear correlation between the assessment of biological communities and ecotoxicological indicators was also detected in historically polluted coastal areas where metal concentrations exceed background levels for the corresponding area (Marín-Guirao et al., 2005; Borja et al., 2008; Cesar et al., 2009). In these studies, the response of bioassays based on invertebrate (namely sea urchin embryo bioassay, amphipod bioassay) and bacterial responses (Microtox® assay), fitted satisfactorily with the ecological assessment performed with ecological indices. More specifically, bioassays supported the findings of the ecological indices, being able to quantify and estimate the cumulative effects of multiple stressors on benthic biota between the low- and high-impacted zones (Marín-Guirao et al., 2005; Cesar et al., 2009). In contrast, indices based on pollution resulting from organic enrichment might not be as successful as in the case of purely toxic pollution (Pinto et al., 2009).

Similarly, Damásio et al. (2011) and Pereira et al. (2012) found some relationships between the ecological status based on invertebrate and biomarker responses in moderately contaminated sites. Through the factor analysis applied to all data sets, biomarkers were able to detect different responses at some of the sites, which allowed discriminating between the contaminated and uncontaminated sites. In addition, the inclusion of more markers belonging to different metabolic pathways substantially increased the discrimination power within sites (Damásio et al., 2011; Pereira et al., 2012).

The environmental quality assessment based on macrobenthic communities alone sometimes led to unclear results in some studies such as, for instance, Pereira et al. (2012), Marín-Guirao et al. (2005), and Cesar et al. (2009). These authors found that despite the fact that some signs of poorer benthic quality can be detected in the respective study areas, differences among sites were not significant, and therefore a biotic index can inform about a good environmental status of the affected sites. Thus the selection of an appropriate ecological index, the one with the greatest ability to correctly characterize the study area, is one of the most difficult tasks. Although ecological indices may provide a good assessment of the status of the biological environments, a universal index efficiency, in all systems (or at least in systems of the same ecological type), appears difficult to achieve (Pinto et al., 2009). In addition, biological communities are complex and not all organisms are equally sensitive to all types of anthropogenic disturbance, likely to respond differently to diverse types of stress (Birk et al., 2012; van Hoey et al., 2013). In this sense, the combination of different approaches considering several components (such as chemical analyses, acute toxicity assessments, and biological elements) at different levels of organization can provide an integrated view of the environmental stresses affecting an ecosystem.

Conversely, Damásio et al. (2007), von der Ohe et al. (2009), Bartzke et al. (2009), Tankoua et al. (2012), and Prat et al. (2013) did not find consistent results between both approaches. Damásio et al. (2007) did not detect effects based on two ecological indices (one based on benthic diatoms and other in benthic macroinvertebrates), after an accidental fuel oil spill in a river, but a clear effect was detected by using fish biomarkers. The lack of a similar response between both approaches could be related to temporal differences between sampling. Diatom and invertebrates were collected a few weeks after the spill, whereas fish were sampled some months after the spill. Although population and community parameters are more ecologically relevant and reflect integrated conditions over a long time period, the ecological community approach often lacks sensitivity since it may be affected by environmental factors other than pollution (such as river morphology; Triebkson et al., 2001). Thus, in the case of Damásio et al. (2007), it is possible that a few weeks after the spill the biological communities had not had the time to react but the responses on fish biomarkers were evident even a few months after the spill, which highlights the need to incorporate ecotoxicological tools in environmental quality assessment monitoring programs (Martinez-Haro et al., 2015).

Similarly, the study carried out by Prat et al. (2013), aiming to evaluate the effect of introducing reclaimed water from a wastewater treatment plant on the river quality in a polluted river basin, showed that most of the metrics selected to assess the biological quality indicated a slight biological impairment after the introduction of the treated water, even though the ecological status remained poor. However, indications of additional stress to the populations were found using several biomarkers, suggesting a potential of further deterioration of the ecological status of the river. Hence, the authors concluded that structural indicators are unable to indicate further impairment in polluted rivers, for which biomarkers may be a useful tool to detect environmental impoverishment.

Finally, in the study developed by Tankoua et al. (2012) intersex (i.e., a variation in sex characteristics including chromosomes, gonads, and/or genitals that do not allow an individual to be distinctly identified as male or female) in a bivalve was detected in all estuaries studied. Although no causal relationship was demonstrated between intersex and population effects, the authors pointed to the subtle effects of endocrine disrupting chemicals, which are active at very low doses, often in the absence of any other sign of toxicity, and therefore suggest the need to incorporate in environmental monitoring early-warning systems that can anticipate the impact at a higher level of biological organization.

9.9 Overview of Applicable Ecological Indicators for EHA

Table 9.4 gives an overview of the classes of ecological indicators, including their applications (see the nine levels in Section 9.3). Level 2 indicators have been widely used for EHA of rivers and may also be used in most other

TABLE 9.4

Overview of Frequently Applied Ecological Indicators

	Indicator Level								
Ecosystem	**1**	**2**	**3**	**4**	**5**	**6**	**7**	**8**	**9**
Coastal, estuary	X							X	X
Lake	X	X	X	X	X	X	X		
Lake					X	X	X	X	X
Coastal	X			X	X	X	X	X	X
Coastal					X	X	X	X	X
Marine	X				X	X	X		
Wetland				X	X	X	X	X	
Pond, lagoon, lake, basin								X	X
Agro-ecosystem									X
Landscape				X	X	X	X	X	X

ecosystems. Similarly, concentrations of chemical compounds are obvious to use for all unhealthy conditions caused by toxic substances. The experience gained by the use of level 7–9 indicators is usually of more general value in the EHA, because the higher level indicators give an overall (holistic) picture of how far a focal ecosystem on the system level is from healthy conditions. The overview is therefore to a higher extent giving information on the applicability of level 7–9 indicators. These indicators should in practical EHA often be supplemented with level 2–6 indicators, which are more specific. The selection of level 1–6 indicators is furthermore obvious in most cases; for instance the use of PCB in birds and zebra mussels in the EHA of the Great North American Lakes, because there has been a major PCB pollution problem and zebra mussels were a threatening invasive species.

9.10 EHA—Procedures

Direct Measurement Method (DMM)

The procedures established for the direct measurement method are as follows:

1. Identify the necessary indicators to be applied in the assessment process; use Table 9.2, Section 9.6.
2. Measure directly or calculate indirectly the selected indicators.
3. Assess ecosystem health based on the resulting indicator values.

9.10.1 Ecological Model Method (EMM)

The procedure established for the ecological modeling method for lake ecological health assessment is shown in Figure 9.3.

Five steps are necessary when assessing lake ecosystem health using the EMM procedure:

1. Determine the model structure and complexity according to the ecosystem structure.
2. Establish an ecological model through designing a conceptual diagram, developing model equations, and estimating model parameters.
3. Calibrate the model as necessary in order to assess its suitability in application to the EHA process.
4. Calculate the ecosystem health indicators.
5. Assess the ecosystem health based on the values of the indicators.

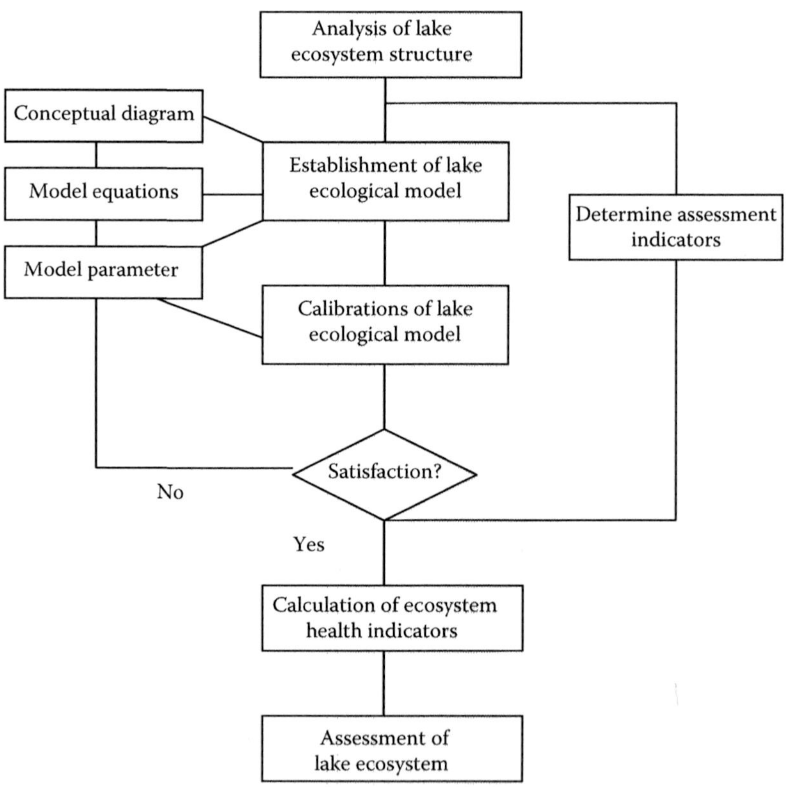

FIGURE 9.3
The procedure of the ecological model method (EMM) for ecological health assessment. (Modified from Xu et al., 2001. *Water Res.* 35(1), 3160. With permission.)

9.10.2 Ecosystem Health Index Method (EHIM)

In order to quantitatively assess the state of ecosystem health, an ecosystem health index on a scale of 0 to 100 was developed. It was assumed that, when the EHI is zero, the healthy state is worst; when the EHI is 100, the healthy state is best. In order to facilitate the description of healthy states, the EHI is equally divided into five segments or ranges as: 0%–20%, 20%–40%, 40%–60%, 60%–80%, and 80%–100%, which correspond with the five health states, respectively.

The EHI can be calculated by the following equation:

$$\text{EHI} = \sum_{i=1}^{n} \omega_i \cdot \text{EHI} \tag{9.2}$$

where EHI is a synthetic ecosystem health index, EHI_i is the *i*th ecosystem health index for the *i*th indicator, ω_i is the weighting factor for the *i*th indicator.

It can be seen from Equation 9.2 that the synthetic EHI depends on subEHIs and weighting factors for each indicator.

Five steps are necessary for the EHI method:

1. Select the basic and additional indicators.
2. Calculate the sub-EHIs for all of the selected indicators.
3. Determine the weighting factors for all of the selected indicators.
4. Calculate the synthetic EHI using sub-EHIs and weighting factors for all of the selected indicators.
5. Assess the ecosystem health based on synthetic EHI values.

EXERCISES AND PROBLEMS

1. An ecosystem consists of about 1 million of individuals and 500 species. The 20 most dominant species have about 20,000 individuals each, while the second 20 most abundant species have about 10,000 individuals. The rest of the species are covered by the remaining 400,000 individuals with approximately the same number of individuals for each of the 460 species. Find Shannon's index and the index of evenness.
2. Why should we expect that the diversity has increased over time?
3. List the advantages of high species diversity.
4. It is shown by modeling and statistics that eco-exergy and the sum of buffer capacities are well correlated. Explain why this correlation is rather obvious.
5. Why is it more difficult to quantify resilience than buffer capacity and resistance (see Section 3.12 for the explanations of these holistic indicators).

References

Allan, I.J., B. Vrana, R. Greenwood, G.A. Mills, B. Roig, and C. Gonzalez. 2006. A "toolbox" for biological and chemical monitoring requirements for the European Union's Water Framework Directive. *Talanta* 69, 302–322.

Amiard-Triquet, C. 2009. Behavioral disturbances: The missing link between sub-organismal and supra-organismal responses to stress? Prospects based on aquatic research. *Human and Ecological Risk Assessment* 15, 87–110.

Azevedo, I., S. Ramos, A.P. Mucha, and A.A. Bordalo. 2013. Applicability of ecological assessment tools for management decision making: A case study from the Lima estuary (NW Portugal). *Ocean and Coastal Management* 72, 54–63.

Baird, D.J., S.S. Brown, L. Lagadic, M. Liess, L. Maltby, M. Moreira-Santos, R. Schulz, and G.I. Scott. 2007. *In situ*-based effects measures: Determining the ecological relevance of measured responses. *Integrated Environmental Assessment and Management* 3, 259–267.

Bartzke, M., V. Delov, P. Stahlschmidt-Allner, B. Allner, and J. Oehlmann. 2010. Integrating the fish embryo toxicity test as triad element for sediment toxicity assessment based on the Water Framework Directive approach. *Journal of Soils and Sediments* 10, 389–399.

Böhmer, J., A. Zenker, B. Ackermann, and B. Kappus. 2001. Macrozoobenthos communities and biocoenotic assessment of ecological status in relation to degree of human impact in small streams in southwest Germany. *Journal of Aquatic Ecosystem Stress and Recovery* 8, 407–419.

Borja, A., J. Franco, and V. Peeresz. 2000. A marine biotic index to establish the ecological quality of Soft-Bottom Benthos within European Estuarine and coastal environments. *Marine Pollution Bulletin* 40(12), 1100–1114.

Borja, Á., I. Tueros, M.J. Belzunce, I. Galparsoro, J.M. Garmendia, M. Revilla, O. Solaun, and V. Valencia. 2008. Investigative monitoring within the European Water Framework Directive: A coastal blast furnace slag disposal, as an example. *Journal of Environmental Monitoring* 10, 453–462.

Boudou, A. and F. Ribeyre. 1997. Aquatic ecotoxicology: From the ecosystem to the cellular and molecular levels. *Environmental Health Perspectives* 105, 21–35.

Bougis, P., M.C. Corre, and M. Etienne. 1979. Sea-urchin larvae as a tool for assessment of the quality of sea water. *Annales du Institute Oceanographique, Paris* 55(1), 21–26.

Brack, W., J. Bakker, E. De Deckere, C. Deerenberg, J. Van Gils, M. Hein, P. Jurajda, et al. 2005. MODELKEY. Models for assessing and forecasting the impact of environmental key pollutants on freshwater and marine ecosystems and biodiversity. *Environmental Science and Pollution Research* 12, 252–256.

Cairns Jr., J. and J.R. Pratt. 1989. The scientific basis of bioassays. *Hydrobiologia* 188–189, 5–20.

Cajaraville, M.P., M.J. Bebianno, J. Blasco, C. Porte, C. Sarasquete, and A. Viarengo. 2000. The use of biomarkers to assess the impact of pollution in coastal environments of the Iberian Peninsula: A practical approach. *Science of the Total Environment* 247, 295–311.

Carson, R. 1962. *Silent Spring*. New York: Houghton Mifflin.

Cesar, A., A. Marín, L. Marin-Guirao, R. Vita, J. Lloret, and T.A. Del Valls. 2009. Integrative ecotoxicological assessment of sediment in Portmán Bay (Southeast Spain). *Ecotoxicology and Environmental Safety* 72, 1832–1841.

Crane, M., G.A. Burton, J.M. Culp, M.S. Greenberg, K.R. Munkittrick, R. Ribeiro, M.H. Salazar, and S.D. St-Jean. 2007. Review of aquatic *in situ* approaches for stressor and effect diagnosis. *Integrated Environmental Assessment and Management* 3, 234–245.

Clements, W.H. 2000. Integrating effects of contaminants across levels of biological organization: An overview. *Journal of Aquatic Ecosystem Stress and Recovery* 7, 113–116.

Costanza, R., B.G. Norton, and B.D. Haskell. 1992. (Eds.). *Ecosystem Health: New Goals for Environmental Management*. Washington, DC: Island Press.

Dale, V.H. and S.C. Beyeler. 2001. Challenges in the development and use of ecological indicators. *Ecological Indicators* 1, 3–10.

Damásio, J.B., C. Barata, A. Munné, A. Ginebreda, H. Guasch, S. Sabater, J. Caixach, and C. Porte. 2007. Comparing the response of biochemical indicators (biomarkers) and biological indices to diagnose the ecological impact of an oil spillage in a Mediterranean river (NE Catalunya, Spain). *Chemosphere* 66, 1206–1216.

Damásio, J., M. Fernández-Sanjuan, J. Sánchez-Avila, S. Lacorte, N. Prat, M. Rieradevall, A.M. V.M. Soares, and C. Barata. 2011. Multi-biochemical responses of benthic macroinvertebrate species as a complementary tool to diagnose the cause of community impairment in polluted rivers. *Water Research* 45, 3599–3613.

Depledge, M.H. 1994. The rational basis for the use of biomarkers as ecotoxicological tools. In Fossi, M.C. and Leonzio, C. (Eds.), *Nondestructive Biomarkers in Vertebrates*. Boca Raton, FL: CRC Press, 368 pp.

Depledge, M.H. and M.C. Fossi. 1994. The role of biomarkers in environmental assessment. (2) Invertebrates. *Ecotoxicology* 3, 161–172.

Depledge, M.H. and T.S. Galloway. 2005. Healthy animals, healthy ecosystems. *Frontiers in Ecology and the Environment* 3, 251–258.

Dietze, U., T. Braunbeck, W. Honnen, H.R. Köhler, J. Schwaiger, H. Segner, R. Triebskorn, and G. Schüürmann. 2001. Chemometric discrimination between streams based on chemical, limnological and biological data taken from freshwater fishes and their interrelationships. *Journal of Aquatic Ecosystem Stress and Recovery* 8, 319–336.

Forrow, D.M. and L. Maltby. 2000. Toward a mechanistic understanding of contaminant-induced changes in detritus processing in streams: Direct and indirect effects on detritivore feeding. *Environmental Toxicology and Chemistry* 19, 2100–2106.

Galloway, T.S., R.J. Brown, M.A. Browne, A. Dissanayake, D. Lowe, M.B. Jones, and M.H. Depledge. 2004. A multibiomarker approach to environmental assessment. *Environmental Science and Technology* 38, 1723–1731.

Hagger, J.A., M.B. Jones, D.R.P. Leonard, R. Owen, and T.S. Galloway. 2006. Biomarkers and integrated environmental risk assessment: Are there more questions than answers? *Integrated Environmental Assessment and Management* 2, 312–329.

Hagger, J.A., M.B. Jones, D. Lowe, D.R.P. Leonard, R. Owen, and T.S. Galloway. 2008. Application of biomarkers for improving risk assessments of chemicals under the Water Framework Directive: A case study. *Marine Pollution Bulletin* 56, 1111–1118.

Hellawell, V.M. 1986. *Biological Monitors of Freshwater Pollution and Environmental Management*. London: Elsevier.

Hynes, H.B.N., 1971. *Ecology of Running Water*. Liverpool, England: Liverpool University Press.

Hyne, R.V. and W.A. Maher. 2003. Invertebrate biomarkers: Links to toxicosis that predict population decline. *Ecotoxicology and Environmental Safety* 54, 366–374.

ICES (International Council for the Exploration of the Sea). 2007. Report of the Working Group on Biological Effects of Contaminants (WGBEC), March 19–23, 2007, Alessandria, Italy. CM 2007/MHC:03.

Jørgensen, S.E. and G. Bendoricchio 2001. *Fundamentals of Ecological Modelling*, Amsterdam: Elsevier, 530 pp.

Jørgensen, S.E., R. Costanza, and F.-L. Xu. 2010. *Handbook of Ecological Indicators for Assessment of Ecosystem Health*. 2nd edition. Boca Raton, FL: CRC Press, 482 pp.

Kay, J. and E.D. Schneider. 1992. Thermodynamics and measures of ecological integrity. In *Proc. "Ecological Indicators."* Amsterdam: Elsevier, pp. 159–182.

Keddy, C.J., J.C. Greene, and M.A. Bonnell. 1995. Review of whole-organism bioassays: Soil, freshwater sediment and freshwater assessment in Canada. *Ecotoxicology and Environmental Safety* 30, 221–251.

Kobayashi, N. 1991. Marine pollution bioassay by using sea urchin eggs in the Tanabe Bay, Wakayama Prefecture, Japan, 1970–1987. *Marine Pollution Bulletin* 23, 709–713.

Lagadic, L., T. Caquet, and F. Ramade. 1994. The role of biomarkers in environmental assessment. (5) Invertebrate populations and communities. *Ecotoxicology* 3, 193–208.

Marques, J.C., F. Salas, J. Patrício, J. Neto, and H. Teixeira. 2009. *Ecological Indicators for Coastal and Estuarine Environmental Assessment—A User Guide*. Southampton, UK: WIT Press, 208 pp.

Martinez-Haro, M., R. Beiras, J. Bellas, R. Capela, J.P. Coelho, I. Lopes, M. Moreira-Santos et al. 2015. A review on the ecological quality status assessment in aquatic systems using community based indicators and ecotoxicological tools: What might be the added value of their combination? *Ecological Indicators*, 48, 8–16 (DOI: 10.1016/j.ecolind.2014.07.024).

Maas, J.L., M.J. van den Heuvel-Greve. 2004. Opportunities for bio-analysis in WFD chemical monitoring using bioassays. RIZA working document 2005.053X.

Marín-Guirao, L., A. Cesar, A. Marín, J. Lloret, and R. Vita. 2005. Establishing the ecological quality status of soft-bottom mining-impacted coastal water bodies in the scope of the Water Framework Directive. *Marine Pollution Bulletin* 50, 374–387.

Mayon, N. and A. Bertrand, D. Leroy, C. Malbrouck, S.N.M. Mandiki, F. Silvestre, A. Goffart, J.P. Thomé, and P. Kestemont. 2006. Multiscale approach of fish responses to different types of environmental contaminations: A case study. *Science of the Total Environment* 367, 715–731.

Micheletti, C., S. Gottardo, A. Critto, S. Chiarato, and A. Marcomini. 2011. Environmental quality of transitional waters: The lagoon of Venice case study. *Environment International* 37, 31–41.

Monserrat, J.M., P.E. Martínez, L.A. Geracitano, L. Lund Amado, C. Martinez Gaspar Martins, G. Lopes Leães Pinho, I. Soares Chaves et al. 2007. Pollution biomarkers in estuarine animals: Critical review and new perspectives. *Comparative Biochemistry and Physiology—C Toxicology and Pharmacology* 146, 221–234.

Niell, X. and J. Pazo 1978. Incidencia de vertidos industriales en la estructura de poblaciones intermareales. II. Distribución de la biomasa y de la diversidad específica de comunidades de macrófitos de facies rocosa. *Investigaciones Pesqueras* 42(2), 231–239.

O'Connor, J.S. and Dewling, R.T. 1986. Indices of marine degradation: Their utility. *Environmental Management* 10, 335–343.

Pereira, P., S. Carvalho, F. Pereira, H. De Pablo, M.B. Gaspar, M. Pacheco, and C. Vale. 2012. Environmental quality assessment combining sediment metal levels, biomarkers and macrobenthic communities: Application to the Obidos coastal lagoon (Portugal). *Environmental Monitoring and Assessment* 184, 7141–7151.

Pinto, R., J. Patrício, A. Baeta, B.D. Fath, J.M. Neto, and J.C. Marques. 2009. Review and evaluation of estuarine biotic indices to assess benthic condition. *Ecological Indicators* 9(1), 1–25.

Prat, N., M. Rieradevall, C. Barata, and A. Munné. 2013. The combined use of metrics of biological quality and biomarkers to detect the effects of reclaimed water on macroinvertebrate assemblages in the lower part of a polluted Mediterranean river (Llobregat River, NE Spain). *Ecological Indicators* 24, 167–176.

Rand, G.M., P.G. Wells, and L.S. McCarty. 1995. Introduction to aquatic toxicology. In Rand, G.M. (Ed.), *Fundamentals of Aquatic Toxicology: Effects, Environmental Fate, and Risk Assessment*. 2nd edition. Washington, DC: Taylor & Francis, pp. 3–67.

Sanchez, W. and J.M. Porcher. 2009. Fish biomarkers for environmental monitoring within the Water Framework Directive of the European Union. *Trends in Analytical Chemistry* 28, 150–158.

Schettino, T., R. Caricato, A. Calisi, M.E. Giordano, and M.G. Lionetto. 2012. Biomarker approach in marine monitoring and assessment: New insights and perspectives. *Open Environmental Sciences* 6, 20–27.

Soares, J., A.M. Coimbra, M.A. Reis-Henriques, N.M. Monteiro, M.N. Vieira, J.M.A. Oliveira, P. Guedes-Dias et al. 2009. Disruption of zebrafish (*Danio rerio*) embryonic development after full life-cycle parental exposure to low levels of ethinylestradiol. *Aquatic Toxicology* 95, 330–338.

Sumpter, J.P. 2005. Endocrine disrupters in the aquatic environment: An overview. *Acta Hydrochimica et Hydrobiologica* 33, 9–16.

Tankoua, O.F., C. Amiard-Triquet, F. Denis, C. Minier, C. Mouneyrac, and B. Berthet. 2012. Physiological status and intersex in the endobenthic bivalve Scrobicularia plana from thirteen estuaries in northwest France. *Environmental Pollution* 167, 70–77.

Triebskorn, R., S. Adam, A. Behrens, S. Beier, J. Böhmer, T. Braunbeck, H. Casper et al. 2003. Establishing causality between pollution and effects at different levels of biological organization: The VALIMAR project. *Human and Ecological Risk Assessment* 9, 171–194.

Triebskorn, R., J. Böhmer, T. Braunbeck, W. Honnen, H.R. Köhler, R. Lehmann, A. Oberemm et al. 2001. The project VALIMAR (VALIdation of bioMARkers for the assessment of small stream pollution): Objectives, experimental design, summary of results, and recommendations for the application of biomarkers in risk assessment. *Journal of Aquatic Ecosystem Stress and Recovery* 8, 161–178.

van der Ohe, P.C., E. De Deckere, A. Prüb, I. Muñoz, G. Wolfram, M. Villagrasa, A. Ginebreda, M. Hein, and W. Brack. 2009. Toward an integrated assessment of the ecological and chemical status of European river basins. *Integrated Environmental Assessment and Management* 5, 50–61.

Valavanidis, A., T. Vlahogianni, M. Dassenakis, and M. Scoullos. 2006. Molecular biomarkers of oxidative stress in aquatic organisms in relation to toxic environmental pollutants. *Ecotoxicology and Environmental Safety* 64, 178–189.

Vashchenko, M.A. and P.M. Zhadan. 1993. Bioassay of bottom sediments of Petr Velikiy Gulf (Sea of Japan) with sexual cells, embryos, and larvae of the sea urchin. *Oceanology* 33, 102–105.

Vidal-Liñán, L. and J. Bellas. 2013. Practical procedures for selected biomarkers in mussels, Mytilus galloprovincialis—Implications for marine pollution monitoring. *Science of the Total Environment* 461–462, 56–64.

Von Bertalanffy, L. 1952. *Problems of Life*. New York: Wiley.

von der Ohe, P.C., A. Prüß, R.B. Schäfer, M. Liess, E. De Deckere, and W. Brack. 2007. Water quality indices across Europe—A comparison of the good ecological status of five river basins. *Journal of Environmental Monitoring* 9, 970–978.

Wharfe, J., W. Adams, S.E. Apitz, R. Barra, T.S. Bridges, C. Hickey, and S. Ireland. 2007. *In situ* methods of measurement—An important line of evidence in the environmental risk framework. *Integrated Environmental Assessment and Management* 3, 268–274.

Weiderholm, T. 1980. Use of benthos in lake monitoring. *Journal of Water Pollution Control Federaction* 52, 537–547.

WHO (World Health Organization). 1993. *Biomarkers and Risk Assessment: Concepts and Principles*. Geneva: World Health Organization.

10

Ecosystem Services as Diagnostic Tools

10.1 Ecosystem Services

Humankind benefits in a number of ways from ecosystems. Collectively, these benefits are defined as ecosystem services. The millennium ecosystem assessment (MA) defines the ecosystem services as benefits people obtain from ecosystems. Ecosystem services can be classified into four broad categories: *provisioning,* such as the production of food and water; *regulating,* such as the control of climate and disease; *supporting,* such as nutrient cycles and crop pollination; *cultural,* such as aesthetic and recreational benefits. To help inform decision makers, many ecosystem services are being assigned economic values, (as mentioned in Chapter 1) will attempt to assess the value of ecosystems services for a number of ecosystems.

The services in the four categories include the following items:

1. *Provisioning services:* "Products obtained from ecosystems":
   - Food (including seafood and game), crops, wild foods, and spices
   - Raw materials (including lumber, skins, fuel wood, organic matter, fodder, and fertilizer)
   - Genetic resources (including crop improvement genes and health care)
   - Water
   - Minerals (including diatomite)
   - Medicinal resources (including drugs, pharmaceuticals, chemical models, and test and assay organisms)
   - Chemical resources (the number of possibly useful chemicals is enormous)
   - Energy (hydropower, biomass fuels)
   - Ornamental resources (including fashion, handicraft, jewelry, pets, worship, decoration, and souvenirs like furs, feathers, ivory, orchids, butterflies, aquarium fish, shells, etc.)

2. *Regulating Services:* "Benefits obtained from the regulation of ecosystem processes":
   - Carbon sequestration and climate regulation
   - Waste decomposition and detoxification
   - Purification of water and air
   - Pest and disease control
   - Buffering water quantities (thereby reducing the number of floods and droughts)
   - Moderating general weather extremes
   - Maintaining diversity
   - Preserving quantities (reduce erosion) and qualities (maintain fertility) of soil
3. *Supporting Services:* Ecosystem services are those that are necessary for the production of all other ecosystem services:
   - Nutrient dispersal and cycling
   - Seed dispersal
   - Primary production
   - Pollinating crops and natural vegetation
4. *Cultural Services:* Nonmaterial benefits people obtain from ecosystems through aesthetic enrichment and experiences, reflection, and recreation:
   - Cultural (including the use of nature as motif in books, film, painting, folklore, national symbols, architecture, advertising, etc.)
   - Historical (including the use of nature for religious or heritage value)
   - Aesthetical (enjoying nature and nature as a source of inspiration in art)
   - Recreational experiences (including ecotourism, outdoor sports, and recreation)
   - Science and education (including the use of natural systems for school excursions and scientific discovery)

10.2 Value of Ecosystem Services

How can we estimate the value of the ecosystem services? Costanza et al. (1997) have found the value by adding the costs of all the services that we utilize by ecosystems: purification of air and water, recycling, recreational services, and natural resources such as timber, fish, drinking water, and so on. All activities require energy that can do work. Services by ecosystems mean activities offered to the user of ecosystems and they could therefore be measured by the work energy. This means as the total amount of eco-exergy (work capacity), that an ecosystem offers. The sustainability of nature (ecosystems) can also be expressed by the work capacity, because

1. Sustainability in the Brundtland's sense means that the same level of ecosystem services must be maintained for the future generation.
2. The amount of energy to be used to break down an ecosystem is equal to the work capacity of the ecosystem (Svirezhev 1998).

This idea to measure ecosystem services and thereby sustainability by eco-exergy or work capacity is pursued in this chapter and the results are compared with Costanza et al.'s results. It is a newly proposed approach, which inevitably will lead to the recommendation to apply eco-exergy or work energy capacity calculations of ecosystems as informative holistic indicators to assess the economic value of ecosystem services. How do we calculate the work capacity of ecosystems?

Work energy is a part of the total energy that can do work. It can therefore be found as the gradient (= difference in potential) x extensive descriptor, dependent on the energy form, for instance (Jørgensen et al. 2007; Jørgensen 2012).

- Chemical work energy = $(\mu 1 - \mu 2) * N$ where = $(\mu 1 - \mu 2)$ is the difference in chemical potential before and after the actual chemical process and N is the number of moles participating in the chemical process.
- Pressure work energy = $(p1 - p2) (-V)$ where $p1 - p2$ is the pressure difference available for the volume V.
- Potential work energy = $(h1 - h2) * m * g$ where h1 and h2 are the high and low altitude respectively, m is the mass and g the gravity constant.
- Electrical work energy = $(V1 - V2) * Q$ where $V1 - V2$ is the difference in voltage and Q is the charge.

We can distinguish between technological work energy also denoted as exergy and ecological work energy also denoted as eco-exergy: It includes in contrast to technological work energy also the work energy of information (see Chapters 3 and 4). Technological exergy uses the environment as a reference state and is useful to find the first class energy (work) that a power plant can produce. Eco-exergy uses as a reference state the same ecosystem with the same temperature and pressure but at thermodynamic–chemical equilibrium; see Chapters 3 and 4 for details.

For calculation of ecosystem services, we will use the work capacity of ecosystems or eco-exergy, which includes the work energy of the information that the organisms use to perform the life processes. As the reference system has the same temperature and pressure and magnetic and electrical work are negligible, only the chemical work energy contributes to the work capacity of ecosystems. The following equation is used for the calculations (see Chapters 3 and 4 for details):

$$\text{Eco-exergy} = \sum_{i=1}^{i=n} \beta i * ci, \tag{10.1}$$

where β is a weighting factor, considering the information carried by the organisms in their genes, that is, how many amino acids in the right sequence are required to make up the organism, or how much information does an organism contain? β-values found based on the genome size and by indirect methods can be seen in Table 3.7.

The value of the annual ecosystem services can be expressed by the annual increase of work capacity or eco-exergy. The annual growth of biomass—included growth of biomass, information, and the ecological networks (see Jørgensen et al., 2000)—for the most common types of ecosystems can be found in general ecology textbooks or on internet. An average β-value for various ecosystems can be estimated based on the organisms that can be found generally in various types of ecosystems. Table 1.2 shows the annual biomass growth in MJ/m^2 using an average energy (exergy) content of biomass of 18.7 kJ/g, the applied average β-values and the work capacities in GJ/ha y found as the product of the biomass growth per m^2 and the β-value.

The work capacity of various ecosystems can easily be compared with Costanza et al.'s (1997) values by using the energy costs:

- 1 MJ has the value of 1 EURO-cent or 1.4 $-cent.
- 1 GJ has therefore the value of 10 EURO or 14 $.

The values found on basis of the results in Table 1.2, are compared with Costanza et al.'s results in Table 10.1. The work capacity values give a much higher money value, which is not surprising, because the work capacities include all the possible services that the ecosystems offer—not only the services that we actually are utilizing. The ratio of the two sets of values is included in the last column of Table 10.1.

10.3 Classification of Ecosystems

We use evidently the ecosystems differently, which can explain the different ratios. The ecosystems can be divided into five classes according to the

TABLE 10.1

Value of Annual Ecosystem Services

Ecosystem	(Based on Eco-Exergy) kEURO/ha Year	According to Costanza et al. (1997) $/ha Year	Ratio kEURO/$
Desert	20.7	–	–
Open sea	23.8	252	94
Coastal zones	48.3	4052	12
Coral reefs, estuaries	960	14,460	66
Lakes, rivers	93.5	8500	11
Coniferous forests	539	969	556
Deciduous forests	1000	969	1032
Temperate rainforests	1580	–	–
Tropical rainforests	3000	2007	1495
Tundra	72.8	–	–
Croplands	400	92	4348
Grasslands	180	232	775
Wetlands	450	14,785	30.4

Source: Jørgensen, S.E. 2010. *Ecol. Complex.* 7, 311–313.

different ratios, starting with the class which is utilized most by man for a series of services:

- A. Coastal zones, lakes, rivers:

  Regulation, water supply, waste treatment, recreation, genetic resources, pollination, nutrient cycles, biological control, food production, refugia, transportation, raw materials, purifying air, moderating weather extremes, cultural; ratio about 10–20.

- B. Wetlands:

  Regulation, water supply, waste treatment, recreation, raw materials, genetic resources, pollination, nutrient cycles, biological control, refugia, cultural, buffering water quantities; ratio about 30.

- C. Open sea, estuaries, coral reefs:

  Only climate and gas regulation, very little waste treatment, much less recreation than A and B, raw materials, genetic resources, pollination, nutrient cycles, (minor) biological control, (minor) refugia, cultural; ratio about 60–100.

- D. Forests, grasslands, and (deserts):

  Mainly as raw materials, too little genetic resources, pollination, nutrient cycles, biological control, (minor) refugia, cultural, recreation; ratio about >500.

- E. Croplands:

  Croplands are only utilized to produce raw materials (mainly food); the ratio is therefore high, 4348.

The results can be summarized in the following conclusions:

- The total value of all services offered by the ecosystems may be estimated from the work capacity (eco-exergy) of the ecosystems.
- It is also a measure of the sustainability.
- We can divide the ecosystems in five classes according to how much we are able to utilize the entire spectrum of possible services.
- The sequence of our utilization of the ecosystem services is: lakes and rivers, coastal zones, wetlands, estuaries, open sea ecosystems, grasslands, forests, and croplands.
- The sequence is understandable and explainable.

The calculations of eco-exergy (work energy with information included) can be used as measure for

- The entire spectrum of all the services that the ecosystems offer—not only the services that we actually utilize.
- The sustainability—at least indirectly.

It is therefore possible to use the calculations of eco-exergy of ecosystems as a very informative holistic ecological indicators.

References

Costanza et al. 1997. The value of the world's ecosystem services and natural capital. *Nature* 387, 252–260.

Jørgensen, S.E. 2010. Ecosystem services, sustainability and thermodynamic indicators. *Ecol. Complex.* 7, 311–313.

Jørgensen, S.E. 2012. *Introduction to Systems Ecology.* (Chinese edition.) Boca Raton, FL: CRC Press, 320 pp.

Jørgensen, S.E. et al. 2007. A *New Ecology Systems Perspectives.* Amsterdam: Elsevier, 275 pp.

Jørgensen, S.E. et al. 2000. Ecosystems emerging: 4 growth. *Ecol. Model.* 126, 249–284.

Svirezhev, Yu. M. 1998. Thermodynamic orientors: How to use thermodynamic concepts in ecology? In Müller, F. and Leupelt, M. (Eds.), *Eco Targets, Goal Functions and Orientors.* Berlin, Heidelberg: Springer, pp. 102–122.

Part V

Tools for Solutions of Environmental Problems

11

Application of Environmental Technology in Environmental Management

11.1 Introduction

Environmental technology was applied more than 100 years ago to solve pollution problems. Biological treatment of wastewater has been applied although on a modest scale in Europe even before the start of the twentieth century. Before Rachel Carson's book, *The Silent Spring*, it was the only tool available for the solution of pollution problems and environmental problems in general. Environmental technology is based on a technological method to reduce the pollution; it means treatment of wastewater, smoke, and solid waste. Environmental technology is best fitted to solve pollution problems from point sources, because it makes it possible to use the technological method directly on the point source, which implies that the amount of wastewater and solid waste is limited and rather concentrated. It is therefore possible to develop a technological method that is able to reduce the concentration of pollutants with a high efficiency in most cases. Environmental technology in many cases requires a relatively high investment and the operational costs are also sometimes high; although there are also many examples of relatively low operational costs. After the development of the other tools to solve the pollution problems—as presented in Chapters 12–15—it is clear that environmental technology often beneficially has to be combined with the other tools to find the most effective and cost-moderate solution, although there are also several cases where environmental technology is the only obvious method to apply. This is particularly the case for industrial pollution problems, where environmental technology even may offer possibilities to recover or recycle raw materials.

This chapter will give a short overview of the most applied environmental technological methods. The overview of water pollution methods is presented in Section 11.2, while Section 11.3 is devoted to air pollution treatment methods, and finally Section 11.4 is focused on the treatment of solid waste problems.

11.2 Application of Environmental Technology on Wastewater Problems

Impact on aquatic ecosystems from point sources originates from the discharge of wastewater. Wastewater discharges into ecosystems are man-controlled forcing functions of crucial importance for the water quality. It is, however, possible in many situations to control it completely, either by water diversion or by wastewater treatment methods. Water diversion, however, results in another downstream water body that has to cope with the pollutant load. Thus, treating the wastewater properly should be considered a generally more acceptable solution to the problem. This gives rise to two questions; namely

i. Is it possible to solve all pertinent wastewater problems?
ii. What is understood by a proper wastewater treatment?

The water pollution problems associated with municipal and industrial wastewaters include their content of (see also Chapter 6 for more details)

- Nutrients causing eutrophication.
- Biodegradable organic matter causing oxygen depletion.
- Bacteria and viruses affecting the sanitary quality of water, which is of particular importance, when the water is used for bathing, swimming, and drinking purposes.
- Heavy metals, namely mainly lead, zinc, and cadmium from gutters, heavy metals from fungicides and other agricultural chemicals, and a wide range of other heavy metals in minor concentrations.
- Refractory organic matter, originating from industries, hospitals and the use of pesticides and even from the use of a wide spectrum of household articles.

Tables 11.1 and 11.2 (Jørgensen, 2000) provide an overview of a wide range of wastewater treatment methods, their efficiencies, and costs. Clearly, there is with good approximations a method available to virtually any of the above mentioned problems.

Industrial wastewaters can cause the same water pollution problems as municipal wastewater plus a few more associated with toxic organic and/ or inorganic compounds (particularly heavy metals and persistent organic pollutants). However, it is necessary to solve at the source the problems associated with industrial wastewater that can hardly be solved with municipal wastewater treatment methods. It is also the general legislation all over the world today, where industries are obliged to treat the wastewater before discharge to the public sewage system. In many countries, the practice of the

TABLE 11.1

Survey of Generally Applied Wastewater Treatment Methods

Method	Pollution Problem	Efficiency	Costs (US$/100 m³)
Mechanical treatment	Suspended matter removal	0.75–0.90	3–5
	BOD_5 reduction	0.20–0.35	
Biological treatment	BOD_5 reduction	0.70–0.95	25–40
Flocculation	Phosphorus removal	0.3–0.6	6–9
	BOD_5 reduction	0.4–0.6	
Chemical precipitation	Phosphorus removal	0.65–0.95	10–15
$Al_2(SO_4)_3$ or $FeCl_3$	Reduction of heavy metals concentrations	0.40–0.80	
	BOD_5 reduction	0.50–0.65	
Chemical precipitation	Phosphorus removal	0.85–0.95	12–18
$Ca(OH)_2$	Reduction of heavy metals concentrations	0.80–0.95	
	BOD_5 reduction	0.50–0.70	
Chemical precipitation and flocculation	Phosphorus removal	0.9–0.98	12–18
	BOD_5 reduction	0.6–0.75	
Ammonia stripping	Ammonia removal	0.70–0.95	25–40
Nitrification	Ammonium is oxidized to nitrate	0.80–0.95	20–30
Active carbon adsorption	COD removal (toxic substances)	0.40–0.95	60–90
	BOD_5 reduction	0.40–0.70	
Denitrification after nitrification	Nitrogen removal	0.70–0.90	15–25
Ion exchange	BOD_5 reduction (e.g., proteins)	0.20–0.40	40–60
	Phosphorus removal	0.80–0.95	70–100
	Nitrogen removal	0.80–0.95	45–60
	Reduction of concentrations	10–25	
Chemical oxidation (e.g., with Cl_2)	Oxidation of toxic compounds	0.90–0.98	60–100
Extraction	Heavy metals and other toxic compounds	0.50–0.95	80–120
Reverse osmosis	Removes pollutants with high efficiency, but is expensive	100–200	
Disinfection methods	Reduction of microorganisms	High, can hardly be indicated	6–10
Ozonation + active carbon adsorption	Removal of refractory compounds	0.5–0.95	100–120

TABLE 11.2

Efficiency Matrix Relating Pollution Parameters and Wastewater Treatment

	Suspended Matter	**BOD_5**	**COD**	**Total Phosphorus**	**Ammonium Nitrogen**
Mechanical treatment	0.75–0.90	0.20–0.35	0.20–0.35	0.05–0.10	~0
Biological treatment[a]	0.75–0.95	0.65–0.90	0.10–0.20	0.05–0.10	~0
Chemical precipitation	0.80–0.95	0.50–0.75	0.50–0.75	0.80–0.95	~0
Ammonia stripping	~0	~0	~0	~0	0.70–0.96
Nitrification	~0	~0	~0	~0	0.80–0.95
Active carbon adsorption[a]	–	0.40–0.70	0.40–0.95	~0.1	High[b]
Denitrification after nitrification	~0	–	–	~0	–
Ion exchange	–0.40	0.20–0.50	0.20–0.95	0.80–0.95	0.80–0.95
Chemical oxidation	–	Corresponding to oxidation	~0	~0	~0
Extraction	–	Corresponding extraction of toxic compounds	~0	~0	~0
Reverse osmosis[a]					
Disinfection methods	–	Much corresp. to appl. of chlorine, ozone, etc.			

	Total Nitrogen	Heavy Metals	*E. coli*	Color	Turbidity
Mechanical treatment	0.10–0.25	0.20–0.40	–	0.80–0.98	
Biological treatment[a]	0.10–0.25	0.30–0.65	Fair	~0	–
Chemical precipitation	0.10–0.60	0.80–0.98	Good	0.30–0.70	0.80–0.98
Ammonia stripping	0.60–0.90	~0	~0	~0	~0
Nitrification	0.80–0.95	~0	Fair	~0	~0
Active carbon ads.[a]	High[b]	0.10–0.70	Good	0.70–0.90	0.60–0.90
Denitrification after nitrification	0.70–0.90	~0	Good	~0	–
Ion exchange	0.80–0.95	0.80–0.95	Very good	0.60–0.90	0.70–0.90
Chemical oxidation	~0	~0	~0	0.60–0.90	0.50–0.80
Extraction	~0	0.50–0.95	~0	~0	~0
Reverse osmosis[a]					
Disinfection methods	Very high	0.50–0.90	0.30–0.60		

[a] Depends on the composition.
[b] As chloramines.

"polluter pays" principle, has forced the industries to solve their pollution problems to keep the production costs low. The major portion of toxic substances is therefore today removed by the industries, at least in the industrialized countries. They would only have been partially removed, if at all, at municipal wastewater treatment plants, and/or could contaminate the sludge produced at municipal wastewater treatment plants, thereby eliminating the possibility of the use of the sludge as a soil conditioner.

The removal of high concentrations of biodegradable organic matter at the source is strongly recommended, since it is usually much more cost-effective to remove these components, at least partially, when they are present in high concentrations. High concentrations of biodegradable organic matter are found in wastewater from slaughterhouses, starch factories, fish industries, dairies, and canned food industries. The removal of x% of biological oxygen demand (BOD_5) costs as a rule of thumb the same independent on the level of BOD_5. It means that 1 kg of BOD_5 can be removed at a much lower price from wastewater with a high BOD_5 as for instance wastewater from slaughterhouses, starch factories, fish industries, dairies, and canned food industries.

The listed methods often are used in combinations of two or more steps to obtain the overall removal efficiency required by the most cost-moderate solution. The methods can also be applied in combination with ecotechnology (Chapter 12) and cleaner technology (Chapter 13). Because wastewater treatment often is costly, it is recommendable in the planning phase to examine *all* possible combinations of treatment options in order to identify the most feasible and appropriate one.

Many existing municipal wastewater treatment plants were constructed years or decades ago, and may not meet today's higher standards. Nevertheless, upgrading an existing wastewater treatment plants is possible, and may be more cost-moderate than building new ones (Novotny and Somlyódy, 1995; van Loosdrecht, 1998). Because the funding allocated to pollution abatement often is limited, the overall effect of upgrading wastewater treatment plants that can be upgraded with sufficient efficiency will be to the benefit of the environment. An attractive solution is often to introduce *tertiary treatment* by chemical precipitation and flocculation in an existing mechanical–biological treatment plant, with the addition of chemicals and flocculants before the primary sedimentation phase. The installation costs for this solution are minor, and the additional running costs are limited to the costs of chemicals. The result is 85%–95% removal of phosphorus at low cost. Similarly, nitrification and denitrification, ensuring a 80%–85% removal of nitrogen, can be realized with the installation of additional capacity for biological treatment (the overall water retention time in the plant is increased by 4–12 h ours, depending on the standards and composition of the wastewater), which is considerably less costly than installation of a completely new treatment plant. For details, see Hahn and Muller (1995) and Henze and Ødegaard (1995).

The second question refers to the selection of the right standards for the treated wastewater. Any removal efficiency of any pertinent parameters (BOD_5, nutrients, bacteria, viruses, toxic organic toxic compounds, color, taste, heavy metals) is possible to obtain with a suitable combination of the available treatment methods. However, what removal efficiencies are needed in the focal case? Because wastewater treatment is costly, the maximum allowable concentrations should not be set significantly lower than the ecosystem receiving the effluents can tolerate. The ban of phosphate detergents to decrease phosphorus concentrations in municipal wastewater treatment plant effluents together with increased efficiency of detergents are points to consider in this context, as the treatment costs can be reduced considerably by the introduction of phosphorus-free detergents, which is more or less the situation today in most industrialized countries. On the other hand, it might be even more expensive to install an insufficient treatment plant. Thus, the potential effects of a wide range of possible pollutant inputs on water quality and on the entire lake or reservoir should be assessed, as the basis for selecting an acceptable option. This will require a quantification of the impacts of various possible pollutant inputs, considering a wide range of solutions. All processes and components affected significantly by the impacts should be included in the quantification. It is usually very helpful to develop a water quality or ecosystem model and use it properly to assist in the selection of specific environmental treatment methods. It is important to emphasize that a model has an uncertainty in all its predictions that must be considered in making a final decision. Thus, it is essential to use safety factors to the benefit of the environment, in order to ensure that the selected treatment methods will have the anticipated effects. If the uncertainty is not taken into account for the sake of economy, as it is unfortunately often done, the investment may be wasted because the foreseen recovery of the freshwater ecosystem will not be realised.

Application of the methods identified in Tables 11.1 and 11.2 gives only approximate results, and the indications should therefore be used with caution. However, first estimates, such as those shown in the tables, are useful for evaluations of various alternative solutions to wastewater pollution problems. The biological treatment may either be an activated sludge plant or a trickling filter.

The cost of treating 100 m^3 of wastewater is also based on approximate indications, because they vary from place to place, as the costs of labor are very different in different locations and are highly dependent on the size of the wastewater treatment plant. The costs are calculated as the running costs (electricity, labor, chemicals, and maintenance), plus 10% of the investment to cover interests and annual appreciation. The annual water consumption of one person in an industrialized country corresponds to approximately 100 m^3.

A problem in many developing countries is the relatively high cost of wastewater treatment. Although this cost might justify diversion of the wastewater, the application of "soft technology"–"ecotechnology" should also be considered. Some corresponding methods will be touched upon in the following sections, but proper planning at an early phase, and considering all predictable problems, offers the widest range of cost-effective possibilities, and may allow the prevention of pollution problems before they will occur.

Corrections at a later stage, when pollution has already degraded the water quality and associated ecosystems, are possible, but will always be more expensive than the costs of proper wastewater treatment at an early stage. This is due, in part, to the fact that the accumulation of pollutants in an aquatic ecosystem over time will always cause additional problems and, therefore, result in additional costs. Thus, pollution prevention at an early stage is better than curing pollution at a later stage. Removal of phosphorus from wastewater at an early stage, for example, is always beneficial since the surplus phosphorus will accumulate in the sediments—to a large extent—allow its remobilization back into the water column under certain chemical conditions in the water body.

Model studies are able to reveal how long it may require to restore an ecosystem, or how much higher phosphorus removal efficiency will be required to compensate for each year that implementation of an appropriate phosphorus removal technology is postponed, as phosphorus is accumulated in the ecosystem, mainly in the sediment. However, it is usual that implementation of a phosphorus removal technology a few years later than it was first feasible may delay the restoration of for instance a lake by one or more decades, due to the fact that the additional phosphorus accumulated in the sediments may significantly increase the quantity of phosphorus in the water column.

Important pollution sources may be reduced if the liquid waste and sludge land disposal is avoided or minimized. For municipal areas, two options are used to decrease the hydraulic load of wastewater treatment plants:

i. Decreasing water use, thereby saving water and producing smaller volumes of polluted water.
ii. Separating storm water from municipal domestic waste, with a similar result. One result of these options is that the capacity of wastewater treatment plants can be kept smaller, achieving significant cost savings.

New approaches have emerged with regard to sustainable development. For instance, serious consideration is being given to separation toilets in some locations. They collect urine separately from feces, thereby allowing utilization of the septic urine as fertilizer and the collected fecalia to be used as compost after appropriate treatment.

The selection of proper wastewater treatment methods for point sources of pollution is summarized in the following points:

- Develop models for the impacts of the wastewater on freshwater ecosystems, considering the impacts on the water quality and the entire lake ecosystem.

- Apply the model to identify the maximum allowable pollutant concentration in the treated wastewater. Any uncertainty associated with the model predictions should be reflected in identifying the lower maximum allowable concentrations.
- Select the combination of available treatment methods able to meet the standards at the lowest costs without impacting the proper operation of the plant.
- If the investment needed for a proper solution to a problem cannot be provided, the application of cost-moderate technology that will reduce the accumulation of pollutants in the lake should be considered. Any measures taken at an early stage will reduce the costs at a later stage.

The composition of domestic sewage varies surprisingly little from place to place, although it to a certain extent reflects the economic status of the society.

A typical composition of municipal wastewater including BOD_5 (biological oxygen demand—the concentration of oxygen consumed over 5 days for a microbiological decomposition of the organic matter, COD (chemical oxygen demand—the amount chromate oxygen used to decompose (almost completely) the organic matter) suspended matter and nutrients is shown in Table 11.3.

Most industrialized countries require a treatment of municipal wastewater. The standards vary for the European Union, the United States, Canada, Australia, New Zealand, and Japan, but generally the following are required: a reduction of BOD_5 to about 10 mg/L, a reduction of nitrogen to about 10 mg/L,

TABLE 11.3

Typical Composition of Municipal Wastewater (mg/L)

Constituent	Soluble	Particulate	Total
BOD_5	100–200	50–100	150–300
COD	200–500	100–200	200–700
Ammonium-N	20–40	0	20–40
Nitrate-N	5–20	0	5–20
Organic nitrogen	0	5–20	25–60
Suspended matter	–	40–80	40–80
Carbohydrates	20–40	10–15	30–55
Amino acids	10–15	15–25	25–40
Fatty acids	0	50–80	50–80
Surfactants	10–20	5–10	15–30
Creatinin	3–5	0	3–5
Phosphorus	2–4	4–10	6–14

Source: Jørgensen, S.E. 2000. *Principles of Pollution Abatement*. Amsterdam: Elsevier, 520 pp.

and a reduction of the phosphorus concentration to 1–2 mg/L or even less. The combinations of methods applied for the treatment of municipal wastewater are presented in the next section and it will be demonstrated that it is possible by a suitable combination of methods to meet practically all realistic effluent standards.

Models are used increasingly to design and to optimize wastewater treatment methods. For details about the applied models refer to Jørgensen (2011), *Handbook of Ecological Models Used in Ecosystem and Environmental Management*, which has a comprehensive chapter devoted to models of wastewater treatment systems.

There are a number of different possible designs of biological treatment plants; see Jørgensen (2000). The use of mechanical–biological wastewater treatment is a classical method or rather the combination of the two methods. Figure 11.1 gives the flow chart of a classical mechanical-activated sludge plan as it is generally used all over the world. The processes involve screening, a separation of grease and sand in the grit chamber, a primary

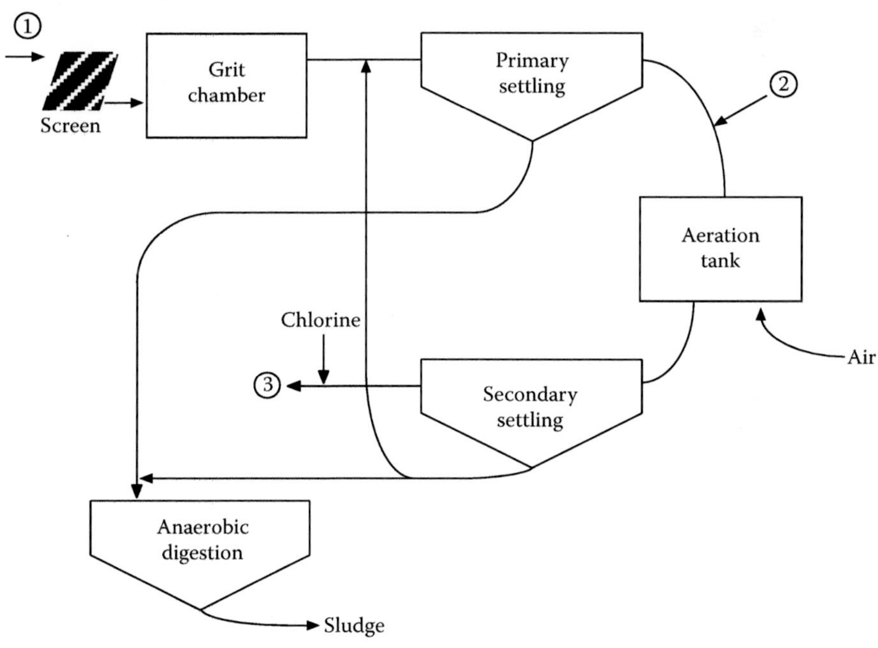

FIGURE 11.1
A flow diagram of a conventional activated sludge plant. It is widely used to treat municipal wastewater. Number 1 is untreated municipal wastewater which would have a BOD_5 of about 150–300 mg/L, total nitrogen of about 25–45 mg/L and total phosphorus of about 6–12 mg/L. The mechanical treatment (sample at point 2) will reduce BOD_5 and total N by about 25%–40%, while the reduction of total phosphorus will be minor. The totally treated wastewater (point 3) will have BOD_5 of about 10–15 mg/L, while the nitrogen only is reduced 35%–50% and the phosphorus concentration only about 10%–20%.

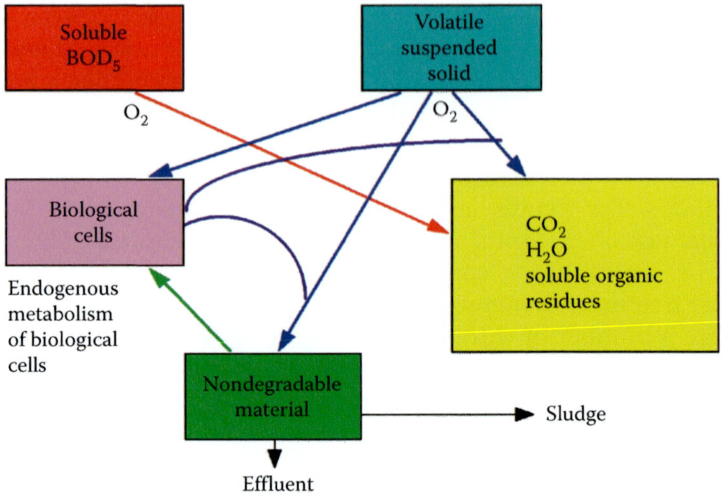

FIGURE 11.2
Processes characteristic for biological treatment.

settling, an aeration step followed by sedimentation. Chlorine is often added before discharge to the receiving water, particularly when it is used for swimming and bathing. The sludge is most often digested anaerobic, which gives a production of biogas. In the activated sludge plant, a rapid adsorption and flocculation of suspended matter take place. Organic matter is oxidized and decomposed. Sludge particles are dispersed and settled by the secondary settling. The processes are conceptualized in Figure 11.2.

ILLUSTRATION 11.1

The BOD_5 (= the oxygen used by microorganisms to decompose the organic matter during a period of 5 days) is often used to express the concentration of organic matter in wastewater. If the BOD_5 is 150 mg/L by 20°C, what would be the BOD_7 (which is used in Sweden) at 15°C? The rate coefficient for decomposition by microorganisms of organic matter in typical municipal wastewater is 0.351/24 h and the Arrhenius constant is 1.05.

Solution

The organic matter is decomposed as a first-order reaction (see Chapter 8) and if we denoted the concentration of organic matter (measured by its oxygen demand) in the wastewater for L, then we know that after 5 days L is reduced 150 mg/L. It means that

$$L(0) - L(5) = 150 \text{ mg/L} = L(0)(1 - \exp(-0.35 * 5)) \text{ or}$$
$$L(0) = 150/(1 - \exp(-0.35 * 5)) = 181 \text{ mg/L}$$

The decomposition rate constant at 15°C is according to the Arrhenius equation = $0.35 \times (1.05^{(15-20)}) = 0.271/24$ h.

$$L(7) \text{ at } 15°C \text{ is therefore } = L(0)(1 - \exp(-0.27 \times 7)) = 154 \text{ mg/L}.$$

Ditches, see Figure 11.3, can replace the activated sludge plan and the secondary settling. The rotor provides the aeration to oxidize the organic matter. The influent and the rotor are usually stopped during the night when the inflow is low to allow for settling. The supernatant is withdrawn through an effluent launder. The retention time is usually 2–5 days.

A trickling filter is a bed packed with rocks, although recently also with plastic media (celite pellets; Sorial et al., 1998) is applied or bioblocks with a high surface area due to the high porosity. It requires substantially less space than the stone packed trickling filter. They usually have a specific surface on 100 square meter per cubic meter or more. It is at least twice the specific surface for trickling filter packed by rocks. The media is covered by a slimy microbiological film. The wastewater is passed through the bed, and oxygen and organic matter diffuse into the film, where oxidation occurs. Recirculation improves the removal efficiency.

Physical–chemical methods have been proposed to replace mechanical–biological treatment, for instance the so-called advanced wastewater treatment or AWT system, which is based on the application of a combination of mechanical treatment, precipitation, adsorption on activated carbon and sand filtration.

It has been considered to recover proteins and grease to at least partially cover the costs of the treatment of wastewater from slaughterhouses, fish industries, starch factories, and other foodstuff industries. Figure 11.4 shows

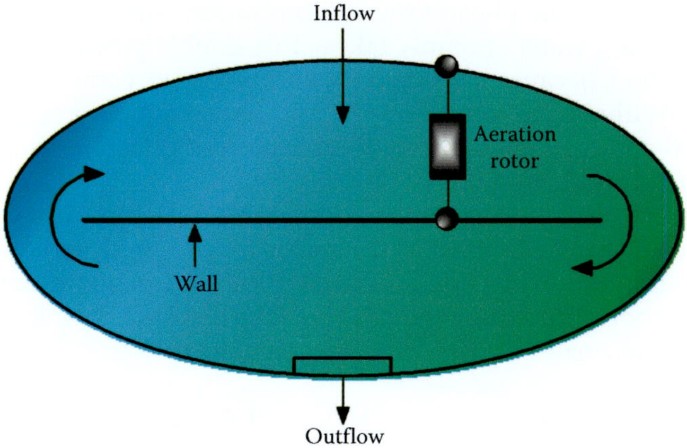

FIGURE 11.3
Oxidation ditch.

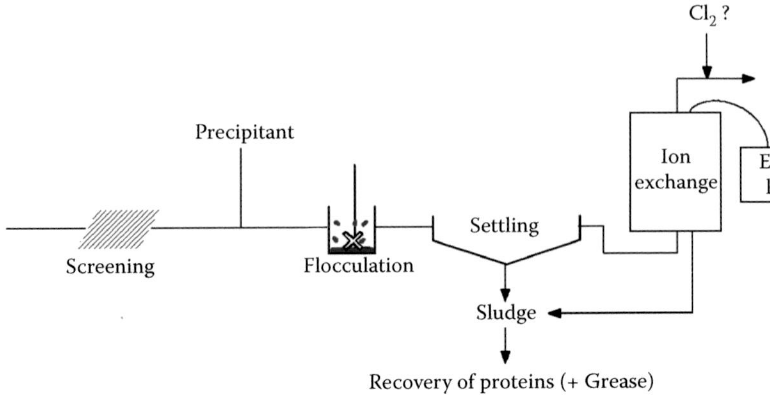

FIGURE 11.4
Recovery of proteins and grease from the wastewater discharge from the foodstuff industries is possible by a combination of precipitation and ion exchange. The recovery partially pays for the treatment of the wastewater by the process.

a method that has been applied in several, but still relatively few cases, but is included in this chapter to present a method that makes it possible to recover valuable components from wastewater. Several other methods for recovery of heavy metals from the metal industry or phosphorus from various types of wastewater are other recovery possibilities.

Nutrient removal is most frequently carried out by chemical precipitation, often combined with mechanical–biological treatment. The chemical precipitation can be applied at three different points in the mechanical–biological plant as shown in Figure 11.5. Sometimes both direct precipitation and post-treatment is applied, which together with sand filtration makes it possible to obtain a concentration of 0.1 mg P/l or less in the effluent. In many cases where such a low phosphorus concentration is needed for particular sensitive receiving waters (mainly lakes), this double precipitation is a very attractive method, because it is relatively cost moderate compared with other alternatives. Without the sand filtration and only a single precipitation, it is usually relatively easy to obtain a concentration of phosphorus in the effluent between 1.0 and 1.5 mg/L.

Aluminum sulfate, various poly-aluminates, calcium hydroxide, and iron(III) chloride can be applied as chemicals for the precipitation.

A combination of nitrification and denitrification can significantly reduce the nitrogen concentration in the effluent. The applied chemical processes are imitating the corresponding processes in nature:

$$NH_4^+ + 2O_2 \rightarrow NO_3^- + H_2O + 2H^+$$

$$4NO_3^- + 5C + H_2O \rightarrow 5HCO_3^- + 2N_2 + H^+$$

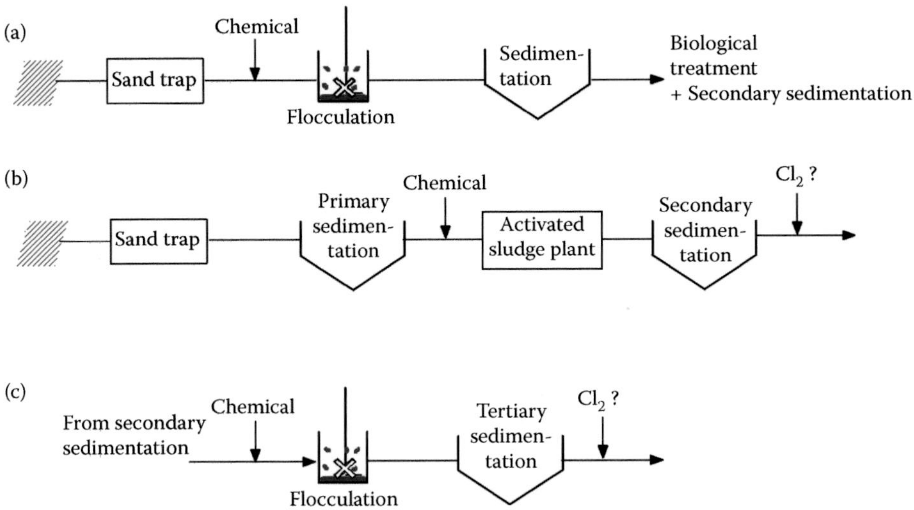

FIGURE 11.5
Precipitation by aluminum sulfate or other aluminum compounds, iron(III) chloride and calcium hydroxide is able to reduce the phosphorus concentration in wastewater significantly. The precipitation is applied often in combination with mechanical–biological treatment and can be carried out after the sand trap: (a) direct precipitation, before the activated sludge plan, (b) simultaneous precipitation or after the mechanical–biological treatment, and (c) post-treatment.

The ammonium is oxidized to nitrate and the nitrate is used to oxidize by anaerobic conditions organic matter—here indicated just as C. Thereby the nitrate is reduced to dinitrogen, which is released to the atmosphere, where there is about 78% dinitrogen and a minor addition of dinitrogen is therefore harmless.

The nitrification process is a two-step process. First, ammonia is oxidized to nitrite by Nitrosomonas. Second, nitrite is oxidized to nitrate by Nitrobacter. The optimum pH range for the nitrification is 6.7–7.0 (Groeneweg et al., 1994). The oxygen concentration for the nitrification has to be at least 2 mg/L. Heavy metal ions are toxic to the nitrification at rather low concentrations. Toxic levels of about 0.2 mg/L are reported for chromium, nickel, and zinc (Halling Sørensen and Jørgensen, 1993).

The denitrification takes place by a number of heterotrophic bacteria present in activated sludge. Anaerobic conditions are absolutely required. The optimum pH for this process is about 7.0. About three times as much BOD_5 as nitrate-N both expressed in mg/L should be applied to ensure adequate denitrification. The process can be realized even at high concentrations of nitrate or at high salinities (Jørgensen, 2000). Addition of organic carbon—for instance methanol—has been proposed to ensure a proper denitrification, but it is of course not an attractive method to apply. Alternatives are

1. To switch between aerobic and anaerobic conditions. It is called an alternative operation (for more details see Diab et al., 1993 and Halling Sørensen and Jørgensen, 1993);
2. To recycle a part of the wastewater containing nitrate after the treatment is completed.

Both methods work properly. Other possibilities are the use of zeolite ion-exchange material to obtain simultaneous nitrification and denitrification (Halling Sørensen and Jørgensen, 1993), the use of electrolysis for denitrification (Islam and Suidan, 1998) and the use of activated carbon to enhance the denitrification (Sison et al., 1996).

Ammonia can furthermore be removed by blowing air through the wastewater. The process is called stripping. The cost of stripping is relatively low, but the process has three crucial limitations:

1. It is practically impossible to work at temperatures lower than 7°C.
2. Deposition of calcium carbonate can reduce the efficiency or even block the tower.
3. After the ammonia removal pH is high and must be adjusted to 8.0 or below.

Moreover, ammonia can be removed by chlorination and adsorption on activated carbon. Chlorine can oxidize ammonia to NH_2Cl, $NHCl_2$, and NCl_3 and the activated carbon is able to adsorb the chloramines formed by these oxidation processes, but the process is expensive due to the high amount of chlorine needed for the oxidation of ammonia.

By a combination of the available treatment methods, it is possible to recycle municipal wastewater—it means produce drinking water of municipal wastewater. The recycling takes about 8–10 days from toilet to water tap so to say. All drinking water is in principle recycled. It takes normally nature a couple of thousands years to recycle water in average, which of course much better to ensure a good water quality than a recycling with a duration of only 8–10 days. Recycling of wastewater has been applied in Pretoria, South Africa and in Windhoek, Namibia due to insufficient supply of natural water for production of portable water. In Pretoria, a treatment consisting of the following steps is applied: mechanical–biological treatment, aeration, lime precipitation, ammonia stripping, sand filtration, chlorination, adsorption on activated carbon, and a second chlorination. The plant in Windhoek has the same steps except no ammonia stripping.

11.3 Abatement of Air Pollution by Environmental Technology

The most important air pollution problems are particulate pollution, pollution by carbon monoxide, carbon hydrides, sulfur dioxide, nitrogenous gases, and heavy metals. Furthermore, a number of gases—carbon dioxide, methane, and nitrogen oxides—have greenhouse effect, that is, that they will be able to change the climate toward higher atmospheric temperature.

When considering particulate pollution, the source should be categorized with regard to contaminant type. Inert particulates are distinctly different from active solids in the nature and type of their potentially harmful human health effects. Inert particulates comprise solid airborne material, which does not react readily with the environment and does not exhibit any morphological changes as a result of combustion or any other process. Active solid matter is defined as particulate material which can be further oxidized or react chemically with the environment or the receptor. Any solid material in this category can generally, but depending on its composition and size, be considerably more harmful than inert matter of similar size.

A closely related group of emissions are from aerosols of liquid droplets, generally below 5 μm. They can be oil or other liquid pollutants (e.g., Freon) or may be formed by condensation in the atmosphere. Fumes are condensed metals, metal oxides, or metal halides, formed by industrial activities, predominantly as a result of pyro-metallurgical processes: melting, casting, or extruding operations. Products of incomplete combustion are often emitted in the form of particulate matter. The most harmful components in this group are particulate polycyclic organic matter (PPOM) which are mainly derivatives of benzo-a-pyrene (see Chapter 6).

Natural sources of particulate pollution are sandstorms, forest fires, and volcanic activity. The major sources in towns are vehicles, combustion of fossil fuel for heating and production of electricity, and industrial activity.

Particulate pollution is an important health factor. The toxicity and the size distribution are the most crucial factors. Many particles are highly toxic, such as asbestos and those of metals such as beryllium, lead, chromium, mercury, nickel, and manganese. In addition, particulate matter is able to absorb gases, which enhances the effects of these components. In this context, the particle size distribution is of particular importance, as particles greater than 10 μm are trapped in the human upper respiratory passage and the specific surface (expressed as meter square per gram of particulate matter) increases with $1/d$, where d is the particle size. The adsorption capacity of particulate matter, expressed as g adsorbed per g of particulate matter, will generally be proportional to the surface area. Table 11.4 lists some typical particle size ranges.

Particulate pollutants have the ability to adsorb gases including sulfur dioxide, nitrogen oxides, carbon monoxide, and so on. The inhalation of these toxic gases is frequently associated with this adsorption, as the gases

TABLE 11.4

Typical Particle Size Ranges

	μm
Tobacco smoke	0.01–1
Oil smoke	0.05–1
Ash	1–500
Ammonium chloride smoke	0.1–4
Powdered activated carbon	3–700
Sulfuric acid aerosols	0.5–5

otherwise would be dissolved in the mouthwash and spittle before entering the lungs.

Particulate pollution may be controlled by modifying the distribution pattern—by building taller chimneys so to say. This method represents, in principle, an obsolete philosophy of pollution abatement: dilution, but it is still widely used to reduce the concentration of pollutants at ground level and thereby minimize the effect of air pollution. Particulate control technology can offer a wide range of methods aimed at the removal of particulate matter from gas. These methods are: settling chambers, cyclones, filters, electrostatic precipitators, wet scrubbers, and modification of particulate characteristics. The methods, their optimum particle size and efficiency are compared in Table 11.5. The cost of the various installations varies of course from country to country and is dependent on several factors (material applied, standard size or not standard size, automatic, and so on). Generally, electrostatic precipitators are the most expensive solution and are mainly applied for large quantities of air. Wet scrubbers also belong among the more expensive installations, while settling chamber and centrifuges are the most cost-effective solutions. (For more technical details reference Jørgensen [2000] could be applied.)

ILLUSTRATION 11.2

The concentration of particulate matter at the ground is inversely proportional to the squared height of the chimney.

If the particulate matter has a concentration at the ground of 20 mg/(24 h m^2) and it is required to be less than 0.2 mg/(24 h m^2). How many times taller do we need to build the chimney to meet this requirement?

Solution

The concentration has to be 100 times smaller, which means that the chimney has to be 10 times taller, because 10 times taller implies that the concentration is $10 \times 10 = 100$ times smaller.

A scrubbing liquid, usually water, is used to assist separation of particles, or a liquid aerosol from the gas phase. The operational range for particle removal includes material less than 0.2 μm in diameter up to the largest particles that can be suspended in air. Gases soluble in water are of course also removed by this process.

All types of fossil fuel will produce carbon dioxide on combustion, which is used in the photosynthetic production of carbohydrates. As such, carbon dioxide is harmless and has no toxic effect, whatever the concentration levels, but increased carbon dioxide concentration in the atmosphere will increase absorption of infrared radiation and the heat balance of the earth will be changed.

Carbon hydrides are the major components of oil and gas, and incomplete combustion will always involve their emission. By regulation of the ratio of oxygen to fuel more complete combustion can be obtained, but the emission of carbon monoxide cannot be totally avoided. Motor vehicles are also the major source of carbon monoxide pollution. On average, 1 L of gasoline (petrol) will produce 200 L of carbon monoxide, while it is possible to minimize the production of this pollutant by using diesel instead of gasoline. In most industrial countries, more than 75% of this pollutant originates from motor vehicles.

Carbon dioxide pollution is inevitable related to the use of fossil fuels. Therefore, it can only be solved by the use of other sources of energy.

Legislation (covered in Chapter 15) is playing a major role in controlling the emissions of carbon hydrides and carbon monoxide. As motor vehicles are the major source of these pollutants, control methods should obviously focus on the possibilities of reducing vehicle emissions. The methods available today are

1. Motor technical methods
2. Afterburners
3. Alternative energy sources

The first method is based on a motor adjustment according to the relationship between the composition of the exhaust gas and the air/fuel ratio. A higher air/fuel ratio results in a decrease in the carbon hydrides and carbon monoxide concentrations, but to achieve this, a better distribution of the fuel in the cylinder is required, which is only possible through construction of another gasification system. This method may be considered cleaner technology. Presently, two types of afterburners are in use—thermal and catalytic afterburners. In the former type, the combustible material is raised above its auto-ignition temperature and held there long enough for complete oxidation of carbon hydrides and carbon monoxide to occur. This method is used on an industrial scale (Waid, 1972, 1974) when low-cost purchased or diverted fuel is available; in vehicles a manifold air injection system is used.

Catalytic oxidation occurs when the contaminant-laden gas stream is passed through a catalyst bed, which initiates and promotes oxidation of the combustible matter at a lower temperature than would be possible in thermal oxidation. The method is used on an industrial scale for the destruction of trace solvents in the chemical coating industry. Vegetable and animal oils can be oxidized at 250–370°C by catalytic oxidation. The exhaust fumes from chemical processes, such as ethylene oxide, methyl methacrylate, propylene, formaldehyde, and carbon monoxide can easily be catalytically incinerated at even lower temperatures. The application of catalytic afterburners in motor vehicles presents some difficulties due to poisoning of the catalyst with lead. With the decreasing lead concentration in gasoline it is becoming easier to solve that problem, and the so-called double catalyst system and even three-way catalysts are now finding a wide application. These systems are able to reduce nitrogen oxides and oxidize carbon monoxide and carbon hydrides simultaneously. New catalysts are currently coming on the market and offer a higher efficiency.

Lead in gasoline has been replaced by various organic compounds to increase the octane number. Benzene has been applied but it is toxic and causes air pollution problems because of its high vapor pressure. Methyl tertiary butyl ether (MTBE) is another possible compound for increasing the octane number. It is, however, very soluble and has been found as groundwater contaminant close to gasoline stations.

Application of alternative energy sources is still at a preliminary stage. Most interest has, however, been devoted to electric and hybrid vehicles.

Fossil fuel contains approximately 2%–5% sulfur (w/w), which is oxidized by combustion to sulfur dioxide. Although fossil fuel is the major source, several industrial processes produce emissions containing sulfur dioxide, for example mining, the treatment of sulfur containing ores and the production of paper from pulp. The total global emissions of sulfur dioxide has been decreasing during the last 25 years due to the installation of pollution abatement equipment, particularly in North America, the European Union, and Japan and due to the use of low-sulfur diesel. The concentration of sulfur dioxide in the air is relatively easy to measure, and sulfur dioxide has been used as an indicator component. High values recorded by inversion are typical.

Sulfur dioxide is oxidized in the atmosphere to sulfur trioxide, which forms sulfuric acid in water. Since sulfuric acid is a strong acid it is easy to understand that sulfur dioxide pollution indirectly causes corrosion of iron and other metals and is able to acidify aquatic ecosystems. See also Jørgensen (2000). Clean air acts have been introduced in all industrialized countries during the 1970s and 1980s. Table 11.6 illustrates some typical sulfur dioxide emission standards, although these may vary slightly from country to country. Notice how the duration is important; see also the discussion in Chapter 7.

The approaches used to meet the requirements of the acts, as embodied in the standards, can be summarized as follows:

1. Fuel switching from high to low sulfur fuels.
2. Modification of the distribution pattern—use of tall stacks.

TABLE 11.5

Characteristics of Particulate Pollution Control Equipment

Device	Optimum Particle Size (μm)	Optimum Concentration (gm^{-3})	Temperature Limitations (°C)	Air Resistance (mm H_2O)	Efficiency (% by Weight)
Settling chambers	>50	>100	−30 to 350	<25	<50
Centrifuges	>10	>30	−30 to 350	<50–100	<70
Multiple centrifuges	>5	>30	−30 to 350	<50–100	<90
Filters	>0.3	>3	−30 to 250	>15–100	>99
Electrostatic precipitators	>0.3	>3	−30 to 500	<20	<99
Wet scrubbers	>2–10	>3–30	0 to 350	>5–25	<95–99

3. Abandonment of very old power plants, which have a particular high emission.
4. Flue gas cleaning.

Desulfurization of liquid and gaseous fuel is a well-known chemical engineering operation. In gaseous and liquid fuels sulfur either occurs as hydrogen sulfide or can react with hydrogen to form hydrogen sulfide. The hydrogen sulfide is usually removed by absorption in a solution of alkanolamine and then converted to elemental sulfur.

Sulfur occurs in coal both as pyritic sulfur and as organic sulfur. Pyritic sulfur is found in small discrete particles within the coal and can be removed by mechanical means, for example, by gravity separation methods. However, 20%–70% of the sulfur content of coal is present as organic sulfur, which can hardly be removed today on an economical basis. Since sulfur recovery from gaseous and liquid fuels is much easier than from solid fuel, which also has other disadvantages, much research has been and is being devoted to the *gasification* or *liquefaction of coal*.

When sulfur is not or cannot be economically removed from fuel oil or coal prior to combustion, removal of sulfur oxides from combustion gases will become necessary for compliance with the stricter air pollution control laws.

TABLE 11.6

SO_2-Emission Standards Applied in the European Union

Duration	Concentration (ppm)	Comments
Month	0.05	
24 h	0.10	Might be exceeded once a month
30 min	0.25	Might be exceeded 15 times/month

The chemistry of sulfur dioxide recovery presents a variety of choices and five methods should be considered:

1. Adsorption of sulfur dioxide on active metal oxides with regeneration to produce sulfur.
2. Catalytic oxidation of sulfur dioxide to produce sulfuric acid.
3. Adsorption of sulfur dioxide on charcoal with regeneration to produce concentrated sulfur dioxide.
4. Reaction of dolomite or limestone with sulfur dioxide by direct injection into the combustion chamber. A lime slurry is injected into the flue gas beyond the boilers.
5. Fluidized bed combustion of granular coal in a bed of finely divided limestone or dolomite maintained in a fluid like condition by air injection. Calcium sulfite is formed as a result of these processes.

The two latter methods have found wide applications, particularly in large industrial installations. It is possible to recover the sulfur dioxide or elemental sulfur from these processes, making it possible to recycle the spent sorbing material.

Seven different compounds of oxygen and nitrogen are known: N_2O, NO, NO_2, NO_3, N_2O_3, N_2O_4, N_2O_5—often summarized as NO_x. From the point of view of air pollution mainly NO (nitrogen oxide) and NO_2 (nitrogen dioxide) are of interest. The major sources of the two gases are: nitrogen oxide—from combustion of gasoline and oil: nitrogen dioxide—from combustion of oil, including diesel oil.

The emissions from motor vehicles can be reduced by the same methods as mentioned for carbon hydrides and carbon monoxide. The air/fuel ratio determines the concentration of pollutants in the exhaust gas. An increase in the ratio will reduce the emission of carbon hydrides and carbon monoxide, but unfortunately will increase the concentration of nitrogenous gases. Consequently, the selected air/fuel ratio will be a compromise. A double catalytic afterburner is applied to day. It is able to reduce nitrogenous gases and simultaneously oxidize carbon hydrides and carbon monoxide. The application of alternative energy sources will, as for carbon hydrides and carbon monoxide, be a very useful control method for nitrogenous gases at a later stage.

Nitrogenous gases in reaction with water form nitrates which are washed away by rainwater. In some cases, this can be a significant source of eutrophication. For a shallow lake, for example, the increase in nitrogen concentration due to the nitrogen input from rainwater will be rather significant. In a lake with a depth of 1.7 m and an annual precipitation of 600 mm, which is normal in many temperate regions, the annual input will be as much as 0.3 mg/L.

The rapid growth in industrial production during recent decades has exacerbated the industrial air pollution problem, but, due to an increased application of continuous processes, recovery methods, air pollution control, use of closed systems and other technological developments, industrial air pollution has, in general, not increased in proportion to production.

Industry displays a wide range of air pollution problems related to a large number of chemical compounds in a wide range of concentrations.

Since industrial air pollution covers a wide range of problems, it is not surprising that all three classes of pollution control methods mentioned previously have found applications: modification of the distribution pattern, alternative (cleaner) production methods and particulate and gas/vapor control technology.

All the methods mentioned earlier are also applied for industrial air pollution control. In gas and vapor technology, a distinction has to be made between condensable and noncondensable gaseous pollutants. The latter must usually be destroyed by incineration, while the condensable gases can be removed from industrial effluents by

Absorption

Adsorption

Condensation

Combustion

Recovery is feasible by the first three methods

Absorption is a diffusion process that involves the mass transfer of molecules from the gas state to the liquid state along a concentration gradient between the two phases. Absorption is a unit operation that is enhanced by all the factors generally affecting mass transfer; that is, high interfacial area, high solubility, high diffusion coefficient, low liquid viscosity, increased residence time, turbulent contact between the two phases, and possibilities for reaction of the gas in the liquid phase. This last factor is often very significant and almost 100% removal of the contaminant is the result of such a reaction. Acidic components can easily be removed from gaseous effluents by absorption in alkaline solutions, and, correspondingly, alkaline gases can easily be removed from effluent by absorption in acidic solutions. Table 11.7 provides an overview of applicable absorber reagents.

Carbon dioxide, phenol, and hydrogen sulfide are readily absorbed in alkaline solutions in accordance with the following processes:

$$CO_2 + 2NaOH \rightarrow 2Na^+ + CO_3^{2-} \tag{11.1}$$

$$H_2S + 2NaOH \rightarrow 2Na^+ + S^{2-} + 2H_2O \tag{11.2}$$

$$C_6H_5OH + NaOH \rightarrow C_6H_5O^- + Na^+ + H_2O \tag{11.3}$$

TABLE 11.7

Absorber Reagents

Reagents	Applications
$KMnO_4$	Rendering, polycyclic organic matter
NaOCl	Protein adhesives
Cl_2	Phenolics, rendering
Na_2SO_3	Aldehydes
NaOH	CO_2, H_2S, phenol, Cl_2, pesticides
$Ca(OH)_2$	Paper sizing and finishing
H_2SO_4	NH_3, nitrogen bases

Ammonia is readily absorbed in acidic solutions:

$$2NH_3 + H_2SO_4 \rightarrow 2NH_4^{+}SO_4^{2-} \tag{11.4}$$

Adsorption is the capture and retention of a component (adsorbate) from the gas phase by the total surface of the adsorbing solid (adsorbent). In principle, the process is the same as deals with wastewater treatment; the theory is equally valid for gas adsorption. Adsorption is used to concentrate (often 20 to 100 times) or store contaminants until they can be recovered or destroyed in the most economical way. These processes are often described as either Langmuir or Freundlich's adsorption isotherms, which are also applied for a quantitative description of this process in nature. Adsorption is dependent on temperature: increased temperature means that the molecules move faster and therefore it is more difficult to adsorb them. There are four major types of gas adsorbents, the most important of which is activated carbon, but also aluminum oxide (activated aluminum), silica gel, and zeolites are used.

The selection of an adsorbent is made according to the following criteria:

1. High selectivity for the component of interest.
2. Easy and economical to regenerate.
3. Availability of the necessary quantity at a reasonable price.
4. High capacity for the particular application, so that the unit size will be economical. Factors affecting capacity include total surface area involved, molecular weight, polarity activity, size, shape, and concentration.
5. Pressure drop, which is dependent on the superficial velocity.
6. Mechanical stability in the resistance of the adsorbent particles to attrition. Any wear and abrasion during use or regeneration will lead to an increase in bed pressure drop.

7. Microstructure of the adsorbent should, if at all possible, be matched to the pollutant that has to be collected.
8. The temperature, which has a profound influence on the adsorption process, as already mentioned. Regeneration of the adsorbents is an important part of the total process. A few procedures are available for regeneration.

Although the regeneration is 100% the capacity of the adsorbent may be reduced 10%–25% after several regeneration cycles, due to the presence of fine particulates and/or high molecular weight substances, which cannot be removed in the regeneration step.

Combustion is defined as rapid, high temperature gas-phase oxidation. The goal is the complete oxidation of the contaminants to carbon dioxide and water, sulfur dioxide, and nitrogen dioxide.

The process is often applied to control odors in rendering plants, paint and varnish factories, rubber tire curing, and petrochemical factories. It is also used to reduce or prevent an explosion hazard by burning any highly flammable gases for which no ultimate use is feasible. The efficiency of the process is highly dependent on temperature and reaction time, but also on turbulence or the mechanically induced mixing of oxygen and combustible material. The relationship between the reaction rate, r and the temperature can be expressed by the Arrhenius equation (see also Chapter 8): r at temperature t = r at reference temperature rt $*k^{(t-tr)}$, k is a constant which in most cases is between 1.02 and 1.12.

Heavy metals, which may be defined as metals with a specific gravity, >5.00 kg/L, comprise 70 elements. Most of them, however, are only rarely found as pollutants. The heavy metals of environmental interest form very heavy soluble compounds with sulfide and phosphate and form very stable complexes with many ligands present in the environment. This means, fortunately, that most of the heavy metals are not very bioavailable in most environments, see also the entry about bioremediation. A number of enzymes activated by metal ions and metallo-enzymes are known. The first mentioned group comprises iron, cobalt, chromium, vanadium and selenium, copper, zinc, iron, cobalt, molybdenum are able with a stronger bond to form metallo-enzymes: metallo-proteins, metallo-porphyrines, and metalloflavines. Pollutants are particularly lead, cadmium, and mercury in focus, because of their extremely high toxicity.

Heavy metals are emitted to the atmosphere by energy production and a number of technological processes (see Table 11.8). It makes the atmospheric deposition of heavy metals, originated from human activities, the dominant pollution source for the vegetation of natural ecosystems—forests, wetlands, peat lands, and so on. The heavy metal content in sludge and fertilizers plays a more important role for agricultural land, which is also where the inputs of heavy metals by irrigation, natural fertilizers, and application of chemicals

TABLE 11.8

Important Atmospheric Pollution Sources of Heavy Metals

Source	Heavy Metals
Incineration of oil	V, Ni
Incineration of coal	Hg, V, Cr, Zn, As
Gasoline	Pb (leaded gasoline)
Metal industry	Fe, Cu, Mn, Zn, Cr, Pb, Ni, Cd, and other
Application of pesticides	Hg, Cr, Cu, As
Incineration of solid waste	Hg, Zn, Cd, and other

including pesticides may add to the overall pollution level. The atmosphere and hydrosphere both have a well-developed ability for "self-purification"—for heavy metals by removal processes, for instance sedimentation. The lithosphere has a high buffer capacity toward the effects of most pollutants, and also has an ability of self-purification, for instance by runoff and uptake by plants, although the rates usually are much lower than in the two other spheres.

11.4 Solution of the Solid Waste Problems by Environmental Technology

Solid wastes include an incredible miscellany of items and materials, which makes it impossible to indicate one simple solution to the problem. It is necessary to apply a wide spectrum of solutions to the problems according to the source and nature of the waste. Table 11.9 shows a classification of solid waste. Typical quantities per inhabitant in a technological society

TABLE 11.9

Classification of Solid Waste

Type of Waste	kg/inhabitant/day (approx.)
Domestic garbage	4.5
Agricultural waste[a]	12
Mining waste[b]	18
Wastes from construction[c]	0.12
Industrial waste	1.35
Junked automobiles	0.15

[a] Not mentioned further in this context. Recycling is recommended and also widely in use.

[b] A substantial part is used for land filling.

[c] Not mentioned further in this context. Recycling is possible and recommended, but a part is also used for land filling.

are included. The numbers represent the situation in EU 2005 and are taken from various sources.

A universal method for treatment of solid waste does not exist, and it is necessary to analyze each individual case to find a relevant solution to the problem. The analysis begins with an examination of mass flows (see Chapter 6) to ascertain whether there is an economical basis for reuse (e.g., bottles), recovery of valuable raw materials (paper and metals) or utilization of organic matter, as a soil conditioner or for the production of energy. These considerations are illustrated in Figure 11.6. Mass balances for important materials, as mentioned in Chapter 6, must therefore be set up.

The analysis of mass flows is the framework for a feasible solution, which takes economic as well as environmental issues into consideration. However, legislation and economical means are required to achieve the management goals. For instance, the use of returnable bottles can be realized either by banning the use of throw-away bottles or by placing a purchase tax on throw-away bottles and not on returnable bottles (more details are presented in Chapter 14). The management problems are highly dependent on the technological methods used for solid waste treatment. Methods based on cleaner technology (see also Chapter 13) and recycling, to reduce the amount of waste produced, often require environmental legislation or economical means (green tax) to guarantee success; see Chapter 14.

Methods based on deposition will require a comprehensive knowledge of environmental effects on the ecosystems involved. Complete decomposition to harmless components is not possible for most solid waste. For instance, incineration will produce a slag, which might cause a deposition problem, and smoke, which involves air pollution.

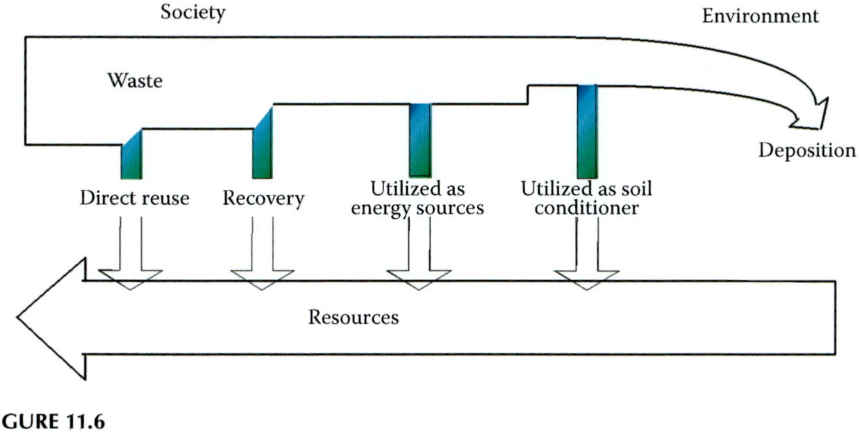

FIGURE 11.6
Principles of recycling illustrated by mass flows.

11.4.1 Methods for Treatment of Solid Waste: An Overview

Table 11.10 gives an overview of the methods applicable to solid waste. The table indicates on which principle the method is based—recycling, deposition, or decomposition to harmless components. It also shows the type of solid waste the method is able to treat.

The solutions of the solid waste problems can often be classified according to the source: (A) sludge, (B) domestic garbage, (C) industrial, mining, and hospital waste, and (D) agricultural waste. The methods used for the four classes are, to a certain extent, different, although there is some overlap, between them.

(A) *Sludge.* In most wastewater treatment the impurities are not actually removed, but rather concentrated in the form of a sludge. Only when a chemical reaction takes place does real removal of the impurities occur, for example, by chemical or biochemical oxidation of organics to CO_2 and H_2O, or denitrification of nitrate to nitrogen gas. Sludge from industrial wastewater treatment units in most cases requires further concentration before its ultimate disposal. In many cases, two- or even three-step processes are used to concentrate the sludge. It is often an advantage to use further thickening by gravity, followed by such treatments as filtration or centrifugation. The characteristics of a sludge are among the factors that influence the selection of the best sludge-treatment method. The sludge characteristics vary with the wastewater and the wastewater treatment methods used. One of the

TABLE 11.10

Methods for Treatment of Solid Waste

Method	Principle	Applicable to
Conditioning and composting	Decomposition before deposition	Sludge, domestic garbage, agricultural waste
Anaerobic treatment	Decomposition, deposition, utilization of biogas	Sludge, agricultural waste including manure
Thickening, filtration, centrifugation, and drying	Dewatering before further treatment and deposition	Sludge
Combustion and incineration	Decomposition and utilization of energy	All types of waste
Separation	Recycling, partially	Domestic garbage
Dumping ground	Decomposition, deposition, domestic garbage	
Pyrolysis	Recycling, utilization of energy, decomposition	Domestic garbage
Precipitation and filtration	Deposition or recycling	Industrial waste
Land filling	Deposition	Industrial, mining, and agricultural waste
Aerobic treatment	Decomposition and deposition	Sludge, agricultural waste

important factors is the concentration of the sludge. The specific gravity of the sludge is another important factor, since the effect of gravity is utilized in the thickening process. The specific gravity of activated sludge increases linearly with the sludge concentration. This corresponds to a specific gravity of 1.08 g/mL for actual solids. The water can be removed from a sludge by such processes as vacuum filtration, centrifugation, and sand-bed drying.

When sludge is being considered for use as a soil conditioner, its chemical properties are of prime importance. The nutrient content (nitrogen, phosphorus, and potassium), in particular, is of interest. Furthermore, knowledge of the heavy metals and organic contaminants in sludge is important because of their toxicity. Standards for sludge applicable as soil conditioner are used in most industrialized countries (see Jørgensen, 2000). Finally, it must be considered, that normal wastewater treatment processes, such as sedimentation, chemical precipitation, and biological treatments, remove considerable amounts of pathogens which are concentrated in the sludge. A significant reduction in the number of pathogenic organisms has been found to occur during anaerobic digestion, but they are not destroyed entirely. Combustion, intensive heat treatment of sludge or treatment with calcium hydroxide would of course eliminate the hazard of pathogenic microorganisms.

Sludge conditioning is a process that alters the properties of the sludge to allow the water to be removed more easily. The aim is to transform the amorphous gel-like sludge into a porous material which will release water. Conditioning of the sludge can be accomplished by either chemical or physical means. Chemical treatment usually involves the addition of coagulants or flocculants to the sludge. Typical doses of inorganic coagulants, such as aluminum sulfate, iron(III) chloride, and calcium hydroxide, are as much as 20% of the weight of the solid, while a typical dose of organic polymer is less than 1% of the weight of the solid. This does not necessarily mean that the cost of using synthetic polymers is lower, since the polymers cost is considerably more per kg than the inorganic chemicals used as conditioners.

The amorphous gel-like structure of the sludge is destroyed by heating. The filtration rate of activated sludge is increased more than a thousand-fold after heat treatment. Typically, the heat-treatment conditions are 30-min treatment at 150–200°C under a pressure of 10–15 atm. A great advantage of heat conditioning is, of course, that the pathogens are destroyed. Conditioning by freezing has also been reported. Thickening of sludge is possible by filtration. The filter area is determined from the specific sludge resistance and the amount to be treated by filtration. As the specific resistance of conditioned sludge may vary considerably, it is recommended to design for a surplus capacity of at least 25%–50%.

Several types of filter are applicable to achieve a high concentration of dry matter (30%–45%) in the sludge which will reduce the cost of transportation significantly. Vacuum filtration is used to remove water from a sludge by applying a vacuum across a porous medium. As the rotary drum passes through the slurry in the slurry tank, a cake of solid is built up on the drum surface and the water is removed by vacuum filtration through the porous medium on the drum surface. The cake is removed from the drum by a knife edge. Often the porous filter is washed with water before it is returned to the slurry tank.

Filter presses (mainly plate and frame types) are more cost-effective than vacuum filters and are more widely used. Calcium hydroxide slurry is often used as filter medium. Filter presses operate by charge basis, as the filter cakes have to be removed when the filtration rate is reduced to a certain level.

Filter belts are also widely used. They have the advantage that they operate continuously, but usually they provide a lower concentration of dry matter than the filter press (about 30%–35% versus 40% for filter presses).

The filtrate from the treatment of municipal sludge has a high ammonium–nitrogen concentration which it may be advantageous to remove by stripping to eliminate the nitrogen loading in the wastewater, as the filtrate must always be recycled.

Centrifugation can also be used in the removal of water from wastewater sludge. Of the various types of centrifuge, the solid-bowl centrifuge is considered to offer the best clarification and water-removal properties. The use of electrolytes will increase the recovery at a given flow rate or increase the flow rate for a given recovery.

If the sludge contains biodegradable organic material it may be advantageous to treat the sludge by aerobic or anaerobic digestion before thickening. Anaerobic digestion is by far the most common method of treating municipal sludge. It creates good conditions for the growth of microorganisms. The end products of anaerobic digestion are carbon dioxide and methane. The temperature is commonly set at about 35°C, in order to maintain optimum conditions in the digester. Unfortunately, anaerobic digestion results in considerable quantities of nutrients going into solution (Dalton et al., 1968), which means that a significant amount of nutrient material will be returned to the treatment plant if the supernatant is separated from the sludge. The principal function of anaerobic digestion is to convert as much as possible of the sludge to end products: liquids and gases. Anaerobic decomposition generally produces less biomass than aerobic processes. Heavy metal ions may slow down the decomposition rate. The presence of ammonia and hydrogen sulfide may have the same effect (Vavilin et al., 1995).

The microorganisms can be divided into two broad groups: acid formers and methane formers. The acid formers consist of facultative and anaerobic bacteria, and soluble products are formed through hydrolysis. The soluble products are then fermented to acids and alcohols of lower molecular weight. The methane formers are strictly anaerobic bacteria that convert the acids and alcohols, along with hydrogen and carbon dioxide, to methane. The biogas produced by anaerobic digestion of sludge from wastewater treatment is often utilized at wastewater treatment plants for the production of heat and for the production of electricity (a gas turbine is used). The amount of gas is usually not sufficient to make the sale of gas possible.

The properties of aerobically digested sludge are similar to those of anaerobically digested sludge. An advantage is that some of the operational problems attending anaerobic digesters are avoided, but the disadvantage compared with anaerobic digestion is that the process is more expensive since oxygen must be provided and energy recovery from methane is not possible. Since aerobic digestion is less used in industrial wastewater processes than in treatment of municipal wastewater, it is not appropriate here to go into further detail. The purpose of drying sludge is to prepare it for use as a soil conditioner or for incineration. Air drying of the sludge on sand beds is often used to reach a moisture content of about 90%. Also, such drying techniques as flash drying and rotary drying are used to remove water from sludge. Often waste heat from the incineration process itself is used in drying. However, the cost of combined drying and combustion is higher than the cost of incineration alone.

(B) *The composition of domestic garbage varies from country to country.* To a certain extent, the environmental legislation and the economic level of the nation is reflected in this composition. The amount of domestic garbage per inhabitant is increasing. In most developed countries, the growth was 2%–4% from the mid-1950s to 1963, while it has been lower in the period from the mid-1970s to 2010 (1%–2%).

It is often advantageous to carry out a comprehensive analysis of the domestic garbage in order to estimate the economic and ecological consequences of various environmental management strategies including the selection of the most relevant treatment methods. The analysis will in this case include the following items: (1) food waste, (2) paper, (3) textiles, (4) leather and rubber, (5) plastics, (6) wood, (7) iron, (8) aluminum, (9) other metals, (10) glass and ceramic products, (11) ash and dust, (12) stone, (13) garden waste, and (14) other types of waste.

Separation of solid waste can be achieved either in a central plant or by the organization of separate collection of paper, glass, metals, and other types of domestic waste.

Dumping grounds (landfills) were previously the most common handling method for solid waste. Today, it is mainly in use in smaller towns, often after grinding or compression which reduce the volume by 60%–80%.

Deposition of solid waste on a dumping ground is an inexpensive method, but it has a number of disadvantages:

1. Possible contamination of groundwater
2. Causes inconveniences due to the smell
3. Attracts noxious animals, such as flies and rats
4. During the deposition several processes take place

5. Decomposition of biodegradable material
6. Chemical oxidation of inorganic compounds
7. Dissolution and wash out of material
8. Diffusion processes

Where the decomposition takes place in the aerobic layers, carbon dioxide, water, nitrates, and sulfates are the major products liberated, while decomposition in anaerobic layers leads to the formation of carbon dioxide, methane, ammonia, hydrogen sulfide, and organic acids.

Water percolating from dumping grounds has a very high concentration of BOD and nutrients and can therefore not be discharged into receiving waters. If it cannot be recycled on the dumping ground, this wastewater must be subject to some form of treatment.

Composting has been applied as a treatment method in agriculture for thousands of years. The method is still widely applied for treatment of agricultural waste. Organic matter from untreated solid waste cannot be utilised by plants, but it is necessary to let it undergo a degree of biological decomposition, through the action of microorganisms.

Again we can distinguish between aerobic and anaerobic processes. A number of factors control these processes:

1. The ratio of aerobic to anaerobic processes, which, of course, is determined by the available oxygen (diffusion process).
2. Temperature. Different classes of microorganisms are active within different temperature ranges (see Table 11.11). Heat is produced by the decomposition processes. The thickness of the layer determines to what extent this heat can be utilized to maintain a temperature of 60–65°C, which is considered to be the optimum. In this context it is also important to mention that composting at a relatively high temperature means a substantial reduction in the number of pathogenic microorganisms and parasites.
3. The water content should be 40%–60%, as this gives the optimum conditions for the processes of decomposition.
4. The C/N ratio should be in accordance with the optimum required by the microorganisms. Domestic garbage has a C/N ratio of 80 or

TABLE 11.11

Classification of Microorganisms

	Temperature Optimum	Temperature Range
Psychrophile	15–20°C	0–30°C
Mesophile	25–35°C	10–40°C
Thermophile	50–55°C	25–80°C

more due to the high content of paper, while the optimum for composting is 30. It is therefore advantageous to add sludge from the municipal sewage plant to adjust the ratio to about 30. Sludge usually has a C/N ratio of 10 or even less. During composting, the C/N ratio is decreased as a result of respiration, which converts some of the organic matter to carbon dioxide and water.

5. The optimum conditions for the microorganisms include a pH around 6 (6–8). Usually, pH increases as a result of the decomposition processes. If the pH is too low, calcium hydroxide should be added and if the pH is too high, sulfur should be added. Sulfur activates sulfur bacteria, which produces sulfuric acid.

A composting plant is generally applying composting in containers followed by composting in stacks. Composting in containers with the addition of *air accelerates* the decomposition processes, but this process might be excluded in smaller plants. The compost is used as soil conditioner when it is stable, that is, the decomposition processes are almost terminated, which may be determined by a respirometric technique. The standards for sludge are applied also for compost.

Incineration of domestic garbage and solid waste in general is very attractive from a sanitary point of view, but it is a very expensive method which has some environmental disadvantages. Valuable material such as paper is not recycled, a slag, which must be deposited, is produced and air pollution problems are involved. Nevertheless, incineration has been the preferred method in many cities and larger towns. This development might be explained by the increasing amount of combustible material in domestic garbage and the possibilities of combining incineration plants with district heating. The optimum combustion temperature is 800–1050°C. If the temperature is below this range incomplete combustion will result (dioxines may be produced), and at a higher temperature the slag will melt and prevent an even air distribution. The composition of the solid waste determines whether this combustion temperature can be achieved without using additional fossil fuel. If the ash content is less than 60%, the water content is less than 50% and the combustible material more than 25%, no additional fossil fuel is required.

The heat value (cal/g) can be calculated from the following equation:

$$Hu = H1\frac{100 - W}{100} - 600\frac{W}{100} \qquad (11.5)$$

where

H1 = the heat value of dry matter (cal/g)
Hu = the heat value
W = the water content (weight%)

The composition of the slag and the fly ash, which is collected in the filter, is, of course, dependent on the composition of the solid waste. It is notable that a minor amount of the slag and the fly ash always will be an unburned material. Slag and fly ash comprise 25%–40% of the weight of the solid waste. The slag will have a high concentration of heavy metals in most cases but they are usually chemically bound as very water insoluble oxides. Although all modern incineration plants have filters, the air pollution problem is not completely solved. Hydrogen chloride, in particular, can cause difficulties, because it is toxic and highly corrosive. It results from the incineration of chlorine compounds among which PVC is the most important.

Pyrolysis is the decomposition of organic matter at elevated temperature without the presence of oxygen. For pyrolysis of solid waste a temperature of 850–1000°C is generally used. The process produces gas and a slag, from which metals can easily be separated due to the anoxic atmosphere. Pyrolysis is a relatively expensive process, although usually slightly less expensive than incineration. The smoke problems are the same as those of incineration, but the easy recovery of metals from the slag is, of course, an advantage. *The gas produced has a composition close to coal gas* and in most cases can be used directly in the gas distribution system without further treatment. Approximately 500 m^3 of gas is produced per ton of solid waste, but 300–400 m^3 is used in the pyrolysis process to maintain the temperature.

Comparison of the methods: Applied for domestic garbage is presented in Table 11.12. Composting is the process which is most acceptable from an ecological point of view for treatment of solid waste including domestic garbage. This is according to the six waste treatment principles that will be mentioned in Chapter 16 where the integrated ecological–environmental management will be discussed. It is furthermore in accordance with Chapter 12, which focuses on ecological engineering methods. Composting may almost be designated an ecological engineering method, as at least to a certain extent

TABLE 11.12

Comparison of Methods for Treatment of Solid Waste, Particularly Domestic Garbage

Method	Resource Friendly	Ecological Evaluation	Area Needed	Costs per kg Waste
1. Dumping ground	Unacceptable	Unacceptable	Large	Low
2. Dumping ground with pretreatment	About 50%	Slightly better than 1	Unacceptable but still unacceptable	More than 1
3. Composting	Acceptable	Acceptable	Medium	70%–85% of 4 and 5
4. Pyrolysis	Almost acceptable	Almost acceptable	Low	High
5. Incineration	Unacceptable, unless heat is recovered	Unacceptable	Low	High

considering the 14 ecological properties presented in Chapter 3. Pyrolysis also gives the opportunity to reuse waste products as raw materials or energy sources. When the waste heat from incineration plants is utilized, this method becomes of course more attractive when evaluated according to ecological principles.

Composting is, however, less area intensive than incineration and pyrolysis, although it will require relatively less area than landfills, especially when compressing and grinding are used to reduce the volume of domestic and industrial garbage. Incineration is the most widely used method for the treatment of solid waste in bigger towns and cities. The relatively small area required per kg of waste is the dominant factor here.

The cost of the treatment of course varies with the size of the plant and the composition of waste. Generally, the treatment costs increases in the following order of treatment methods: dumping ground → dumping ground with compressing and grinding of waste → composting → incineration or pyrolysis.

(C) *Industrial, mining, and hospital waste.* This type of waste causes particular problems because it may contain toxic matters in relatively high concentrations. Hospital wastes are especially suspect because of contamination by pathogens and the special waste products, such as disposable needles and syringes and radioisotopes used for detection and therapy.

The waste from industry and mining varies considerably from place to place, and it is not possible to provide a general picture of its composition. If the composition permits, it can be used for land filling, but if it contains toxic matter special treatment will be required. The composition may be close to that of domestic garbage in which case the treatment methods mentioned above can be applied.

Hospital solid waste is being studied in only a few locations to provide data on current practices and their implications. Most hospital waste is now incinerated and this might, in many cases, be an acceptable solution. However, if the solid waste contains toxic matter it should be treated along the lines given for industrial wastes below. At least, waste from hospital laboratories should be treated the same as other types of chemical waste.

Industrial and mining waste containing toxic substances, such as heavy metals or toxic organic compounds, should be treated separately from other types of solid waste, which means that it cannot be treated by the methods mentioned earlier. A number of countries have built special plants to handle this type of waste, which could be called chemical waste. Such plants may include the following treatment methods:

1. Combustion of toxic organic compounds. The heat produced by this process might be used for district heating. Organic solvents, which are not toxic, should be collected for combustion, because discharge to the sewer might overload the municipal treatment plant. For example, acetone is not toxic to biological treatment plants, but 1 kg

of acetone uses 2.2 kg of oxygen in accordance with the following process:

$$(CH_3)_2CO + 4O_2 \rightarrow 3CO_2 + 3H_2O \quad (11.6)$$

A different combustion system might be used for pumpable and nonpumpable waste.

2. Compounds containing halogens or sulfur should be treated only in a system which washes the smoke to remove the hydrogen halogenideor sulfuric acid formed.
3. Waste oil can often be purified and the oil reused.
4. Waste containing heavy metals requires deposition under safe conditions after a suitable pretreatment. Recovery of precious metals is often economically viable and is essential for mercury because of its high toxicity. The pretreatment consists of a conversion to the most relevant oxidation state for deposition, for example, chromates should be reduced to chromium in oxidation state 3. Furthermore, metals should be precipitated as the very insoluble hydroxides (see also Section 11.1) before deposition.
5. Recovery of solvents by distillation becomes increasingly attractive from an economical viewpoint due to the growing costs of oil products, as most solvents are produced from mineral oil.

It is finally recommended to examine the possibilities of recovering plastics. The main problem with recycling plastics is associated with the wide range of chemically different plastic type. A mixture of different plastic types which may be recovered does not have the same properties as the particular plastic types, and can therefore only be used where the strength and the color is of minor importance, for instance for plastic bags.

(D) *Agricultural waste.* Recycling is used to a high extent for many types of agricultural waste. For instance, manure and dunghills are used as a natural fertilizer. The problem is, however, that manure contains nutrients that are useful to apply as natural fertilizer, but if it is used without sufficient control and too intensively, the nutrients will inevitably pollute rivers and lakes and even coastal areas, due to the organic matter and the nutrients of the manure. Therefore, there are limits in many countries on how much manure that is possible to apply per ha. If it is a problem, however, the use of ecotechnology should be considered as it is fitted to nonpoint sources of pollution. Recently, manure has been used more and more for the production of biogas, which is practically the same method as the above described anaerobic digestion of sludge.

Straw left on the field after the harvest of grains was previously incinerated on the field, but is not allowed any longer in many countries due to the fire danger. Straw is, however, being increasingly used as fuel for heating.

A possibility is of course also to plow the straw in the soil and thereby be able to utilize the carbon and nutrients of the straw. Agricultural waste will be discussed further in the next chapter, which discusses ecotechnology.

EXERCISES AND PROBLEMS

1. Calculate the total oxygen demand (BOD_5 + nitrification) for domestic wastewater with $BOD_2 = 160$ and 38 mg/L ammonium (as ammonium). The rate constant is 0.351/24 h.
2. Calculate BOD_2, BOD_{10}, and BOD_7 when BOD_5 is 200 mg/L. The rate constant is 0.351/24 h.
3. Mention the advantages of disadvantages by application of iron(III) chloride, calcium hydroxide, and aluminum sulfate for precipitation of phosphorus in municipal wastewater.
4. Nitrification (99%) of ammonium in a biological treatment plant requires 6 h retention time at 20°C, how many hours are needed for 99% nitrification at 0°C?
5. Compare the influence of SO_2 on the global sulfur cycle and CO_2 on the global carbon cycle.
6. Indicate a method to remove the following air pollutants: (a) ammonia, (b) hydrogen sulfide, (c) phenol, and (d) chlorine.
7. Discuss the possibilities to recycle iron, other metals, glass, paper, and cardboard in domestic garbage.

References

Dalton, F., E. Stein, and H.T. Lyman. 1968. Land reclamation—A complete solution to the sludge and solids disposal problem. *J. Water Pollut. Control. Fed.* 40, 789–801.

Diab, S. et al. 1993. Nitrification pattern in a fluctuating anerobic-aerobic pond environment. *Water Res.* 27, 1469–1475.

Groeneweg, J. et al. 1994. Ammonia oxidation by nitrosomonas at ammonia oncentrations near Km: Effects of pH and temperature. *Water Res.* 28, 2561–2566.

Hahn, H.H. and N. Muller. 1995. Factors affecting water quality of (large) rivers: Past experiences and future outlook. In Novotny, V. and Somlyódy, L. (Eds.), *Remediation and Management of Degraded River Basins with Emphasis on Central and Eastern Europe*. Berlin: Springer-Verlag, pp. 385–426.

Halling Sørensen, B. and S.E. Jørgensen. 1993. *The Removal of Nitrogen Compounds from Wastewater*. Amsterdam: Elsevier. 444 pp.

Henze, M. and H. Ødegaard. 1995. Wastewater treatment process development in Central and Eastern Europe—Strategies for a stepwise development involving chemical and biological treatment. In Novotny, V. and Somlyódy, L. (Eds.),

Remediation and Management of Degraded River Basins with Emphasis on Central and Eastern Europe. Berlin: Springer-Verlag, pp. 357–384.

Islam, S. and M.T. Suidan. 1998. Electrolytic denitrification. *Water Res.* 32, 528–536.

Jørgensen, S.E. 2000. *Principles of Pollution Abatement*. Amsterdam: Elsevier. 520 pp.

Jørgensen, S.E. (Ed.). 2011. *Handbook of Ecological Models Used in Ecosystem and Environmental Management*. Boca Raton, FL: CRC Press, 620 pp.

Novotny, V. and L. Somlyódy (Eds.). 1995. *Remediation and Management of Degraded River Basins with Emphasis on Central and Eastern Europe*. Berlin: Springer-Verlag.

Sison, N.F. et al. 1996. Denitrification with external carbon source utilizing adsorption and desorption capability of activated carbon. *Water Res.* 30, 217–227.

Sorial, G.A. et al. 1998. Evaluation of trickle-bed air biofilter performance for styrene removal. *Water Res.* 32, 1593–1603.

van Loosdrecht, M. 1998. Upgrading of wastewater treatment process for integrated nutrient removal—The BCFS© process. *Wat. Sci. Technol.*, 37, 234–256.

Vavilin, V.A. et al. 1995. Modelling ammonia and hydrogen sulfide inhibition in anaerobic digestion. *Water Res.* 29(3), 827–835.

Waid, D.E. 1972. Controlling pollutants via thermal incineration. *Chem. Eng. Prog.* 68, 57–58.

Waid, D.E. 1974. *Thermal Oxidation or Incineration*. Pittsburgh, PA: *Pollut. Control Assoc.* pp. 62–79.

12

Application of Ecotechnology and Ecological Engineering in Environmental Management

12.1 What Is Ecotechnology and Ecological Engineering?

H.T. Odum (1982, 1983) was among the first to define ecological engineering as the "environmental manipulation by man using small amounts of supplementary energy to control systems in which the main energy drives are coming from natural sources." Odum (1983) further developed the concept of ecological engineering as follows: ecological engineering, the engineering of new ecosystems designs, is a field that uses systems that are mainly self-organizing. Straskraba (1985) has defined ecological engineering, or as he calls it ecotechnology, more broadly, as the use of technological means for ecosystem management, based on a deep ecological understanding, to minimize the costs of measures and their harm to the environment. Ecological engineering and ecotechnology may be considered synonymous.

Mitsch and Jørgensen (1989) give a slightly different definition which, however, covers the same basic concept as the definition given by Straskraba and also encompasses the definition given by H.T. Odum. They define ecological engineering and ecotechnology as the design of human society with its natural environment for the benefit of both. It is engineering in the sense that it involves the design of man-made or natural ecosystems or parts of ecosystems. Like all engineering disciplines it is based on basic science, in this case ecology and systems ecology. The biological species are the components applied in ecological engineering. Ecological engineering therefore represents a clear application of ecosystem theory.

Ecotechnic is another often applied word, but it encompasses in addition to ecotechnology or ecological engineering also the development of technology applied in society, based upon ecological principles, for instance all types of cleaner technology, particularly if they are applied to solve an environmental problem.

Ecological engineering should not be confused with bioengineering or biotechnology. Biotechnology involves the manipulation of the genetic structure of the cells to produce new organisms capable of performing certain functions. Ecotechnology does not manipulate at the genetic level, but operates at several steps higher in the ecological hierarchy. The manipulation takes place on an assemblage of species and/or their abiotic environment as a self-designing system that can adapt to changes brought about by outside forces, controlled by human or by natural forcing functions.

Ecological engineering is *not* the same as environmental engineering which is involved in cleaning up processes to prevent pollution problems—it is engineering applying environmental technology, which is covered in Chapter 11. It uses settling tanks, filters, scrubbers, and man-made components which have nothing to do with the biological and ecological components that are applied in ecological engineering, although the use of environmental engineering aims toward reducing man-made forcing functions on ecosystems. Ecotechnic mentioned earlier may be considered to include in addition to ecological engineering, environmental technology based on ecological principles such as technology using recirculation. The tool boxes of the two types of engineering are completely different. Ecological engineering uses ecosystems, communities, organisms, and their immediate abiotic environment.

All applications of technologies are based on quantification—it means that a diagnosis is developed before the selection of the solution to the environmental problem is made. Ecosystems are very complex systems and the quantification of their reactions to impacts or manipulations therefore becomes complex. Fortunately, ecological modeling and the use of ecological indicators represent a well-developed tool to survey ecosystems, their reactions and the linkage of their components. Ecological modeling is able to synthesize our knowledge about an ecosystem and makes it possible to a certain extent to quantify any changes in ecosystems resulting from the use of both environmental engineering and ecological engineering (see Chapter 8). Ecological engineering may also be used directly to design constructed ecosystems. Consequently, ecological modeling and ecological engineering are two closely cooperating fields. Research in ecological engineering was originally covered by the *Journal of Ecological Modeling* which was initially named *Ecological Modeling—International Journal on Ecological Modeling and Engineering and Systems Ecology* to emphasize the close relationship between the three fields: ecological modeling (Chapter 8), ecological engineering, and systems ecology (Chapters 3–5). *Ecological Engineering* was launched as an independent journal in 1992, and the name of *Ecological Modeling* was changed to *Ecological Modeling—An International Journal on Ecological Modeling and Systems Ecology*. Meanwhile, the journal *Ecological Engineering* has been successful in covering the field of ecological engineering which grew rapidly during the 1990s due to the increasing acknowledgement of the need for technologies other than environmental technology in our efforts to solve pollution problems the best way and simultaneously be cost moderate. This development does not imply that ecological modeling and ecological engineering are moving in different directions. On the contrary, ecological engineering is increasingly using models to perform design of constructed ecosystems or to quantify the results of application of specific ecological engineering methods for comparison with alternative, applicable methods.

Section 12.2 is devoted to the classification of ecotechnology that will be able to reveal in more detail the applicability of ecotechnology. Section 12.3 shows how ecotechnology is rooted in systems ecology and how the ecosystem properties can be applied to perform a holistic integrated ecological–environmental management. The following sections cover illustrative examples on how to apply ecotechnology for restoration of lakes, for treatment of agricultural waste, for treatment of wastewater by the use of waste stabilization ponds and constructed wetlands and for soil remediation.

12.2 Classification of Ecological Engineering and Ecotechnology

Ecotechnology may be based on one or more of the following four types or classes:

1. Ecosystems are used to reduce or solve a pollution problem that otherwise would be (more) harmful to other ecosystems. A typical example is the use of wetlands for wastewater treatment.
2. Ecosystems are imitated or copied to reduce or solve a pollution problem, leading to constructed ecosystems. Examples are fishponds and constructed wetlands for the treatment of wastewater or diffuse pollution sources.
3. The recovery of ecosystems after significant disturbances. Examples are coal mine reclamation and restoration of lakes and rivers.
4. The use of ecosystems for the benefit of humankind without destroying the ecological balance, that is, utilization of ecosystems on an ecologically sound basis. Typical examples are the use of integrated agriculture and development of organic agriculture. This type of ecotechnology finds wide applications in the ecological management of renewable resources.

The idea behind these four classes of ecotechnology is illustrated in Figure 12.1. Notice that ecotechnology operates in the environment and the ecosystems. It is here that ecological engineering has its tool box.

Illustrative examples of all four classes of ecological engineering may be found in situations where ecological engineering is applied to replace environmental engineering because the ecological engineering methods offer an ecologically more acceptable solution and where ecological engineering is the only method that can offer a proper solution to the problem. Examples

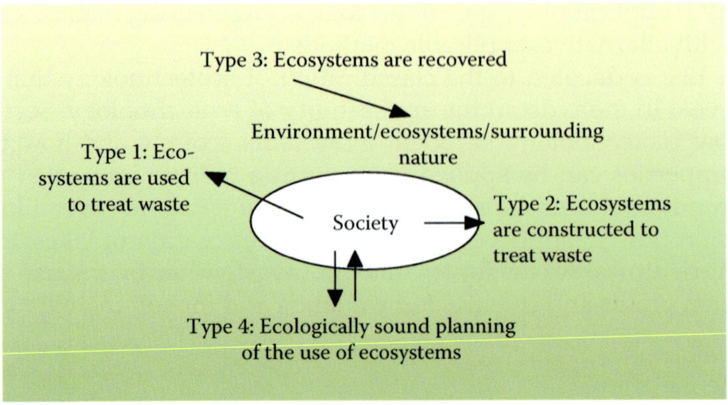

FIGURE 12.1
An illustration of the four types of ecological engineering.

are shown in Table 12.1, which is also where the alternative environmental technological solution is indicated. This does not imply that ecological engineering consequently should replace environmental engineering. On the contrary, the two technologies should work hand in hand to solve environmental management problems better than they could do alone. This is illustrated in Figure 12.2, where control of lake eutrophication has required both ecological engineering and environmental technology.

Type 1 ecological engineering, the application of ecosystems to reduce or solve pollution problems, may be illustrated by wetlands utilized to reduce the diffuse nutrient loadings of lakes. This problem could not be solved by environmental technology. Treatment of sludge could be solved

TABLE 12.1
Ecological Engineering Examples

	Example of Ecological Engineering		
Type of Ecological Engineering	**Without Environmental Engineering Alternative**	**With Environmental Engineering Alternative**	**Environmental Engineering Alternative**
1	Wetlands utilized to reduce diffuse pollution	Sludge disposal on agricultural	Sludge incineration
2	Constructed wetland to reduce diffuse pollution	Root zone plant (subsurface wetland)	Traditional wastewater treatment
3	Recovery of lakes	Recovery of contaminated land *in situ*	Transport and treatment *in situ*
4	Agroforestry	Ecologically sound planning of harvest rates of resources	

Note: Alternative environmental engineering methods are given, if possible.

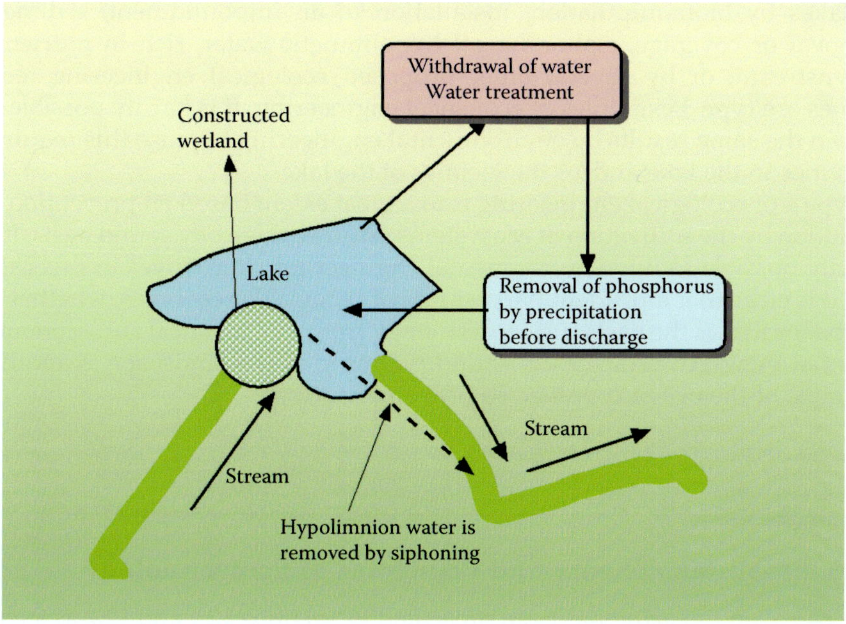

FIGURE 12.2
Control of lake eutrophication with a combination of chemical precipitation for phosphorus removal from wastewater (environmental technology), a wetland to remove nutrients from the inflow (ecotechnology type 1 or 2), and siphoning off of hypolimnetic water, rich in nutrients, downstream (ecotechnology type 3).

by environmental technology, namely by incineration, but the ecological engineering solution, sludge disposal on agricultural land which involves a utilization of the organic material and nutrients in the sludge, is a considerably more sound method from ecological perspectives. The application of constructed wetlands to cope with the diffuse pollution is a good example of ecological engineering type 2. Again, this problem cannot be solved by environmental technology. The application of root zone plants (constructed subsurface wetland) for the treatment of small amounts of wastewater is an example of ecological engineering, type 2, where the environmental engineering alternative, a mechanical–biological–chemical treatment cannot compete, because it will have excessive costs relative to the amount of wastewater (sewage system, pumping stations, and so on). A solution requiring less resources will always be ecologically more sound.

Recovery of land contaminated by toxic chemicals is possible using environmental technology, but it requires transportation of soil to a soil treatment plant, where biological biodegradation of the contaminants takes place. Ecological engineering will propose a treatment *in situ* with adapted microorganisms or plants. The latter method will be much more cost-effective and the pollution related to the transport of soil will be omitted. Restoration of lakes by biomanipulation, installation of an impoundment, sediment removal or coverage, siphoning off hypolimnetic water, rich in nutrients, downstream, or by several other proposed ecological engineering techniques are type 3 examples of ecological engineering. It is hardly possible to obtain the same results by environmental engineering because this requires activities in the lake and/or the vicinity of the lake.

Type 4 of ecological engineering is to a great extent based on prevention of pollution by the utilization of ecosystems on an ecologically sound basis. It is hardly possible to find environmental engineering alternatives in this case, but it is clear that a prudent harvest rate of renewable resources, whether of timber or fish, is the best long-term strategy from an ecological and economic point of view. An ecologically sound planning of the landscape is another example of the use of type 4 ecological engineering.

12.3 Ecosystem Theory and Principles of Ecotechnology

Ecological engineering has been presented in Sections 12.1 and 12.2 as a useful technological discipline, but it is important that it is based on systems ecology presented in Chapters 3–5, because ecotechnology is operating in the ecosystems and has therefore to consider the properties of ecosystems. Mitsch and Jørgensen (1989, 2003) apply 19 system ecological principles to elucidate the basic concepts of ecotechnology and to ensure proper application of the practical use of this approach. These 19 principles and their basis in systems ecology are given one by one as follows.

Principle 1: The external factors (in ecology denoted the forcing functions) determine the reactions of ecosystems is rooted in the openness of ecosystem. It is important when we discuss the possibilities to make man-made system more sustainable that we realize that we by regulation of the forcing functions can change the reactions of the man-made systems.

Principle 2: Available storage of matter and energy is limited. It is explained by the thermodynamic laws and in the ability of ecosystems to recycle (Chapter 3).

Principle 3: Ecosystems are open systems and have to be considered in all management plans. This is of course completely consistent with the openness of ecosystems in general.

Principle 4: Ecosystems are homeostatic systems with many feedbacks complies with the system properties of ecosystems as presented in Chapters 4 and 5.

Principle 5: Homeostasis of ecosystems requires accordance between the biological function and chemical composition of the forcing functions and is rooted in the biochemistry that ecosystems have to follow; see Chapter 3.

Principle 6: Ecosystems recycle all matter and energy is explained by the network properties of ecosystems; see Chapters 3 and 5.

Principle 7: Ecosystems are self-designing systems that are an emergent holistic property of ecosystems; see Chapter 3.

Principle 8: Ecosystems have characteristic scales in space and time, which is clearly in accordance with the hierarchical organization of ecosystems; see Chapter 3.

Principle 9: Ecosystems have a high diversity and complexity. The biodiversity should be championed to maintain an ecosystem self-design capacity and is presented directly in Chapter 3. The high complexity is furthermore a holistic emerging property.

Principle 10: Ecotones, transition zones, are as important for ecosystems as membranes are for cells. This principle is consistent with the openness of ecosystems, which of course makes ecosystems vulnerable; see Chapters 3 and 4.

Principle 11: Due to their openness, ecosystems are coupled with other ecosystems, which is also rooted in the openness, Chapter 3.

Principle 12: Ecosystems form networks. The components of ecosystems are connected.

Principle 13: Ecosystems have the history of development. The application of ecotechnology should respect that a part of the history is explained by the growth and development of ecosystems (Chapter 4). Another part is of course due to the change in the external factors, which means the openness (Chapter 3). The development of ecosystems has ensured the maintenance of the functions—primary production, secondary production, and so on—not certain species. The maintenance of the function can actually be considered a consequence of the high buffer capacity and resilience of ecosystems; see Chapter 3.

Principle 14: Ecosystems and their species are most vulnerable at their geographical edge, which is due to the lower buffer capacity and resilience, when the conditions are unfavorable. This is touched upon in Chapter 3.

Principle 15: Ecosystems are hierarchical systems. This is directly covered in Chapter 3. This form of organization is important to consider in ecological management.

Principle 16: Ecosystems are pulsing systems. This is due to the pulsing of the forcing functions and is therefore due to the openness.

Principle 17: Physical and biological processes are interactive, which is rooted in the enormous variability of the forcing functions, which are presented in Chapter 3.

Principle 18: Ecosystems are integrated systems with holistic properties. This is directly presented in Chapter 3 about the holistic emerging properties.

Principle 19: Ecosystems store their information in the structure. This is directly treated and discussed in Chapters 3 and 4 and is considered by the work energy of the information.

Straskraba (1993) has presented 7 principles of ecosystems which he uses to set up 17 rules on how to use ecotechnology. The rules may, however, be expanded to be applied in environmental management in general. They are based upon basic scientific principles of systems ecology and should therefore be respected by all means whenever an environmental management decision is to be taken. The principles (taken from Straskraba, 1993) are presented below with a short explanation, followed by the rules (Straskraba, 1993) with comments on their application in ecological technology and environmental management in general.

Principle 1: Energy inputs to the ecosystems and available storage of matter are limited. This principle is based on the conservation of matter and energy. Nuclear processes may convert matter to energy, but these processes are insignificant in ecosystems.

The dominant energy input to life on the Earth is solar energy. We may be able to supplement this energy source with other forms of energy, for instance fossil fuel, but the only sustainable energy source is solar energy.

Favorable conditions for vegetation growth are represented by temperature, humidity (water), and sufficient concentrations of nutrients. Mass conservation leads to the concept of limiting factors for growth and the Michaelis–Menten equation: growth = $v \cdot S/(k + S)$, where S represents the concentration of the substrate or limiting nutrient, and v and k are constants. A maximum growth rate is attained when $S \gg k$, which implies that another nutrient or factor becomes limiting (see also Chapter 3).

Principle 2: Information is stored in ecological systems in structures. Structures are a result of the input of energy which is utilized to move away from thermodynamic equilibrium, gain work energy (see Chapters 3 and 4), or build structures. Such structures include not only organisms but also the physical structure of the landscape. Size is an important characteristic of structures. Organism size determines many important features of life, like the rate of development, speed of movement, and the range of areas they inhabit. A certain minimum size of structures surrounding the organisms is necessary to satisfy their needs.

Principle 3: The genetic code of single organisms stores information about the development of the Earth's environment, and the process of adaptation to the changing conditions of existence. For organisms information about the past evolution is stored in the genes. The genetic code reflects not only the past of an individual organism, but also its surroundings, included other coexisting organisms. It is the consequence of tight coupling between each organism and its environment. The evolution of the organisms is highly dependent on the conditions of life (Straskraba, 1993).

In Chapter 3, use of work energy was proposed to describe the propensity for the development of an ecosystem and to assess ecosystem health. These proposals are in accordance with the concepts behind Principle 3.

Principle 4: Ecosystems are open and dissipative systems. They are dependent on a steady input of energy from outside. This principle is an application to ecosystems of the second law of thermodynamics. As ecosystems of course must obey the conservation principle, they must also obey basic scientific laws, including the basic laws of thermodynamics. The energy input is utilized to cover the energy needs for maintenance, respiration, and evapotranspiration. If the work energy (exergy) input exceeds these needs, the surplus work energy is utilized to build more structure.

Principle 5: Ecosystems are multimediated feedback systems. Feedback systems are characterized by influences on one part of the system being fed back to other components of the system. Feedback creates many completely unexpected effects and the entire network describing the interrelations of the organisms may also work as feedback. In Chapter 5, it is shown that indirect effects may often exceed direct effects due to feedback and the relationships determined by the network.

Principle 6: Ecosystems have homeostatic capabilities which result in smoothing out and depressing the effects of strongly variable inputs. However, this capability is limited. Once it is exceeded, the system breaks down. Several homeostatic mechanisms are known in biology, for instance the maintenance of pH in our blood and the maintenance of the body temperature of warm-blooded animals. The homeostatic capability may be expressed by means of the concept of ecological buffer capacity (see Chapter 3). As the homeostatic capability is limited, so are the buffer capacities. It is obviously important in environmental management to know and respect the buffer capacities, as the ecosystem may otherwise change radically and even collapse.

Principle 7: Ecosystems are adaptive and self-organizing systems. We may distinguish between biochemical and biological adaptation. Biochemical adaptation takes place when the individual organism adapt to changed conditions. Phytoplankton, for instance, can adjust its content of chlorophyll a to the intensity of solar radiation. Biological adaptation implies adaptation at the ecosystem level. The properties of the species in the ecosystem are changed currently in accordance with the prevailing conditions, either for the next generation by dominant heritage of the properties best fitted to the prevailing conditions, or by complete or partial replacement of species with different tolerances. Other species waiting in the wings take over, if they have a combination of properties better fitted to the (new) emergent conditions.

Self-organization of the ecosystem may be understood as a directional change of the species composition of the ecosystem when its surroundings change, for example, as a result of human activities (Straskraba, 1993).

These principles give occasion to introduce 17 rules (Straskraba, 1993) for the application of ecotechnology and environmental management in general. The rules are presented in this context, but they are also very useful for the development of integrated ecological–environmental management and therefore should also be applied in Chapter 16. The 17 rules with reference to the principles are presented as follows.

Rule 1: Minimize energy waste. This rule is based on the conservation principle; see Principle 1. Energy is limited and therefore a resource and in addition all uses of energy cause pollution: carbon dioxide for fossil fuel, nuclear waste for nuclear energy, noise and landscape pollution for windmills, etc. The selection of environmental technology should also be based on this rule and rule number two, so not only are we reducing the use of resources, but we are also reducing the pollution correspondingly, as energy not wasted and matter recycled are not discharged to the environment. Generally, the three Rs should be considered for all environmental management plans: reduce, reuse, and recycle.

Rule 2: Recycle. This rule is also based on the conservation principle, Principle 1. This rule is a core principle in cleaner technology (see Chapter 6) and is an imitation of ecosystems. For all essential elements, ecosystems use recycling which will ensure that the resources can be used again and again. The development of the green audit is based upon the conservation principle. The purpose is to increase recycling; see Chapters 3 and 5.

Rule 3: Retain all kinds of structures. This rule is based on Principle 2. It is in accordance with the use of ecotechnology which prescribe that hedges, littoral zones, a landscape pattern of agriculture, wetlands, forests, hedges, and hills are essential to maintain.

Rule 4: Consider long-term horizons. Sustainable development, or sustainable life, may be realized only when we switch from short-term evaluation of our economic practices to long-term horizons (Straskraba, 1993). This implies that we need to consider the long-term consequences of changes in the ecological structures and extinction of species due to our activities (Principles 2 and 3).

Rule 5: Do not neglect that mankind is dependent on ecosystem services and many organisms and that their loss may lessen our ability to survive in the steadily changing environment. This rule is based on Principle 3. Our dependence on many species is not necessarily known today. We may find medicine for curing cancer in a rainforest plant which has not yet been discovered. The presence of important indirect effects makes it very difficult to evaluate the importance of many species. It seems therefore to be a prudent strategy to maintain as many species as possible and maintain the networks in which they are operating to the greatest possible extent.

Rule 6: Consider ecosystem dynamics. This rule is directly translated from Principle 4. Open and dissipative systems have complex dynamics which should not be neglected in environmental practices. Dynamic and structurally dynamic models appear to be tools needed for decisions concerning ecosystem management, as they consider these important dynamics of ecosystems.

Rule 7: Understand that nature is a teaching ground for handling complex systems. In particular the ability to survive by adapting to changing conditions can be learned only from nature. This rule is based on Principles 4 and 5. The study of adaptation and evolution of individual organisms and ecosystems seems to be the only way we can recognize our own possibilities and limitations of survival under conditions of the environment subject to unexpected major changes created by our activity. The recent development of cleaner technology and organic agriculture is based on imitation of nature or learning from nature.

Rule 8: People are part of the ecosystems, which are directly dependent on solar energy: we need it in the form of light to see and in the form of heat for warmth. We are also indirectly dependent on energy stored in fossil fuels as well as in vegetable and animal food—we are dissipative structures, too. This rule acknowledges that Principle 4 is also valid for us—we are ourselves open and dissipative systems and we are part of open and dissipative systems. We are therefore also victims of Principle 4.

Rule 9: Understand that we are not only dependent on and sensitive to external inputs, in particular solar radiation, but also to the supply of minerals from the Earth. This rule is based upon Principles 1, 2, and 4. It is also associated with Rules 1, 2, and 8. Our activity should not alter the inputs by solar radiation by destruction of the ozone layer as that may cause skin cancer. On the other hand, the modern postindustrialized society still needs a supply of minerals and energy—it demonstrates that we are indeed a part of the global dissipative structure. Therefore we should minimize energy waste and recycle.

Rule 10: Manage the environment as an interconnected system, not as isolated subsystems. This rule is based on Principle 4. You cannot close an ecosystem. It has to be open which means that we have to consider inputs and outputs, but outputs from one ecosystem may be inputs for another ecosystem. Pollution problems can therefore not be solved by discharge/emission/dilution, but only by degradation and recycling.

Rule 11: Evaluate available management options simultaneously. Any technological goal can be achieved in different ways using different options. The options are, however, different in expenditure, quality of product, material needs, waste, etc. The options available to solve an environmental problem have to be recognized and evaluated with respect to best satisfying the goal while respecting all the principles and rules presented in this section.

Rule 12: Includes secondary effects elsewhere. This rule is based on Principle 5 and partly on Principle 6. It is very difficult to survey the consequences of our activities for ecosystems due to the complexity and openness of these systems and due to the often surprising indirect effects. As wide an evaluation as possible—if possible a global system evaluation—of the consequences of our activities is therefore needed.

Rule 13: Do not exceed the ecosystem homeostatic capacity. This rule is directly derived from Principle 6. It is important to know the buffer capacities of focal ecosystems, possibly by use of models to consider the indirect effects and the ability of ecological networks to absorb disturbances.

Rule 14: Consider self-adaptation in management strategies. This rule is directly based on Principle 7. The self-adaptation ability of natural components in ecosystems should be reflected in our environmental management strategy. The best illustration may be taken from pest control. The use of pesticides is problematic because the organisms become adapted to the pesticide, while biological methods do not violate Principle 7. There are unfortunately numerous examples of environmental management where we have omitted to use this rule and therefore failed. To mention a few: the introduction of rabbits to Australia and introduction of the Nile perch to Lake Victoria.

Rule 15: Evaluate the socioeconomic environment. This rule is an acknowledgement of the self-organization (Principle 7) of human society, which is another complex, open, dissipative system with many properties similar to an ecosystem. As society and ecosystems are both open systems; they are also interchanging matter and energy. The ultimate environmental management should therefore consider both types of systems which makes the problem even more complex. It has been attempted to develop ecological–sociological–economic models but this type of model is in its very early infancy.

Rule 16: Evaluate all possible human uses of the environment. The same environmental area or object, like a body of water or a forest, can be used for different purposes. When decisions are made, all of these possible purposes have to be taken into account to find the most environmentally friendly use.

Rule 17: Base measures on ecosystem principles. In other words, use the principles presented in this section and in Chapters 3–5.

12.4 Restoration of Lakes

It has been pointed out throughout this book that a close to optimum solution of an environmental problem can only be found by quantification of the problem and the source, combined with a quantitative diagnosis and correct selection among the wide spectrum of applicable methods. This principle is even more important to use when the solution requires interference with the ecosystem. Without quantification of the problem, we are unable to measure and compare the effects of the various alternatives. Ecological engineering is obviously more applicable when the pollutants are threshold rather than nonthreshold agents, although it has also been applied for nonthreshold pollutants. In this section, a short overview of ecological engineering methods applied to lake restoration will be given as an illustration of the possibilities offered by ecotechnology. Those who seek a more comprehensive introduction into the application of these methods are referred to Mitsch and Jørgensen (1989, 2003). The most important methods are listed as follows and a brief description of the application, advantages, and disadvantages is given.

1. *Diversion* of wastewater has been extensively used, often to replace wastewater treatment. Discharge of effluents into an ecosystem which is less susceptible than the one used at present is, as such, a sound principle, which under all circumstances should be considered, but quantification of all the consequences has often been omitted. Diversion might reduce the number of steps in the treatment,

but cannot replace wastewater treatment totally, as discharge of effluents, even to the sea, always should require at least mechanical treatment to eliminate suspended matter. Diversion has often been used with a positive effect when eutrophication of a lake has been the dominant problem. Canalization, either to the sea or to a lake outlet, has been used as a solution in many cases of eutrophication. However, effluents must be considered as a freshwater resource. If it is discharged into the sea, the effluents cannot be recovered; if it is stored in a lake, after sufficient treatment of course, it is still a potential water resource, and it is far cheaper to purify eutrophic lake water to an acceptable drinking-water standard than to desalinate seawater. Diversion is often in a few cases the only possibility when a *massive* discharge of effluents goes into a susceptible aquatic ecosystem (a lake, a river, a fjord, or a bay). The general trend has been towards the construction of larger and larger wastewater plants, but this is quite often an ecologically unsound solution. Even though the wastewater has received multistep treatment, it will still have a large amount of pollutants relative to the ecosystem, and the more massive the discharge is at one point, the greater the environmental impact will be. If it is considered that canalization is often a significant part of the overall cost of handling wastewater, it might often turn out to be both a better and cheaper solution to have smaller treatment units with individual discharge points.

2. *Removal of superficial sediment* can be used to support the recovery process of very eutrophic lakes and of areas contaminated by toxic substances (for instance, harbors). This method can only be applied with great care in small ecosystems. Sediments have a high concentration of nutrients and many toxic substances, including trace metals. If a wastewater treatment scheme is initiated, the storage of nutrients and toxic substances in the sediment might prevent recovery of the ecosystem due to exchange processes between sediment and water. Anaerobic conditions might even accelerate these exchange processes; this is often observed for phosphorus, as iron(III) phosphate reacts with sulfide and forms iron(II)-sulfide by release of phosphate. The amount of pollutants stored in the sediment is often very significant, as it reflects the discharge of untreated wastewater for the period prior to the introduction of a treatment scheme. Thus, even though the retention time of the water is moderate, it might still take a very long time for the ecosystem to recover.

The method is, however, costly to implement, and has therefore been limited in use to smaller systems. Maybe the best-known case of removal of superficial sediment is Lake Trummen in Sweden. The transparency of the lake was improved considerably, but decreased again due to the phosphorus in overflows from storm water basins.

Probably, treatment of the overflow after the removal of superficial sediment would have given a better result.

3. *Uprooting and removal of macrophytes* have been widely used in streams and also to a certain extent in reservoirs, where macrophytes have caused problems in the turbines. The method can, in principle, be used wherever macrophytes are a significant result of eutrophication. A mass balance should always be set up to evaluate the significance of the method compared with the total nutrient input. A simultaneous removal of nutrients from the effluent should also be considered.
4. *Coverage of sediment by an inert material* is an alternative to removal of superficial sediment. The idea is to prevent the exchange of nutrients (or maybe toxic substances) between sediment and water. The general applicability of the method is limited due to the high costs, even though it might be more moderate in cost than removal of superficial sediment. It has only been used in a few cases and a more general evaluation of the method is still lacking.
5. *Siphoning of hypolimnetic water* is more moderate in cost than methods 2 and 4. It can be used over a longer period of time and thereby gives a pronounced overall effect. However, the effect is dependent on a significant difference between the nutrient concentrations in the epilimnion and the hypolimnion, which, however, is often the case if the lake or the reservoir has a pronounced thermocline. This implies, on the other hand, that the method will only have an effect during the period of the year when a thermocline is present (in many temperate lakes from May to October/November), but as the hypolimnetic water might have a concentration fivefold or higher than the epilimnetic water, it might have a significant influence on the nutrients' budget to apply the method.

   As the hypolimnetic water is colder and poorer in oxygen, the thermocline will move downward and the possibility of anaerobic zones will be reduced. This might have an indirect effect on the release of nutrients from the sediment.

   If there are lakes or reservoirs downstream, the method cannot be used, as it only removes, but does not solve the problem. A possibility in such cases would be to remove phosphorus from the hypolimnetic water before it is discharged downstream. The low concentration of phosphorus in hypolimnetic water (maybe 0.5–1.0 mg L^{-1}) compared with wastewater makes it almost impossible to apply chemical precipitation. However, it will be feasible to use ion exchange or adsorption on activated alumina oxide, because the capacity of an adsorbent and ion exchange is more dependent on the total amount of phosphorus removed and the flow than on the total volume of water treated. Figure 12.3 illustrates the use of siphoning and ion exchange of hypolimnetic water.

6. *Flocculation* of phosphorus in a lake or reservoir is another alternative. Either aluminum sulfate or iron(III)-chloride can be used. Calcium hydroxide cannot be used, even though it is an excellent precipitant for wastewater, as its effect is pH-dependent and a pH of 9.5 or higher is required. The method is not generally recommended as (1) it is not certain that all flocs will settle and thereby incorporate the phosphorus in the sediment, and (2) the phosphorus might be released from the sediment again at a later stage.
7. *Circulation* of water can be used to break down the thermocline. This might prevent the formation of anaerobic zones, and thereby the release of phosphorus from sediment.
8. *Aeration* of lakes and reservoirs is a more direct method to prevent anaerobic conditions from occurring. Aeration of highly polluted rivers and streams has also been used to avoid anaerobic conditions.
9. *Regulation of hydrology* has been used extensively to prevent floods. Lately, it has also been considered as a workable method to change the ecology of lakes, reservoirs, and wetlands. If the retention time in a lake or a reservoir is reduced with the same annual input of nutrients, eutrophication will decrease due to decreased nutrient concentrations. It should also be considered to change the seasonal variation of the retention time. Particularly, a short retention time during the spring and maybe also during the summer is often able to reduce the eutrophication, because the nutrient concentration is reduced when the blooms occur. This form of ecotechnology is named ecohydrology. The role of the depth, which can be regulated by use of a dam, is more complex. Increased depth has a positive effect on the reduction of eutrophication, but if the retention time is increased simultaneously, the overall effect cannot generally be quantified without the use of a model. The productivity of wetlands is highly dependent on the water level, which makes it highly feasible to control a wetland ecosystem by this method (Figure 12.3).
10. *Fertilizer control* can be used in agriculture and forestry to reduce nutrient loss to the environment. Utilization of nutrients by plants is dependent on a number of factors (temperature, humidity of soil, soil composition, growth rate of plant (which again is dependent on a number of factors), chemical speciation of nutrients, etc.). Models of all these processes are available today on computers.
11. *Application of wetlands or impoundments as nutrient traps* could be considered as an applicable method, wherever the nonpoint sources are significant; see also Figure 12.2. The use of wetlands has also been

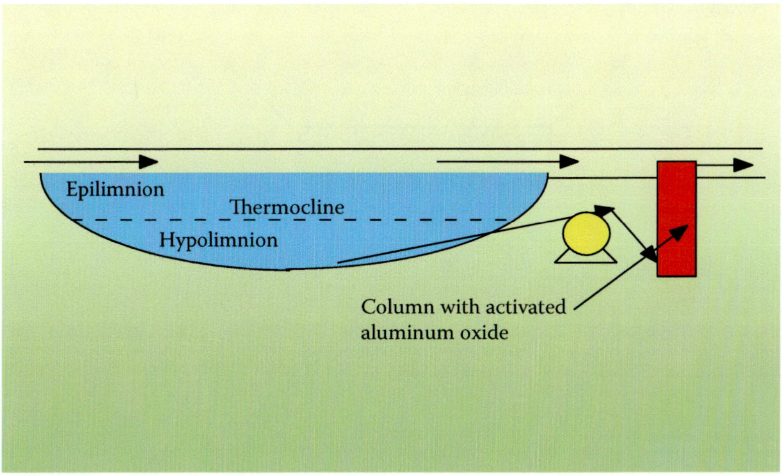

FIGURE 12.3
Application of siphoning and adsorption on activated alumina oxide, which removes by adsorption the phosphorus before it is discharged downstream from the lake.

applied as a direct wastewater treatment method, for instance in Florida, in Tanzania, and in a number of countries particularly in the tropic zone, but it is probably most effective in dealing with nutrient losses from agricultural areas. Inputs of nutrients into a wetland will to a great extent be denitrified, adsorbed on the sediment or used for growth of algae and macrophytes (phragmites, etc.). Management of a wetland or an impoundment as a nutrient trap obviously requires that a major part of the nutrient input is removed by denitrification (nitrogen only) and stored in sediment and plants. The storage capacity is often large, but of course limited. This implies that nutrients must be removed by harvest of macrophytes, which is quite feasible mechanically. However, if the retention time of a lake or reservoir in the temperate zone is short, it might be favorable to let the in flowing water pass through a wetland or an impoundment. In the wintertime, when almost no growth takes place in the wetland, the nutrients are washed out, but due to the short retention time, the water with high nutrient concentration will have passed the lake or reservoir before the spring bloom starts. The nutrients of the water which passes the wetland during the spring and summer will, on the other hand, be used to a large extent in the wetland or the impoundment for growth of algae and macrophytes. The water flowing to the lake or reservoir will therefore have a significantly lower nutrient concentration at the time of the year when eutrophication may appear. Figure 12.4 shows the removal processes. Indications show that

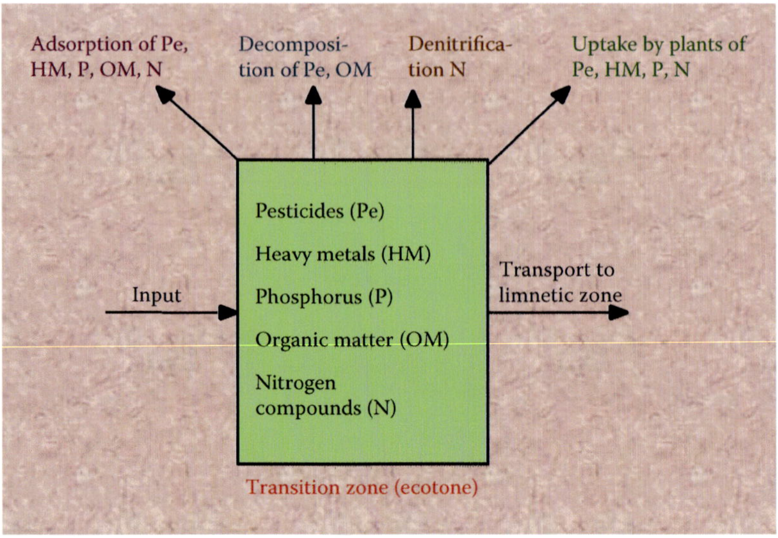

FIGURE 12.4
The figure shows that pesticides, heavy metals, phosphorus, organic matter, and nitrogen compounds can be removed by wetlands, which are often transition areas (ecotones) between land and aquatic ecosystems.

wetlands are important not only as nutrient traps but also for maintenance of species diversity and for an ecologically sound hydrology in the region.

12. *Calcium hydroxide* is widely used to neutralize low pH values in streams and lakes in those areas where acidic rain has a significant impact. Sweden is using on the order of $100 million per year to neutralize acid streams and lakes.
13. *Algicides*, for instance various copper salts, have previously been used widely for relatively small lakes, but are now hardly in use due to the general toxicity of copper which is accumulated in the sediment and thereby can contaminate the lake for a very long time.
14. *Shading by use of trees* at the shore line is a cost-effective method which, however, only can give an acceptable result for small lakes due to the low area/circumference ratio.
15. *Biomanipulation* can only be used in the phosphorus concentration range from about 50 to about 150 μg/L, depending on the lake. In this range, two ecological structures are possible. This is illustrated in Figure 12.5. When the phosphorus concentration initially is low and increases, zooplankton is able to maintain a relatively low phytoplankton concentration by grazing. Carnivorous fish are also able to maintain a low concentration of planktivorous fish which implies relatively low predation on zooplankton. At a certain phosphorus

concentration (above 120–150 μg/L), zooplankton is not any longer able to control the phytoplankton concentration by grazing and as the carnivorous fish (for instance Nile perch or pike) is hunting by the sight, and the turbidity increases, the planktivorous fish become more abundant which involves more pronounced predation on zooplankton. In other words, the structure changed from control by zooplankton and carnivorous fish to control by phytoplankton and planktivorous fish. When the phosphorus concentration decreases from a high concentration the ecological structure is initially dominated by phytoplankton and planktivorous fish. This structure can, however, be maintained until the phosphorus concentration is reduced to about 50 μg/L.

There are therefore two possible ecological structures in the phosphorus range of approximately 50–150 μg/L. Biomanipulation (de Bernardi and Giussani, 1995) can be used in this range—and only in this range—to make a "short cut" by removal of planktivorous fish and release carnivorous fish. If biomanipulation is used above 150 μgP/L, some intermediate improvement of the water quality will usually be observed, but the lake will later get the ecological structure corresponding to the high phosphorus concentration, i.e., a structure controlled by phytoplankton and planktivorous fish. Biomanipulation is a relatively cheap and effective method provided that it is applied in the phosphorus range where two ecological structures are possible. De Bernardi and Giussani (1995) gave a comprehensive presentation of various aspects of biomanipulation. Simultaneously, biomanipulation makes it possible to maintain relatively high biodiversity which does change the stability of the system, but a higher biodiversity gives the ecosystem a greater ability to meet future, unforeseen changes without changes in the ecosystem function (Figure 12.5).

From this brief survey of restoration methods, it is possible to conclude that ecological engineering methods 4, 5, 8, 9, 10, 12, and 15 offer a cost-moderate restoration of lakes and sometimes also rivers. The methods should however in most cases work hand in hand with other methods reducing the nutrient loadings, which is mainly possible by environmental technology.

ILLUSTRATION 12.1

A lake with a total phosphorus concentration of 0.2 mg/L is very eutrophic. Seventy-five percent of the phosphorus is coming from wastewater and the remaining 25% from other sources. Give a solution to the problem, where environmental technology and ecotechnology are working together.

Solution

Since the phosphorus concentration is too high for application of biomanipulation, it would be beneficial to remove at least 90% of the

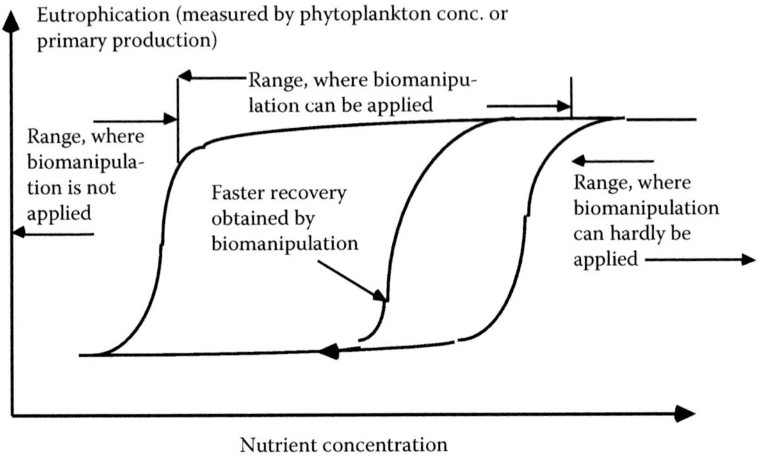

FIGURE 12.5
The hysteresis relation between the nutrient level and eutrophication measured by the phytoplankton concentration is shown. The possible effect of biomanipulation is shown. An effect of biomanipulation can only be expected in the range approximately 50–150 μgP/L. Biomanipulation can hardly be applied successfully above 130–150 μg/L (see also de Bernardi and Giussani, 1995 and Jørgensen and de Bernardi, 1998).

phosphorus from the wastewater by chemical precipitation, which would be able to reduce the total phosphorus concentration to less than 0.1 mg/L and combine this treatment with biomanipulation. Of course, the retention time of the water in the lake determines how long a time it will take to reduce the total phosphorus concentration below about 0.125 mg/L.

12.5 Agricultural Waste

Particular animal waste causes great problems in intensive farming. The productions of chickens, pigs, and cattle are in many industrialized countries concentrated in rather large units, which implies that the waste from such production units requires hundreds of hectares for suitable distribution and feasible use of its value as fertilizer. Animal waste has a high nitrogen concentration, 5%–10% based on dry matter (2%–4% dry matter). If the nitrogen is not used as fertilizer, it may

1. Either evaporate as ammonia,
2. Be lost to deeper layers, where it contaminates the groundwater, or
3. Be lost by surface runoff to lakes and streams.

Agricultural waste has therefore become a crucial pollution problem in many countries with intensive agriculture. The following possibilities give a summary of the available methods, based on ecological engineering or interface between environmental technology and ecotechnology, to reduce the pollution originating from agricultural waste.

1. Storage capacity for animal waste to avoid spreading on bare fields.
2. Green fields in the winter to ensure the use of the fertilizing value of agricultural waste.
3. Composting of agricultural waste assure conditioning before it is used. The composting heat may be utilized. It is an intermediate method between ecological engineering and environmental engineering.
4. Anaerobic treatment of agricultural waste for production of biogas and conditioning before use as fertilizer. The biogas can be used to heat the farm buildings. It is preferable to mix cattle and swine manure to obtain the highest efficiency with this process, as an excessive free ammonia concentration (more than 1.1 g/L) results in a decreased rate of gas formation. This method may also be considered an intermediate method between ecological engineering and environmental engineering
5. Chemical precipitation of agricultural (animal) waste with activated benthonite is applied to bind ammonium and obtain a solid concentration of 6%–10%. This process provides two advantages:
   a. The required storage capacity is reduced by a factor of 2–4 (the solid concentration is increased from 2%–4% to 6%–10%).
   b. The loss of nitrogen by evaporation of ammonia is reduced by a factor of 3–8 corresponding to an adsorption of about 60%–90% of the ammonium on the added benthonite.
6. In China, animal waste is applied in fishponds. Zooplankton eats the detritus and fish feed on the zooplankton.

12.6 Wastewater Treatment by Ecotechnology

Traditionally, waste stabilization ponds (WSPs) are built as flow-through systems with an anaerobic, a facultative and one or more maturation ponds. The following processes are utilized in the pond system: settling (mainly in the first ponds), anaerobic decomposition of organic matter (mainly in the first ponds), aerobic decomposition of organic matter (mainly in the last ponds, where algae are present and produce oxygen), uptake of phosphorus and nitrogen by algae (facultative and maturation ponds), evaporation of ammonia (mainly where pH is high, i.e., in the last ponds), and settling of algae, denitrification (in the anaerobic zones). High removal efficiencies of BOD_5, COD, microorganisms, nitrogen, and phosphorus may be obtained provided that the guidelines for design and maintenance are followed. The method has mainly been applied in the tropical region where the temperature ensure a high rate of the processes, mentioned above.

One of the major maintenance problems is associated with the removal of sludge. It is absolutely necessary to remove sludge at least once a year because the sludge reduces the volume of particularly the anaerobic ponds and causes an undesired internal loading. Generally, an increase in the sediment depth of 0.2 cm per month can be expected. Ammonia volatilization is the main mechanism of nitrogen removal. Removal of phosphorus by WSPs is in the range of 20%–50% but depends highly on the possibility of removing algae from the effluent before discharge. Enhanced removal of phosphorus which is often required for discharge of wastewater to lakes and reservoirs can be achieved by addition of precipitants such as calcium compounds and clay minerals. It is recommended to examine the possibilities to use addition of cheap local and natural compounds. Addition of the chemicals to the influent may increase the removal efficiency to about 80%–95%. Overall high efficiency is, however, best obtained in most cases by using a constructed wetland after the WSPs (see below), preferably with the harvest of the wetland two to four times per year.

The uptake of nutrients by algae can be utilized as a wastewater treatment process, too. Per 100 g of dry matter phytoplankton will, on average, contain 4–10 g nitrogen and 0.5–2.5 g phosphorus. This process is known from the eutrophication problem. It can be considered a form of biological nutrient removal. Algal ponds are, under favorable conditions able to produce 100 g of phytoplankton per m^3 during 24 h, which means that as much as 25% of the nitrogen and phosphorus can be removed from the solution into phytoplankton (i.e., into suspension) during 24 hours. If a retention time of 3–5 days is applied, as much as 40%–60% of the nutrient in municipal wastewater will be removed from solution, but it is hardly possible to increase this efficiency as the uptake rate decreases with decreasing concentration. If chemical precipitation follows the treatment in algal ponds the suspended phytoplankton can easily be removed. The cost of the method is moderate, and the removal of nitrogen especially is often a great advantage. Particularly for the last ponds in the series, waste stabilization ponds will react as algal ponds. Lately, it has been shown that it is possible to produce biodiesel from phytoplankton harvest from algal ponds. It requires that the N:C ratio in the water is adjusted to a relatively low value, as it will imply that the carbon and lipid concentration in the algae is high.

The method, however, has some pronounced disadvantages that limit its application:

1. Only in tropical or subtropical regions is solar radiation sufficient throughout the year to ensure an acceptable efficiency.

2. The areas needed are large—retention time is a minimum of 3–5 days.
3. The growth of phytoplankton is sensitive to many organic compounds, for example, mineral oil.

The ecotones between lakes and terrestrial ecosystems are crucial for the protection of the lake ecosystem against anthropogenic impacts (see Figure 12.4). It is therefore crucial to preserve the shore ecotones around an aquatic ecosystem and the wetlands in the catchment area, independent of the management strategy implemented. Any man-made construction should be omitted in a zone 50–100 m from the lake shoreline to keep the ecotones intact. Nonpoint or diffuse pollutants from the environment will inevitably flow toward the lake, but the transition zone is able to transform and/or adsorb the pollutants entirely or partially. It will thereby significantly reduce the overall irreversible effects on the lake ecosystems. The most important processes are summarized in Figure 12.4.

The denitrification potential of wetlands is often surprisingly high. As much as 2–3000 kg of nitrate-nitrogen can be denitrified per ha of wetland per year, dependent on the hydraulic conditions. It is of great importance for the protection of lakes, as significant amounts of nitrate are released from agriculture. As much as 100 kg nitrate-N/ha may be found in the drainage water from intensive agriculture. As the denitrification is accompanied by a stoichiometric oxidation of organic matter, significant amounts of organic matter are also removed by this process. The phosphorus bound as organic matter or adsorbed to the organic matter may, however, be released by these processes (Jørgensen, 2000, 2009) and Jørgensen et al. (2012) and Palmeri et al. (2013). These processes should be examined carefully and quantitatively in each case, including whether the released phosphorus will flow toward the lake or toward the groundwater. These possibilities should be included in the development of management strategies. The adsorption capacity of the transition zone offers significant protection against pollution by toxic substances, both heavy metals and toxic organic substances (primarily pesticides originating from agriculture). The ratio of the concentration of heavy metals or pesticides in organic matter to the concentration in water at equilibrium is strongly dependent on the composition of the organic matter, but is usually between 50 and 5000, which indicates that the transition zone has an enormous binding capacity for these pollutants.

Ecotones serve as a buffer zone not only for pollutants, but also for the species present in the adjacent ecosystems. Preservation of wetlands at the lake shore line may therefore be crucial for maintenance of the biodiversity in the aquatic ecosystem—a function, which the manager must not overlook in development of an appropriate only management strategy. Table 12.2 gives an overview of different types of wetlands, which are found adjacent to aquatic ecosystems, wet meadows, forested wetlands, marshes, bogs, and shore wetlands. The table provides the characteristics of the seven types

TABLE 12.2

Characteristics of Wetlands Adjacent to Lakes

Type of Wetland	Characteristics	Ability to Retain Nonpoint Pollutants
Wet meadows	Grasslands with waterlogged soil. Standing water for a part of the year	Denitrification only in standing water. Removal of N and P by harvest
Freshwater	Reed-grass dominated marshes often with peat formation	High potential for denitrification, which is limited by the hydraulic conductivity
Forested wetlands	Dominated by trees and shrubs. Standing water part of the year	High potential for denitrification and of pollutants, provided that accumulation. Standing water is present
Salt water marshes	Herbaceous vegetation possible	Medium potential for denitrification. Harvest usually with mineral soil
Bogs	Peat-accumulating wetlands with minor flows	High potential for denitrification but limited by small hydraulic conductivity
Shore wetlands	Littoral vegetation, of great importance for lakes	High potential for denitrification and accumulation of pollutants

of wetlands and their different ability to cope with the nonpoint pollution problems. The importance of wetlands adjacent to aquatic ecosystems has caused drainage of wetlands to cease in many countries, and the restoration of previously drained wetlands. The construction of artificial wetlands offers a solution to diffuse pollution originating from agriculture, septic tanks, and other sources (examples are given in Mitsch and Gosselink, 2007; Jørgensen 2009). U.S. legislation does not allow wetlands to be drained, unless another wetland of the same size is installed somewhere else.

Construction of artificial wetlands is an attractive and cost-effective solution to solve the pollution problems of diffuse sources and wastewater. Wetlands are first of all able to cope with the nitrogen and heavy metal pollution from these sources, but it is essential to make a proper planning on where to place the artificial wetlands, as their effects are dependent on the hydrology (they should be covered by water most of the year and should have sufficient retention time to allow them to solve the key pollution problems) and on the landscape pattern (they should protect the most vulnerable ecosystems which often are lakes and reservoirs). It is furthermore important to ensure that the wetlands are not releasing other components such as phosphorus as mentioned above. The following emergent types of plants are proposed for use in constructed wetlands: cattails, bulrush, reeds, rushes, papyrus, and sedges. Submerged species can be applied in deep water zones. Species that have been used for this purpose include coon tail or horn wart, redhead grass, widgeon grass, wild celery and watermilfoil. It should also be kept in mind that a wetland in most cases will reduce the water budget due to evapotranspiration (Mitsch and Gosselink, 2007). Wetlands, however, reduce the wind speed at the water surface and may thereby also reduce the evaporation. It is important to consider these factors in the planning of artificial wetlands. Finally, it should not be forgotten that an artificial wetland is not fully developed overnight. In most cases, it will require 2 to 4 years for an artificial wetland to obtain sufficient plant coverage and biodiversity to be fully operational. It is, however, clear from the experience gained in the relatively few constructed wetlands that the application of models for the wetland, encompassing all the processes reviewed above, as well as for the lake is mandatory, if positive results are to be expected.

Wetlands encompassing the so-called root zone plants (subsurface wetlands) may also be utilized as wastewater treatment facilities. This application of soft technology seems particularly advantageous for developing countries due to its moderate cost. The self-purification ability of wetlands has found wide application as a wastewater treatment method in several developing countries (China, Philippines, Burma, India, and Thailand). Generally, the constructed wetlands in the tropic regions show higher efficiency than in the temperate regions due to higher photosynthetic and microbiological activity. In addition, the seasonal variations of the removal capacity of constructed wetlands are in the tropic regions much smaller than in the temperate regions. Constructed wetlands in the northern part of Europe have in the winter period (December–February) a capacity expressed for instance as kg N or kg COD removed/ha 24 h, which may be less than 30% of the summer capacity. This implies that constructed wetlands in the temperate zone have to be constructed with an overcapacity relative to the average situation to be able to meet the required efficiency in the winter period. These differences between constructed wetlands in tropic regions and in temperate regions explain why it is usually more expensive to use constructed wetlands in the temperate zone.

Much attention, for instance, has been devoted to the stocking of fish cultivated in biological sewage stabilization ponds. These studies show that nutrients such as nitrogen and phosphorus are retrieved from the sewage by fish, through the intermediate activities of bacteria, algae and other types of plankton. In order to mobilize the self-purification of a body of water for sewage treatment, the main principle is to utilize coordination among four chief components:

a. Producers, aquatic vegetation
b. Consumers including fish
c. Decomposers (mainly bacteria)
d. Abiotic factors such as solar radiation and water exchange

Bacteria decompose organic materials, including organo-phosphates and purify the water in the preliminary phase. Carbon dioxide, ammonia, phosphates are produced and utilized by algae and other aquatic vegetation. The oxygen released by photosynthesis is supplied for the needs of oxidizing organic matter by bacteria. A great quantity of algae and other types of plankton are utilized by rearing fries. The energy fixed by algae appears finally in the form of fish harvesting and the water is thereby further purified.

Wastewater treatment using floating species has also been proposed. Different types of duck weed and water hyacinths (*Eichhornia crassipes)* have been applied as an alternative to waste stabilization ponds. Use of water hyacinths requires, however, strict control, as they easily become widely spread as a weed which may get completely out of control. The inorganic nitrogen and phosphorus brought in by sewage and decomposed from organic pollutants by microorganisms are absorbed by water hyacinths.

Experiments in China show that an average yield of up to about 10 kg/m^2 may be obtained during the growing period from May to November. Such production is able to absorb 1500–3600 kg of nitrogen, 150–500 kg of phosphorus and 100–250 kg of sulfur per ha. The water hyacinths with microorganisms and organic pollutants attached or coagulated on root surfaces are harvested to serve as feed in fish culture ponds, duck farms, pig farms and oxen farms. From May to November the concentrations of COD, total nitrogen, ammonium, total phosphorus and ortho phosphate were less than half the concentrations in the inlet. On the other side, when the water hyacinths were absent from December to April the differences in concentrations of COD, nitrogen, and phosphorus between the inlet and outlet were very small. In tropical areas, it may be possible to obtain the removal efficiencies indicated above for a wetland with floating plants.

The treatment results of constructed and natural wetlands can be estimated by a downloadable software named Sub-Wet 2.0. A first estimation is also possible by use of relatively simple equations, but because the various processes in wetlands are influencing each other the use a model as for instance Sub Wet 2.0 gives a more accurate result and is therefore recommended. The set of simple equations is however presented below to be able to give a first estimation and also to give an understanding of the removal processes. The nitrogen removal can be estimated from the following equation for aquatic treatment systems in tropical regions:

$$N/N_o = \exp(-k\ t) \tag{12.1}$$

where N is the nitrogen concentration in the effluent and No in the in fluent; k is the rate constant which is 0.2–0.6 1/24 h dependent on the plant density (0.2 at 3 t/ha, 0.5 at 10 t/ha and 0.6 at 20 t/ha) and t is the retention time in days. Equation 12.1 assumes harvest of the plants on at least an annual basis. The BOD_5 removal can be estimated at 50 kg–250 kg/ha, dependent on the plant density (from 3–20 t/ha; see Wolverton, 1982). The phosphorus removal is about 5–10 times lower than the nitrogen removal, based on at least an annual harvest.

The root zone plant has found its application for the treatment of small volumes of municipal wastewater in industrialized countries, particularly, where construction of a sewage system to an adjacent wastewater treatment plant would be prohibitively expensive. The decomposition of organic matter and denitrification usually do not cause any problem, provided that the plant is 5–10 m^2 per person equivalent, dependent on the climatic conditions. Phosphorus is mostly not removed by an efficiency of more than 10%–20% by a root zone plant, but with addition of iron chloride the efficiency may be increased to 80% or more due to precipitation of iron phosphate. The application of this method seems attractive to recreational areas, where the density of population is low, but where the loading from the wastewater has a significant impact.

Constructed wetlands or root zone plants are generally designed for BOD-removal in accordance with the following equations (Jørgensen, 2000):

$$C/C_o = \exp(-K * V/Q), \tag{12.2}$$

$$Q_{w+p} = b * d * HC * i \tag{12.3}$$

where C and C_o are the BOD_5 concentration or COD concentration of the inflowing water and the treated water, respectively, while K is a rate constant (1/24 h), V is the volume (m^3) of the plant (depth 0.4–1.0 m), Q is the waste flow expressed in m^3/24 h. Q_{w+p} is the flow rate of wastewater plus rain often estimated to be 3–4 times Q, b is the width of the plant, d is the depth (0.5–1.0 m), HC is the hydraulic conductivity of the soil and i is the slope in cm/m. K is estimated to be 0.2–0.5 for root zone plants and constructed wetlands at approximately 20°C. K at temperature t can be found from the following equation:

$$K_t = K_{20} * 1.06(t - 20) \tag{12.4}$$

EXAMPLE 12.2

Design a root zone plant for the treatment of the wastewater from 1000 inhabitants (water consumption 0.2 m^3/24 h). The BOD_5 of the untreated water is 200 mg/L and it must be reduced to 20 mg/L by the planned treatment. 20°C is presumed. The plot to be applied for construction of the root zone plant has a slope of 5 cm/m and the soil has a hydraulic conductivity of 4 m/24h.

Solution

K = the average = 0.35 (range 0.2–0.5). It is presumed that 0.5 m depth can be applied. By use of Equation 12.2, we get:

$$20/200 = \exp(-0.35 * V/200)$$

This gives V = 1315. We choose an area of 2800 m^2 (0.5 m depth).
Equation 12.3 is used to control the hydraulic loading:

$$800 = 4 \cdot 200 = b \cdot 0.5 \cdot 4 \cdot 5$$

This leads to b = 80 m. The length is therefore 35 m, which seems to be a reasonable dimension.

The recommended management of natural or artificial wetlands and of the lake shore ecotones may be summarized in the following points:

1. Maintenance and preservation of the transition zone between the terrestrial ecosystems and the lake ecosystem should be considered mandatory. The ecotone between the two types of ecosystems functions as a buffer zone for preservation of species diversity in both ecosystems (Jørgensen, 2012).
2. The different types of wetlands forming the transition zone have a high adsorption capacity for many pollutants such as heavy metals and toxic organic compounds.
3. The different types of wetlands forming the transition zone are able to denitrify up to several tons of nitrate-nitrogen per ha and year.
4. Recovery and construction of artificial wetlands are crucial abatement methods for diffuse pollutants originating from agriculture.
5. Application of artificial wetlands, including waste stabilization ponds and root zone plants may be an attractive wastewater treatment method for developing countries, for recreational areas adjacent a lake ecosystems and generally for areas with a low population density.
6. The application of quantitatively based management of all the mentioned types of wetlands, including development of models for these ecosystems, is recommended.

The combination of WSPs and constructed or reclaimed wetlands are particularly attractive (Jørgensen et al., 2012), because:

- Wetlands offer by use of gravel media a significant reduction of suspended matter from maturation pond effluents.
- Wetlands buffer the pH value of the effluent from WSPs.
- Effluents from WSPs need most often a polishing step as a post treatment. Wetlands offer an excellent and cost-effective solution to this problem.
- The two steps, WSPs and wetlands, are to a great extent independent, and it therefore provides a much greater certainty to play on both steps.

- Constructed wetlands may also be used to treat dairy farm wastewaters, mine water pollutants, textile wastewater, and pulp mill wastewater.

12.7 Soil Remediation

Today, there is a wide range of available methods to recover contaminated soil. We distinguish between methods based on removal of the contaminated soil and *in situ* treatment methods. The first group of methods is usually more costly, as they entail relatively expensive transportation which in itself produces pollution. These methods are partially environmental technological methods, but they are included in this chapter as they may be considered partially ecological engineering methods. From an ecological point of view, *in situ* methods must be preferred, but they cannot be used in all possible situations

Removal and treatment of contaminated soil treatment facilities to handle contaminated soil have been constructed in many industrialized countries where the problem is most relevant. They are based on one of three principal treatment options: incineration, bioremediation by a composting like process or extraction.

12.7.1 Incineration or Thermal Treatment

Incineration or thermal treatment of the contaminated soil can be used to eliminate organic contaminants susceptible to destruction. This process is energy intensive and expensive, as the transportation costs add to the incineration costs. The method is not applicable to inorganic-contaminated soil for instance, soil contaminated by heavy metals.

12.7.2 Biological Degradation

Biological degradation of organic contaminants is a cheaper alternative to incineration. Dedicated bioreactors may be used or just open composting sites. The latter possibility is of course more cost-effective, but entails that the drainage water must be treated as it will inevitably contain the contaminants in an unacceptably high concentration. It is sometimes necessary to add water to the process to ensure rapid biological decomposition of the organic contaminants, particularly in dry seasons or arid climates. If the contaminant is difficult to decompose, because it has a very slow biodegradation rate (see Section 7.2 in Chapter 7), addition of adapted microorganisms may offer a solution.

12.7.3 Extraction

Extraction may also be used to reclaim soil, but the method has several technical and ecological disadvantages. The extraction liquid must be treated and recovered which is sometimes technically difficult to accomplish with high efficiency. If even a minor emission of the extraction liquid, takes place, unless water is used, it will cause an additional pollution problem to be solved. Only if the contaminant can be recovered or is soluble in water, will this method offer some perspectives.

12.7.4 *In Situ* Treatment of Contaminated Soil

These methods seem more attractive from both an economic and an ecological view point, because the elimination of the transportation step entails that the cost and the pollution associated with the transportation are eliminated too. In addition, *in situ* treatment has the advantage that the mixture of contaminant is known which makes it possible to tailor a treatment method to this specific mixture. If the contaminants are water soluble, it may be advantageous to dissolve them and thereby remove them from the soil. Addition of surfactants to the water may increase the solubility of the pollutants. Above the water table, extraction is normally conducted by creating a vacuum to remove vapor—the contaminants in volatilized form—from the unsaturated zone. This method has proven to be an especially useful process for the hazardous components of gasoline and other hydrocarbons. Heavy metal ions may be removed by use of a solution of EDTA or similar ligands which considerably enhance the solubility of heavy metal ions. As the solubility of heavy metal ions in EDTA-solutions is highly dependent on pH, the pH is adjusted to the optimum value (for most heavy metals around pH = 5.0) is applied.

The most desirable method of all treatment processes is *in situ* biodegradation to render the soil harmless. The rate of recovery with these processes should of course be sufficiently high to prevent the contaminants travelling far off-site, where they might pose a risk to humans or ecosystems, before they are decomposed. A promising possibility is to apply compressed air. Air is pumped down pipes to below the water table to accelerate the decomposition in the saturated and unsaturated zone. *In situ* biodegradation involves a conflict between biodegradation and mobility which obviously is dependent on the polluting compounds. The addition of nutrients or oxygen may provide the enhancement needed to achieve acceptable rates of degradation. Also the use of adapted microorganisms may accelerate the decomposition. The presence of easily degradable compounds can lead to cometabolism and the destruction of other refractory compounds. This is the case with high molecular weight PAHs in the presence of 2–3 ring polycyclic aromatics.

Recovery of soil by phytoremediation is a recently developed technique, which is clearly an ecotechnological method. Heavy metals may also be removed by application of electrolysis. This process is applicable entirely to high concentrations which in addition makes a recovery of the metals possible. A combination of electrolysis and plant bioaccumulators seems to be an attractive solution for soil even highly contaminated with heavy metals.

Recently, it has been shown that a few plant species (for instance the alpine pennycress) can be used as bioaccumulators (hyperaccumulators). When they are growing in contaminated soil, they are able to achieve a many times higher (10–100 times) heavy metal concentration than normal plant species. If the plants are harvested, the corresponding amount of heavy metal is removed from the soil. If the soil is only slightly contaminated with heavy metal, this method may be able to remediate the soil in a few years which makes the method very attractive due to its low costs. For very contaminated soil, the number of years needed for complete recovery of the soil will probably be too high. By watering the plants carefully with a diluted EDTA solution, it is possible to enhance the uptake of heavy metals in the plants. Jørgensen (2009) and Mitsch and Jørgensen (2003) reports a removal of 11.5% lead per harvest of plants watered by 0.02 M EDTA solution at an initial concentration of 380 mg Pb/kg soil (dry matter). Several plants species release chelating ligands and enzymes into the soil and thereby accelerate the removal rate. Chelating agents will also reduce the toxicity of heavy metal ions.

Hyperaccumulators are usually plants that are slow to grow and therefore slow to remove heavy metals as mg/m^2 year. Fast growing poplar trees are not such good hyperaccumulators as many other species, but due to their fast growth they are promising phytoremediators.

Aquatic plants, *Salvinia* and *Spriodela,* have been used to remove chromium and nickel from wastewater. A significant removal efficiency in the concentration range 1–8 ppm, although with fluctuations, has been reported (Mitsch and Jørgensen, 2003). Fungal biomass has recently been proposed for removal of heavy metals. The amounts of metal adsorbed per unit weight are several mg/g, for cadmium, copper, lead and nickel. Plants can efficiently take up organic substances that are moderately hydrophobic, with a log K_{ow} from about 0.5–3.0. More hydrophobic substances exceeding a log K_{ow} (the ratio of the solubility in octanole and water) of 3.0 approximately bind so strongly to the soil and the roots that they are not easily taken up within the plants.

EXERCISES AND PROBLEMS

1. A wetland has a hydraulic conductivity of 0.25 m/24h. It is presumed that the upper 1 m layer can be applied for denitrification of 2000 m^3 of agricultural drainage water which contains 12 mg/L nitrogen compounds. How much wetland would you recommend to use for a removal of 90% of the nitrogen, when it can be anticipated that the plant density is 20 t/ha? What is the expected BOD_5 removal?

2. Design a root zone plant for the treatment of the wastewater from 500 inhabitants (water consumption 200 L/24 h). The BOD_5 of the untreated water is 250 mg/L and it must be reduced to 10 mg/L by the planned treatment. 15°C is presumed. The plot to be applied for construction of the root zone plant has a slope of 6 cm/m and the soil has a hydraulic conductivity of 2.8 m/24 h. It has been recommended to use a depth of 0.8 m.
3. Which of the 7 ecological principles and the 17 rules given in the text are violated in conventional agriculture? In organic agriculture? In integrated agriculture?
4. How many ha waste stabilization ponds are needed for the reduction of COD from 700 to 100 mg/L in domestic wastewater mixed with industrial wastewater. The volume to be treated is 1000 m^3/24 h.

References

De Bernard, R. and G. Giussani (Eds.). 1995. Biomanipulation in lakes and reservoirs management. *ILEC. UNEP*, Otsu, Japan, 208 pp.

Jørgensen, S.E. 2000. *Principles of Pollution Abatement*. Amsterdam, Oxford: Elsevier, 520 pp.

Jørgensen, S.E. (Ed.). 2009. *Applications in Ecological Engineering*. New York: Elsevier, 380 pp.

Jørgensen, S.E. 2012. *Introduction to Systems Ecology*. (Chinese edition.) Boca Raton, FL: CRC Press, 320 pp.

Jørgensen, S.E. and R. de Bernardi. 1998. The use of structurally dynamic models to explain the success and failure of biomanipulation. *Hydrobiologia* 379, 147–158.

Jørgensen, S.E., J.G. Tundisi, and T.M. Tundisi. 2012. *Handbook of Inland Aquatic Ecosystem Management*. Boca Raton, FL: CRC Press, 422 pp.

Mitsch W.J. and L.G. Gosselink. 2007. *Wetlands*. 4th edition. New York: Van Nordstrand Reinhold, 580 pp.

Mitsch, W.J. and S.E. Jørgensen (Eds.). 1989. *Ecological Engineering. An Introduction to Ecotechnology*. New York: John Wiley & Sons, 430 pp.

Mitsch, W.J. and S.E. Jørgensen. 2003. *Ecological Engineering and Ecosystem Restoration*. New York: John Wiley & Sons, 412 pp.

Odum, H.T. 1962. Man in the ecosystem. In *Proceedings Lockwood Conference on the Sururban Forest and Ecology*. Bull. Conn. Agr. Station 652, Storrs, CT, pp. 57–75.

Odum, H.T. 1983. *Systems Ecology*. New York: Wiley Interscience, 520 pp.

Palmeri, L., A. Barausse, and S.E. Jørgensen. 2013. *Ecological Processes Handbook*, Boca Raton, FL: CRC Press, 386 pp.

Straskraba, M. 1985. *Simulation Models as Tools in Ecotechnology Systems. Analysis and Simulation*, Vol. II. Berlin: Academic Verlag, pp. 362–372.

Straskraba, M. 1993. Ecotechnology as a new means for environmental management. *Ecol. Eng.* 2, 311–332.

13

Application of Cleaner Production in Environmental Management

13.1 Introduction

All human activities cause an impact on the environment. Particularly agriculture, industry, and their products together with the consumption and following waste disposal have caused many pollution problems. We can, in principle, solve these problems by use of *environmental technology* and *ecotechnology*. This, however, obviously raises also the question: Could we not change the production, the production method, or even the products in such a way that we would be able to reduce the impact on the environment? Such a change in production, for instance, by recycling inside the production unit may even be much cheaper to realize than the treatment of the subsequent emissions. It may even result in a cost reduction as recycling means less use of "virgin" materials or resources. There are numerous examples of simultaneous reductions in costs and environmental impact due to changes in a production method. At the same time, it is interesting to note that it was earlier believed that it would only pay to implement cleaner technology and production on major production units or factories but experience has shown that even smaller- and medium-sized enterprise units may benefit from such considerations (Klewitz and Hansen, 2014).

The various technological approaches (and environmental legislation) can work hand in hand with the basic idea to find the optimum solution based on the diagnosis by playing on the entire spectrum of available methods. The improvement of production processes with the considerations on how to save the burden on the environment is today widely known under the name of *cleaner production*. In this context, it should be realized that concerns of the environment not only deals with the relations of a factory to its externals but also relates to its employees by considering the internal environment. In fact, improving production by reducing the load on the surrounding environment takes on various forms that may be organized in a hierarchical manner ranging from hard core technological considerations via the change of the production process and addressing design and innovation the so-called *industrial ecology* where several production units are (re-) organized or restructured for the benefits of the environment. Both the areas of cleaner production and industrial ecology today have their own dedicated journals and both serve the purpose of increasing the sustainability of necessary productions of our society (Almeida et al., 2015, Korhonen, 2004).

It is difficult to find the optimum solution to an environmental problem, because of the many possible solutions including those based on combinations of methods, which often in practice give the optimum solution. Not only technological but also social and organizational issues such as LEANing activities may be taken into consideration (Medeiros et al. 2014; Neto and Jabbour, 2010). This also means that a concerted action in the area is bound to be multidisciplinary. In many industries it has at least been close to the optimum solution to enhance recycling, reduce water consumption, and to introduce many minor changes in production which would together reduce emissions. So the improved environmental management of productions can be achieved by mainly four initiatives:

1. Introduction of an environmental management system (EMS) or better, an environmental management and audit scheme, EMAS (EU, 2009, 2013) requires rather detailed knowledge about the flows of a factory or production site which unfortunately is not always readily available. Thus, the first step would be to acquire this information and examine all the production steps from an environmental view point: Could we not reduce the emission here and there? How do we do it? This step often leads to environmental certification (e.g., EMAS) of the industry and for instance the use ecolabeling (Hale, 1996) which may be utilized in negotiation with contractors and customers. Many global and regional firms require that their subsuppliers have environmental certification in order to ensure that their final product is "green." The following ISO standards support the enterprises in their effort to improve their green image and reduce emissions by internal changes in production: ISO 14001–14004 and partly ISO 140014. This is the topic of Section 13.2.
2. The establishment of environmental auditing requires that the enterprise set up a balance for energy usage and mass of all components. The quantities of consumed raw materials, produced products, and material lost as pollutants to the environment at various points of production must be indicated. Such knowledge may have been gained through establishment of Life Cycle Inventories, so called LCIs (e.g., Weidema and Wesnæs, 1996) or full Life Cycle Assessment, LCA of products (Reap et al. 2008a, b; Azapagic, 1999). Such studies investigate the environmental impacts from cradle to grave or even from cradle to cradle (Braungart and McDonough, 2009). The latter approach puts attention and focus on the possibilities of a change in design of products and maybe a total redesign with the purpose of

closing cycles further by reusing and in this context touches upon the ISO 9000 standards. The approaches typically use what is known as a *functional unit* such the production of a certain mass of product or waste as a base for evaluation of the environmental load of a specific production. The ISO standards 14010–14013 provide guidance for environmental auditing. This is the topic of Section 13.3.

3. The implementation of cleaner technology and production covers the actual changes in the production frequently resulting from points 1 and 2. It is an attempt to give a full answer to the pertinent question: How can we produce our products by a more environmental-friendly method? Which changes in the production should be made? The ISO standards mentioned under 1 and 2 are supplied in ISO 14014, 14015, and 14031. Section 13.4 is devoted to cleaner technology.
4. Not only the production but also the product may be changed for instance by changing design and change of materials. Environmental risk assessment which was introduced in Chapter 7, may be used to change a product with a high environmental risk to a product with a low environmental risk. A life cycle assessment, LCA, examines all the emissions associated with the use of the product from the cradle to the grave. The idea is to examine the possibilities for reductions in emissions by minor or major changes in the product, including possibilities of recycling the final discarded waste product. This topic will be discussed further in Section 13.5.

13.2 Environmental Management Systems

The regulation, control, and requirements of documentation of the impact of industrial activities on health and the environment has undergone a significant transition since the late-1970s and early-1980s in the European Union, North America, and Japan. Since the mid-1980s, industries themselves have taken a more proactive stance in recognizing that sound environmental management on a voluntary basis can enhance their corporate image, increase profits, reduce costs, and obviate the need for further environmental legislation in the area. The commercial values of green products have been fully acknowledged. In some countries, governments have even set up what may be named a "green purchase policy" which is to be followed not only by the state but also by counties and municipalities, fiscal authorities, and governmental institutions. It is therefore not surprising that there has been an enormous interest among many enterprises. Introduction of an environmental management system can help companies to approach environmental issues systematically in order to integrate environmental concern as a normal part of their operation and business strategy. The major triggers for the companies to initiate such environmental changes are rooted in legislation, in pressure from the stakeholders, in the importance of image and reputation, in the competition (environmental certification is required for sub-suppliers and the commercial value of a green product) and in economies (sales potential, recycling means lower consumption of resources, green taxes).

ISO 14001 uses the following definition of the term environmental management system, EMS: the part of the overall management system which includes organizational structure, planning activities, responsibility, practices, procedures, processes, and resources for developing, implementing, achieving, reviewing, and maintaining the environmental policy.

EMS is intended to

- Identify and control the environmental aspects, impacts, and risks relevant to the organization.
- Develop and achieve its environmental policy, objectives, and targets, including compliance with environmental legislation.
- Establish short-, medium-, and long-term goals for environmental performance, ensuring a balance of costs and benefits, for the organization and its stakeholders.
- Determine what resources are needed to achieve these goals, assign responsibility for them, and commit the necessary resources.
- Define and document specific tasks, responsibilities, authorities, and procedures to ensure that every employee in the course of their daily work helps to achieve the defined goals.
- Communicate these throughout the organization and train people to effectively fulfill their responsibilities.
- Measure performance against pre-agreed standards and goals, and modify the approach if necessary.

EMS has benefited mainly from two developments:

a. The rising cost of environmental liabilities, which has led companies to develop environmental auditing as a management tool.
b. Total quality management concepts, which were adopted to reduce defects = noncompliance with specifications.

The development of EMS may be described in several phases (Dyndgaard and Kryger, 1999):

1. Set up an *environmental policy*, expressing the commitment of the enterprise to specific environmental goals. These are most clear if they are quantified for instance to reduce the emission of a specific

pollutant by 90%. This point also includes decisions on disclosure of production and results of LCAs, and corporate social responsibility (CSR) issues and initiatives.

2. An environmental *program of action plan* describing the measures to be taken over a given period of time. The action plan should identify the steps to be taken by each department to reduce emissions, commit the necessary resources, and provide monitoring and coordination of progress toward the achievement of the defined goals.
3. *Organizational structures* delegating authority and assigning responsibility for actions. In many cases a whole new structure has to be built within the organization.
4. *Integration* of environmental management into business operations and internal communication including risk assessment to identify potential accidents.
5. *Monitoring* and measurements and record-keeping procedures to document and monitor the results as well as the overall effects of the environmental improvements.
6. *Corrective actions* to eliminate potential nonconformance to objectives, targets, and specifications.
7. EMS *audits* to check the adequacy of efficiency of the implementation.
8. *Management reviews* to assess the status and adequacy of the EMS in light of changing circumstances.
9. *Internal information and training* to ensure that all employees understand that they need to fulfill their environmental responsibility, and why they should.
10. *External communications* and community relations to communicate the enterprise's environmental goals and performance to interested persons outside the enterprise to obtain the advantage of a green image.

All 10 steps are covered in detail by ISO 14001 and are consistent with the quality management in ISO 9000 which discusses four phases: the planning phase, action phase, evaluation phase, and correction phase. Performance measures are needed in EMS to provide indication of how well the goals have been reached and to serve as an incentive for the employers as their efforts are recognized and appreciated through the results. The reward to the enterprise can be substantial because even a limited EMS often demonstrates that the economic gains can outweigh the costs and that inexpensive measures may yield important environmental and financial returns.

The effect of EMS introduced into different enterprises including several industries with high pollution potential (manufacturers of dyestuff, cables, enzymes, and printed matter) have been examined. The results of the examination are summarized in the following list of advantages including indirect advantages gained by the introduction of EMS:

1. A direct measurable reduction of emissions with all the benefits that may entail.
2. A general increased engagement by all employees not only with EMS but with all the enterprise's areas of activities.
3. Improved (green) image.
4. The economic gains outweigh all environmental investments and associated costs from within a few months to a maximum 3.8 years.
5. In several cases, new orders have been obtained directly due to the introduction of EMS.
6. Internal reductions in noise, dust, and odors were obtained.
7. Optimization of processes and better utilization of resources.
8. Better cooperation with environmental agencies.

13.3 Environmental Audit, LCI, and LCA

In order to implement an EMS or EMAS it is necessary to create an overview of the inputs and outputs of the production system. This can be done by establishing a *green accounting system* or *environmental audit* covering a mass and energy balance of the raw materials and energy used in a production unit. Chapter 3 presented the concept for ecosystems. Through a quantification of the mass and energy flows in a production unit (e.g., a factory), often presented in the form of one or more flow charts, it becomes possible to identify how much materials and energy are lost and where this takes place. Hence, the established balances are at the crux of LCIs and LCAs. These results enable us to answer the following pertinent questions:

- Could the losses (=emissions) be reduced? Which are quantitatively, qualitatively significant? What may be achieved immediately by implementation of already existing technology?
- Could the loss of energy eventually be recycled?
- How would that influence the mass and energy balances and the quality of the product?
- How could the mass and energy flows be optimized with relation to the total costs, including energy, raw materials, environmental, labor, and miscellaneous expenditures?

Mass and energy balances are in this context called *green accountings* or *environmental audits*, as they are comparable with accounts of income (useful outputs), expenses (inputs), profit (value of useful outputs—inputs), loss (nonuseful outputs). An account is settled by looking at the money flow in the same manner, using the principles of conservation: the input–output corresponds to the accumulation in the firm = an increased asset. This does not entail that the environmental audit is made by an economic auditor which has been the tendency. Mass balances, particularly in the chemical industry, are often based on stoichiometric calculations which require a good knowledge of chemistry and the involved chemical processes. Similarly, energy balances can hardly be set up without the use of thermodynamics and studies of energy transfer processes. The environmental issues are, however, in most environmental audits covered at least partially, while 8%–10% of the audits give very incomplete information about the environmental impact. This examination has also uncovered the economy behind the use of green audit. In several cases, the enterprises show gains from the results of an environmental audit.

It is often surprising how little enterprises know about their losses (emissions to the environment) before they have introduced a green audit. Most enterprises are focusing almost entirely on the two P's: product (quality and price) and profit. By introduction of a green audit many enterprises have, however, acknowledged they can increase profit by reduction of the losses (emissions) and decrease the price and increase the quality of the product from a commercial point of view by developing a green image. A few examples are presented below to illustrate how green audit has been able to elucidate the value of waste.

A well-managed Danish farm bought waste proteins used as animal feed from a fermentation industry. The manure was utilized as natural fertilizer. The composition of the waste proteins was, however, not at the optimum for animal feed as the potassium content was too high. The manure therefore also had too high a potassium content which meant that the fertilizer effect of the potassium content was not fully utilized, but was emitted to the environment. By an adjustment of the composition of the animal feed obtained mainly by use of a mixture of soya beans, barley, and the waste proteins, it was possible to get an optimum composition which brought the manure closer to a fertilization optimum. The result of the changes was that the farmer reduced the emissions to the environment caused by the manure and could simultaneously increase his profit.

Fibrous material is lost to the wastewater in paper and pulp production. During the 1970s, it was realized on the basis of mass balance that a significant amount of fibrous material was lost to the environment. It is, however, possible using chemical precipitation with bentonite and a polyflocculant as precipitant to collect the fibers as a sludge (5%–10% dry matter) which can be used to produce lower quality paper. Mass balances also revealed 25–40 years ago that the amount of fish oil lost by fish filleting plants and the amount of chromium lost by tanneries were very significant, particularly from an economic point of view. Today, a fish filleting plant which is not recovering fish oil or a tannery which is not recycling chromium cannot survive the tight competition.

13.4 Cleaner Production and Technology

Cleaner production is in accordance with UNEP defined as "the continuous application of an integrated preventative environmental strategy to processes, products, and services so as to increase efficiency and reduce the risks to humans and the environment." The main emphasis is on all aspects of the manufacturing processes and products but nowadays the attention has been drawn to the importance of resource efficiency. Cleaner production therefore includes the efficient use of raw materials and energy, the elimination of toxic and dangerous materials, and the reduction of emissions and wastes at the source. In addition, it focuses on reduction of the impacts along the entire life cycle of the products and services—from design to use and to ultimate disposal (for further details; see Section 13.5). Table 13.1 shows

TABLE 13.1

Comparison of Pollution Control Approaches and Cleaner Production Approaches

Pollution Control Approaches	Cleaner Production Approaches
Pollutants are controlled with filters and through integrated measures	Pollutants are prevented at their sources, wastewater treatment methods
Pollution control is evaluated when processes and products have been developed and when the problems arise	Pollution prevention is an integrated part of product and process development
Pollution control is considered to be a cost factor	Pollutants and waste are considered to be potential resources
Environmental challenges are to be addressed by environmental experts	Environmental challenges are the responsibility of people throughout the company
Environmental improvements are to be accomplished with technology	Environmental improvements include nontechnical and technical approaches
Environmental improvements should fulfill authorized standards	Environmental improvement involves a process of working continuously to achieve higher standards
Quality is defined as meeting the customer's requirements	Quality means the products that meet the customer's needs and have a minimal impact upon human health and the environment

a comparison of pollution control based mainly on environmental technology and cleaner production based on all the approaches presented in this chapter: EMS, environmental audit, changes of processes, and technologies applied in production and LCA.

Cleaner production is especially important for developing countries (it also explains UNEP interest in this topic), because it provides industries in these countries for the first time with an opportunity to bypass the more established industries which are still stocked with the costly environmental technology to solve the problems at the end of the pipeline. Table 13.2 is mainly based on Kryger and Dyndgaard (1999), but partially on the examples given earlier in the context of the environmental audit and UNEP (1994), which gives several examples of how sometimes relatively simple changes or introduction of low cost equipment can provide cleaner production, with reduced emissions and lower costs.

Kryger and Dyndgaard (1999) attempted to review the barriers to the introduction of cleaner production. The reviews are in short: bureaucratic resistance, conservatism, uncoordinated legislation, misinformation, scarce money, and lack of technical reliable information.

Introduction of cleaner production often follows the procedure presented in Figure 13.1. An option generating process is often used extensively from the planning stage to the implementation. It is often beneficial to consider the following elements in this process (Kryger and Dyndgaard, 1999):

TABLE 13.2

Examples of Cleaner Production

Industry	Country	Method Reduction	Waste and Emission	Payback Period
Fish filleting	Denmark	Removal of fish oil from wastewater	75% reduction of BOD_5	3 months
Tannery	Denmark	Recycling of chromium	95% reduction of chromium in wastewater	<1 year
Paper	Switzerland	Precipitation of fibers	70% reduction of BOD_5	2–3 years
Fruit juice	The Philippines	Collection of fruit drops	55 L organic waste/h	9 months
Wood	Malaysia	Waste segregation finishing	54,000 kg of hazardous waste + 5700 m^3 w.w.	3 months
Lead oxide	India	New insulation in furnace	Fuel consumption reduced 50%, increased production 3%	3 months
Cement	Indonesia	Control system to ensure optimum temperature	3% reduced energy use, 9% higher production, reduced emissions of NO_x	<1 year

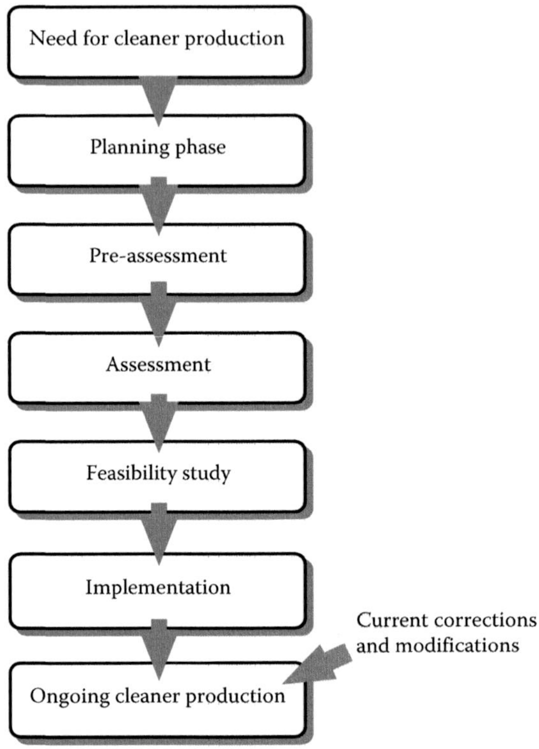

FIGURE 13.1
Introduction of cleaner production by a 7-step procedure.

ILLUSTRATION 13.1

Compare a solution of an environmental problem based on (1) environmental technology, (2) ecological engineering, (3) cleaner technology and production considering: (a) the investment, (b) the running costs, (c) the environmental aspects (solution of the pollution problem), and (d) the ecological aspects (impacts on ecosystems).

Solution

	1	2	3
(a)	High	Medium	Medium
(b)	High	Low	Low to medium
(c)	Good	Good but may f(time)	Good
(d)	Good, depends on the ecological aspects	Usually good ecological aspects are incl.	Good, depends on if they are included in the solution

The focus for (1) and (3) are often only on the pollution problems and not on the ecological consequences, which explains the answers in the table.

Environmental technology is usually more expensive to apply as well the investment costs and the running costs. Because nature has a high variability, the use of ecotechnology will inevitably have a high variation in the efficiency as a function of time.

1. *Change in raw materials* either by changing to another (more pure) grade of the same raw material, eventually by purification of the raw material or by substituting the present raw material with a less hazardous material. Example: Detergents containing polyphosphates are replaced with detergents based on zeolites.
2. *Technological changes* covering changes in the production processes, modification of equipment, use of automation, and changes of process conditions (flow rates, temperatures, pressures, residence times, and so on). The aim with these changes is to reduce the emissions and/or improve the quality of the product. Example: Use of mechanical cleaning to replace cleaning by chemicals.
3. *Good operating practices* can often be implemented with a little cost. They include management and personnel practices, material handling and inventory practices, training of employees, loss prevention, waste segregation, cost accounting practices, and production scheduling. Example: Reduce loss due to leaks.
4. *Product changes* include changes in quality standards, product composition, and product durability or product substitution. Example: Replace CFCs with ammonia in refrigerators.
5. *Reuse and recycling* involves the return of a waste material to the originating or another process to substitute input material. Examples: Recover nickel plating solution with an ion exchanger, recover dyestuffs from wastewater in the textile industry by ultrafiltration.

These five possibilities for cleaner production are summarized in Figure 13.2.

The range of tools to catalyze the industry to adopt cleaner production is large (Kryger and Dyndgaard, 1999). Governmental agencies may use the following instruments:

a. Regulations, standards, and legislation
b. Green taxes
c. Economic support
d. Provide external assistance

The two principles to solve pollution problems. (A) to reduce emissions by cleaner production and (B) reuse and recycling, work hand in hand and can hardly be separated. (B) is however not always a consequence of cleaner

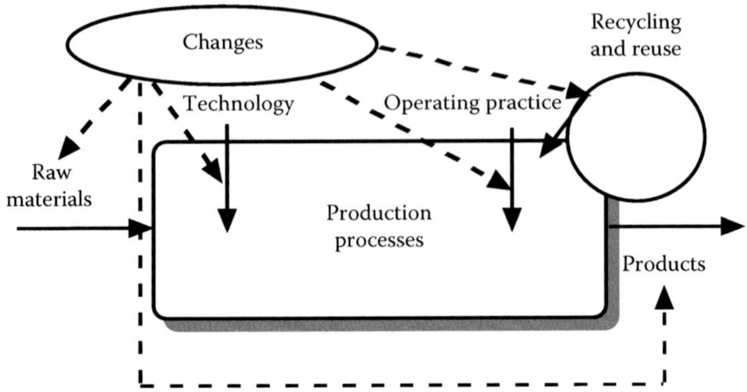

FIGURE 13.2
Cleaner production can be introduced by changes (dotted arrows) in raw materials, technology, operating practices, products, and reuse or recycling.

technology but may be an obvious solution after environmental technology has been introduced.

On the other hand, cleaner technology/production may apply environmental technology as a tool to achieve the goals set up in a cleaner production scheme. This very clearly illustrates that integrated environmental management is needed to achieve an optimum solution to a pollution problem. Up-to-date integrated environmental management requires that the entire spectrum of considerations and tools presented in this volume is applied.

13.5 Life Cycle Inventory and Assessment

Life cycle assessment (LCA) is a methodology for assessing the environmental impacts and resource consumption associated with a product throughout its entire life—from cradle to grave, or from raw material to production to product to use of product to disposal. It is a holistic approach, using system analysis. Several industrialized countries are discussing or have already implemented LCA. The following initiatives have enhanced the interest of industries for the introduction of LCA:

1. Eco-labeling or environmental declarations about products based on LCA.
2. Green purchase guidance taking environmental considerations into account.
3. Take back responsibility for certain products, for example, cars and electronics.

LCA consists of four steps (Hautschild and Wenzel, 1999):

1. *Definition of the goals of the assessment* and the scope of the study. The goals are often formulated as questions:

   Which product on the market causes the least impact on the environment?

   Which are the most important environmental impacts of the products during its life cycle? What would happen if the product was not produced at all?

   How does the impact of a specific product during its life cycle compare with alternative products (comparative studies which should be based on functional units)?

   What influence would a specific proposed change in the product or production processes have on the results of the LCA?

   Which properties of the product cause the undesirable impacts? Can we change them? How?
2. *Preparation of an inventory of inputs and outputs* for the processes that occur during the life cycle of the product. Traditionally, emissions are divided into air pollutants, wastewater, and solid waste. The emissions are expressed relative to the function of the product in order to make a proper comparison.
3. *Impact assessment* where the results under point 2 are translated into environmental impacts. We distinguish the impact on human health, ecosystem health, and the resource base. Various techniques are used to compare two or more emissions with the same or different types of effects. When the emissions have the same type of effects normalization may be applied, while different effects require the introduction of weighting factors. The selection of weighting factors is at least to certain extent a political question, as it is impossible by objective criteria to compare effects on human health and on ecosystems or the depletion of two different resources. Table 13.3 shows an example of the proposed weighting factors.
4. *Interpretation of the impact assessment according to the defined goals and scope of the study* including sensitivity analysis of key elements of the assessment. This phase attempts to answer the questions raised in the first phase. The outcome may be a series of recommendations to the decision maker, but may also lead to revision of the scope of the study. The result of the sensitivity analysis is used to identify which product properties and processes are most significant for the total result. The significance of the uncertainty of the key figures is examined by letting them vary within their estimated range and investigating how this affects the total result of the LCA and the conclusions that can be drawn.

TABLE 13.3

Weighting Factors Proposed

Impact	Weighting Factor
Global warming	1.3
Ozone depletion	23
Photochemical smog formation	1.2
Acidification	1.3
Nutrient enrichment	1.2
Persistent toxicants	2.5
Human toxicity	2.8
Ecotoxicity	2.3

Source: Wenzel, H. et al. 1997. *Environmental Assessment of Products. Volume 1. Methodology. Tools and Case Studies in Product Development*. London: Chapman & Hall, 378 pp.

The following stages of the life cycle are considered in a complete LCA:

a. Extraction of raw materials
b. Transportation of raw materials to the production site
c. Manufacture of the product
d. Transportation from production to the user (this may consist of several stages)
e. Use of product
f. Disposal

In all these stages, emissions may occur and the following treatment processes may be considered as the most obvious alternatives: reuse (after a suitable treatment), recycling, incineration, composting, wastewater treatment, and dumping. The result of the LCA is changed in accordance with any decision on treatment of the waste.

13.6 Industrial Ecology by Network Cooperation

As stated above, cleaner technology and production can be taken to a higher level by integrating several companies in the continuous effort to increase efficiency (Kunz et al. 2013; Tschiggerl and Wolf, 2012; van Berkel, 2007). Network cooperation offers a series of benefits, and it can be strongly recommended to adopt it in society on many levels and in many contexts and may be seen as an extended form of industrial ecology (Korhonen, 2004). Ecosystems utilize the synergistic effect of network cooperation, as depicted in Chapters 3 and 5. The establishment of networks entails a better efficiency in using resources. A few possibilities mentioned below show how these benefits clearly can be applied in society to reduce the pollution and the environmental problems.

Industrial ecological symbiosis can be achieved by the deliberate formation of a network similar to the ecological network of ecosystems, that is, shaping a network consisting of flows of energy, matter, and information from structures and production units which already exist. This idea has been tested in the Danish town, Kalundborg, see also Jørgensen (2006). The network is shown in Figure 13.3. The heart of the system is the Asnaes Power Station (APS). It is the largest power plant in Denmark. One-half of the plant is fueled by coal and the other half by a new fuel called orimulsion, a bituminous product produced from Venezuelan tar sand. The use of orimulsion was introduced in 1998 and gave an 18% reduction in carbon dioxide emissions compared with the use of 100% coal, but gave, due to the sulfur content of the Venezuelan tar, a higher production of gypsum by the sulfur dioxide scrubber.

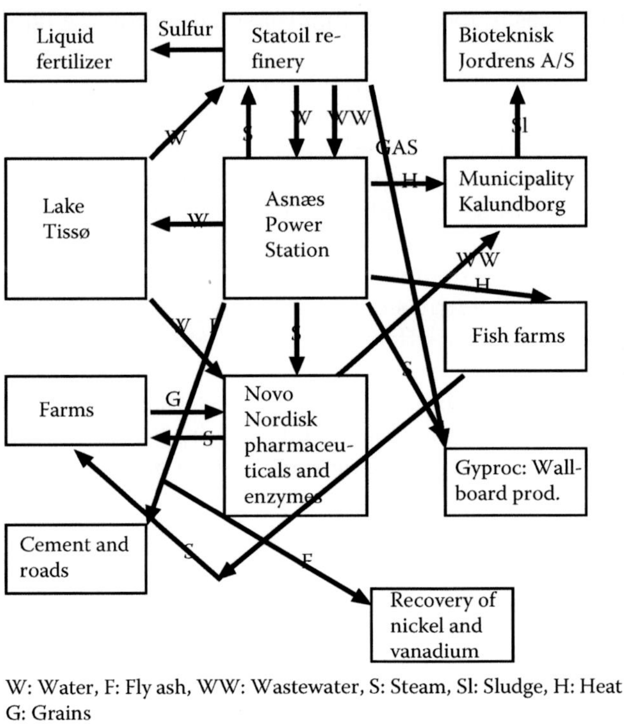

FIGURE 13.3
The industrial symbiosis at Kalundborg, Denmark. The flows of energy and matter make up an "ecological" network that implies that the overall efficiency of the utilization of the exergy input to the network is high.

The efficiency in the use of work energy (chemical work energy in the fossil fuel used) has increased to almost 80% by these uses of waste work energy. APS has reduced the fraction of available energy directly discharged by about 80%. Since 1981, the municipality of Kalundborg has eliminated the use of 3500 oil-fired residential furnaces by distributing heat from the power plant through a network of well-insulated underground pipes. The home owners have paid for the pipeline but now receive cheap and reliable heat supply in return. The power plant also supplies cooling water that is warmed 7–8°C to an on-site fish farm producing about 200 t of trout per year. In addition, APS also delivers process steam to its neighbor Novo Nordisk (a pharmaceutical and biotechnological industry) and Statoil (an oil refinery). APS produces 70,000 tons of ash that is sold for road building and cement production.

In 1993, APS installed a sulfur dioxide scrubber that produces calcium sulfate or industrial gypsum. Gypsum is the primary ingredient of wallboard and ASP is the primary supplier of gypsum to the wallboard factory Gyproc. The refinery has refinery gas as waste product. This is sold after desulfurization as natural gas to be used in the town or elsewhere. The desulfurization process produces liquid sulfur that is used to produce a liquid fertilizer, ammonium thio-sulfate.

Novo Nordisk is a world leader in the production of insulin and enzymes. The plant employs 1000 people. The products are produced by fermentation processes. Based upon agricultural crops, valuable products are produced by microorganisms. The waste is a nutrient-rich sludge that can be used by the nearby farmers as animal feed (the yeast sludge) or fertilizers (the microorganisms sludge). More than 3000 t of the two types of sludge are produced daily. The yeast sludge is sold and the microorganisms sludge is given away due to the firm's concern for disposal security. Distribution of the sludge was the cost-effective way to comply with environmental regulations.

A new partner, A/S Bioteknisk Jordrens (means Biotechnical Soil Purification), joined the industrial symbiosis in 1999. The company uses municipal sludge as a nutrient in a bioremediation process to decompose the pollution of toxic substances in contaminated soils.

It is clear from this review of the industrial ecological network in Kalundborg that there is an enormous work energy-saving potential designing an integrated network that promotes by-product synergies. In this context, it could also be mentioned the Nordiske Affaldsbørs (Nordic Waste Exchange) maintains an Internet page with information about industrial waste produced in the Nordic countries: Finland, Island, Norway, Sweden, and Denmark. The idea is that what may be waste for one industry could be used as a raw material for another industry. For instance, used solvents in the pharmaceutical industry may be applicable for production of paint or dyestuff. It is very important to see such opportunities if we want to reduce the loss of work energy or of raw material waste to the environment.

13.7 Integrated Agriculture and Organic Farming

The idea behind industrial symbiosis—an ecological-industrial network—has also been applied in agriculture. The calculations of work energy for various agricultural systems have revealed that the more an agricultural system imitates nature—high diversity, many interactive components, many feedback loops and recycling of matter and thereby also of energy—the higher work energy capacity the system has. Biological diversity in agriculture provides buffer capacity and resilience, while monocultures increase vulnerability.

The structure of modern industrial agriculture has been heavily discussed during the last decades. Organic farming focuses on maintaining healthy soils such that plants can flourish and be productive without the need for synthetic subsidies (pesticides, artificial fertilizers, etc.). It solves most of the pollution problems associated with industrial agriculture; and organic farming in practice often means that a wider spectrum of agricultural products are produced at the same farm with less resources, but not necessarily more profit.

Industrial agriculture is in crisis. It has imitated the structure of industries: Production of one or at the most a few products with as little labor as possible because labor is an expensive resource in the industrialized countries, which has lead to a very energy intensive form of agriculture and farms units becoming bigger and bigger; but while the industrial products have been sold in an increasing amount, the industrialized countries have had the need for food of higher quality but no need for more food due to a very low population growth. The result of the industrialization of agriculture has been an overproduction of agricultural products in the industrialized countries. Modern industrial agriculture is furthermore heavily subsidized because it has not been possible in spite of the industrialization of the agricultural production to be able to offer farmers the same increase in the life standard in the form of higher revenues as the rest of the population has got due to the industrialization. The farmers have tried to produce more and more to increase their income, but as the size of the market for agricultural products is almost unchanged in the industrialized countries, this strategy is bound to fail and only given a higher and higher overproduction. The result is that an average farm in Europe or North America today receives annually approximately $35,000 in economic support. When it is considered that agriculture at the same time is a massive polluter, mainly by nonpoint pollution, it is clear that structural changes are urgently needed in North America and Europe to stop this negative spiral for the industrialized agriculture.

The implementation of organic agriculture using network synergism has been demonstrated at die Hermannsdorfer Landwerkstätten (abbreviated below as LHL) in Glonn close to München, Germany, and the results are presented below. The practical application of the principles is of major interest in this context.

LHL was founded in 1986. It is a farm based on sustainability principles. The characteristics of the farm and calculation of the total energy and work energy of this agricultural system are presented below. Figure 13.4 shows a conceptual diagram of the material and energy flows of this integrated agriculture farm.

The characteristics of LHL can be summarized in the following points for the application of cleaner technology approach (Schweisfurth et al., 2002).

The principles of organic farming are applied and a principle of more gentle soil handling has been introduced.

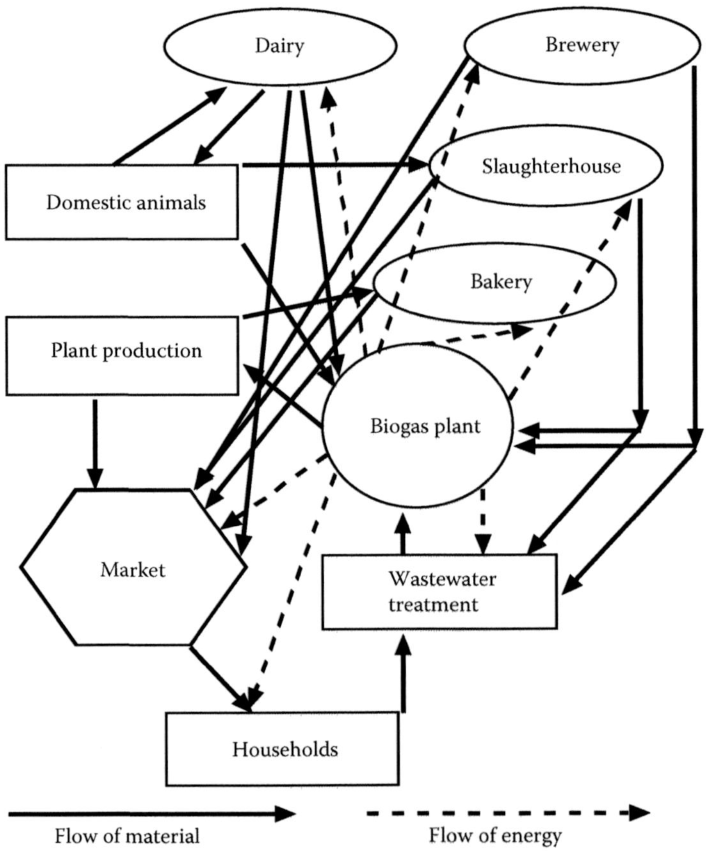

FIGURE 13.4
The material flows (full line arrows) and energy flows (dashed lines arrows) in the LHL. The figure illustrates that LHL forms a symbiotic ecological network, which facilitates the possibilities to obtain high efficiencies for the use of material and direct work energy, that is, it yields a high overall work energy efficiency.

A. All distribution is locally ensuring that waste of energy for transportation and conservation is avoided and all products are tasteful and fresh, when delivered to the customers.
B. The local network of skilled workers, customers, and others is utilized. Modern technology (provided nonlocally) is utilized but only if it has clear advantages. The various production units are linked in a symbiotic ecological networks, which facilitate the possibilities to obtain a high efficiency of material and work energy use; see Figure 4.6 in Chapter 4. The selection of products and their quality is a result of a dialog between LHL and the consumers.
C. A wide spectrum of animal and vegetable products is produced.
D. LHL encompasses a slaughterhouse, a bakery, a dairy, a brewery, and a restaurant.
E. *LHL works with nature not against nature* because it uses nitrogen fixation, hedgerows, and a natural mosaic of the landscape.
F. The energy supply is based on biogas, wind energy, and solar energy. The water supply is to a high extent covered by collecting rainwater and the wastewater is treated by constructed wetlands and reused to a high extent.

Ecological management of farms as illustrated by LHL requires a more comprehensive and complicated control. For example, it is possible to reduce fertilizer use by developing a fertilizer model that considers all the possible forms of nitrogen and phosphorus, how and with which rate the forms are transformed to each other. Figure 13.5 shows such a model for nitrogen.

If the available ammonium and nitrate is sufficient to obtain a close to optimum growth, then there is no reason to fertilize the soil further. Currently, new ammonium and nitrate are formed by the transformation processes, and if these rates are known it is of course possible to calculate when and whether the soil needs more ammonium and nitrate and how long a time there will be a sufficient concentration of ammonium and nitrate for the applied crop.

It is also possible to develop a better weed control by considering:

1. The normal emergence of different types of weeds that are known for the considered fields.
2. Expected crop yield losses due to the weed density based upon recent and anticipated weather conditions.
3. The development of different weed types that are followed by a rough quantification of the amount of weeds of different types.

The LHL agricultural system has been compared with a normal West European heavily subsidized agricultural system (abbreviated below LDK). The work energy of the inputs and the products have been found by the

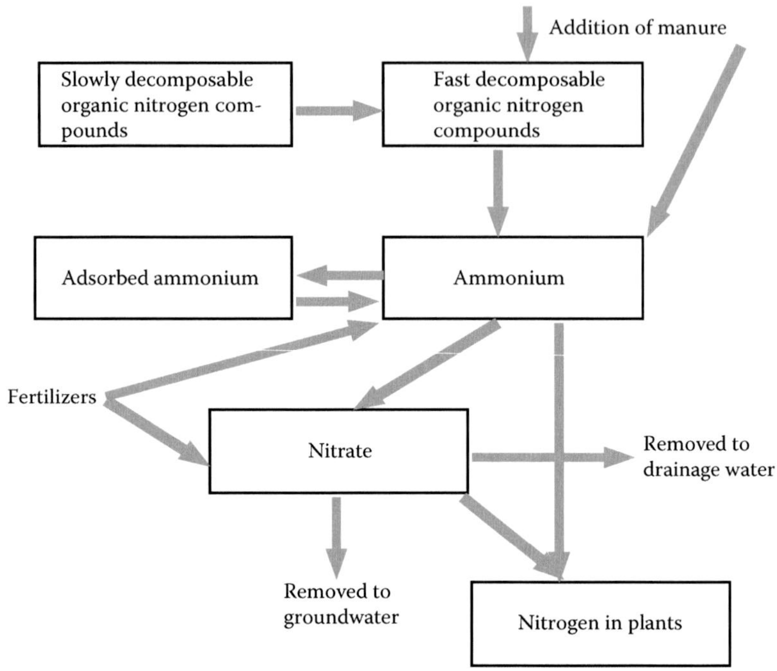

FIGURE 13.5
The model shows the nitrogen cycle in agricultural soil. It is possible to optimize the addition of manure and/or fertilizers by the model. It is not necessary to add more manure and/or fertilizer than corresponding to the amount of ammonium and nitrate which is sufficient for the uptake by plants. Nitrogen shall of course not be limiting, but there is no reason that the nitrogen is more abundant than needed. That would only increase the loss of nitrogen to the environment (drainage water and groundwater).

usually applied equations for work energy capacity. The work energy efficiency has been calculated as = work energy of products/work energy of inputs (not considering the inputs from solar energy). The results are shown in Table 13.4. The table also contains information about the percentage of animal production because the production of animal products is not surprisingly more work energy-expensive than plant products. LDK has a slightly higher animal production than LHL, but far from sufficient to explain the enormous difference in work energy efficiencies.

TABLE 13.4
The Work Energy Efficiency and Percentage of Animal Products of an Organic, Integrated Farm System (LHL) as Compared to a Conventional Farm System (LDK)

Factor	LHL	LDK
Work energy efficiency	2.64	0.3
% Animal products	29.2	33.6

The difference in work energy efficiency between LHL and LDK (almost a factor 9!) is explained by the use of artificial fertilizers, pesticides, import of animal food, use of groundwater, and a significant transportation of the products for LDK. About 46% of the total work energy consumption by LDK is applied for transportation, while the need for transportation by LHL only is 7%. Nearly 16% of the resources are applied for transportation by LDK, while only 6% by LHL. So, the difference in the need for transportation between LDK and LHL explains partly the surprisingly high difference in work energy efficiencies between the two types of agriculture.

It can be concluded on the basis of this analysis that introducing system analysis and ecosystem considerations in agriculture can lead to a more sustainable agriculture, and what is maybe more important: it *is probably* possible to develop a sustainable agriculture, if we use network considerations.

A possible explanation of the present economic crisis could be a decoupling between parts of the financial sector from their system—it means the society that the financial sector should serve. The financial sector thought that it could profit from a decoupling, but the profit was only short term and not sustainable, as the outcome of the crisis has shown. If the financial sector would have worked fully and unconditional with the society, it could have benefitted from the higher utilization efficiency that characterizes systems compared with single components or processes. The society—the system—would of course also have benefited from open and unconditional cooperation with the financial sector. It would have been a win–win situation.

QUESTIONS AND PROBLEMS

1. Explain how an LCA can reduce the waste originating from disposal of a used car.
2. Give examples on the use of cleaner technology in dairies and abattoirs.
3. A specific resource A is used at a rate of 1000 t/y, increasing by 2% per year. The known amount of resource A is 25,000 t, increasing with new discoveries linearly by 1000 t/y. Calculate the lifetime of the resource under these circumstances. How much longer would the lifetime of resource A be by introduction of A. 50% and B. 75% recycling?

References

Almeida, C.M.V.B., F. Agostinho, B.F. Gianetti, and D. Huising. 2015. Integrating cleaner production into sustainability strategies: An introduction to this special volume. *J. Clean. Prod.* 96, 1–9.

Azapagic, A. 1999. Life cycle assessment and its application to process selection, design and optimisation. *Chem. Eng. J.* 73, 1–21.

Braungart, M. and W. McDonough. 2009. *Cradle to Cradle. Remaking the Way We Make Things*. New York/London: Vintage, 208 pp.

Dyndgaard, R. and J. Kryger.1999. Environmental management system. In Jørgensen S.E. (Ed.), *A Systems Approach to the Environmental Analysis of Pollution Minimization*. Boca Raton, FL: Lewis Publishing, pp. 67–87.

EU, 2009. Regulation (EC) No 1221/2009 of the European Parliament and of the Council of 25 November 2009 on the voluntary participation in a Community Eco-Management and Audit Scheme (EMAS).

EU, 2013. Commission decision of 4 March 2013 establishing the user's guide setting out the steps needed to participate in EMAS, under Regulation (EC) No. 1221/2009 of the European Parliament and of the Council on the voluntary participation by organisations in a Community Eco-Management and Audit Scheme (EMAS).

Hale, M. 1996. Ecolabelling and cleaner production: Principles, problems, education and training in relation to the adoption of environmentally sound production processes. *J. Clean. Prod.* 4(2), 85–95.

Hauschild, M. and H. Wenzel. 1999. LCA. In Jørgensen S.E. (Ed.), *A Systems Approach to the Environmental Analysis of Pollution Minimization*. Boca Raton, FL: Lewis Publishing, pp. 155–191.

Jørgensen, S.E. Eco-exergy as sustainability. 2006. Southampton: WIT, 220 pp.

Klewitz, J. and E.G. Hansen.2014. Sustainability-oriented innovation of SMEs: A systematic review. *J. Clean. Prod.* 65, 57–75.

Korhonen, J. 2004. Industrial ecology in the strategic sustainable development model: Strategic applications of industrial ecology. *J. Clean. Prod.* 12, 809–823.

Kryger, J. and R. Dyndgaard, S.E. Jørgensen.1999. Introduction to cleaner production. In Jørgensen S.E. (Ed.), *A Systems Approach to the Environmental Analysis of Pollution Minimization*. Boca Raton, FL: Lewis Publishing, pp. 87–101.

Kunz, N.C., C.J. Moran, and T. Kastelle. 2013. Conceptualising "coupling" for sustainability implementation in the industrial sector: A review of the field and projection of future research opportunities. *J. Clean. Prod.* 53, 69–80.

Medeiros, J.F. J.L.D. de Ribeiro, and M. Nogueira. 2014. Success factors for environmentally sustainable product innovation: A systematic literature review. *J. Clean. Prod.* 65, 76–86.

Neto, A.S. and C.J.C. Jabbour. 2010. Guidelines for improving the adoption of cleaner production in companies through attention to non-technical factors: A literature review. *Afr. J. Bus. Manage.* 4(19), 4217–4299.

Reap, J.F. Roman, S. Duncan, and B. Bras. 2008a. A survey of unresolved problems in life cycle assessment. Part 1: Goal and scope and inventory analysis. *Int. J. Life Cycle Assess.* 13, 290–300.

Reap, J., F. Roman, S. Duncan, and B. Bras. 2008b. A survey of unresolved problems in life cycle assessment. Part 1: Impact assessment and interpretation. *Int. J. Life Cycle Assess.* 13, 374–388.

Schweisfurth, K.L, F.T. Gottwald, and M. Dierkes. 2002. *Wege zu einer Nachhaltigen Agrarund Ernährungskultur*. München: Schweifurth-Stiftung.

Tschiggerl, K. and P. Wolf. 2012. Innovative CP networks: The case of the ÖKOPROFIT® network promoting innovative clean production solutions for 20 years. *Clean. Technol. Environ. Policy* 14, 1029–1035.

van Berkel, R. 2007. Eco-efficiency in primary metals production: Context, perspectives and methods. *Res. Cons. Recycl.* 51, 511–540.

Weidema, B.P. and M.S. Wesnæs. 1996. Data quality management for life cycle inventories—An example of using data quality indicators. *J. Clean. Prod.* 4(3–4), 167–174.

Wenzel, H. et al., 1997. *Environmental Assessment of Products. Volume 1. Methodology. Tools and Case Studies in Product Development*. London: Chapman & Hall, 378 pp.

14

Resource Management

14.1 Introduction

The three Rs: reduce, reuse, and recycle, which have found important applications in cleaner technology. The idea is if waste is reduced, reused, or recycled, the environmental problems of waste can at least partially be solved. To solve the waste problems by use of the three Rs is a very important consideration when we have to develop holistic, integrated environmental management, but it is furthermore important to integrate environmental management with resource management, because we use the same method to solve a waste problem and a resource problem, since when the waste is reduced, reused, or recycled, we reduce the use of resources. Thus, environmental management and resource management become two sides of the same coin. So, the results of the methods mentioned in Chapter 13 can be viewed from two different angles: pollution abatement and longer duration of resources. The methods were in the chapter presented as environmental management tools, because it is probably the primary aim by using the presented methods, but the reduced use of resources is an additional very important advantage.

This chapter will not repeat the methods from Chapter 13 from a resource management point of view, because it is obvious how the methods described are able to prolong the duration of the resources. The chapter will, however, discuss the introduction of a green tax to reduce the use of resources, which of course also on the other side will reduce pollution. Furthermore, in Section 14.3, we will present the so-called cradle-to-cradle method, which was primarily introduced with the objective to reduce the depletion of resources but also has a significant effect on pollution reduction.

Our present economic strategy of a continuous growth and expansion of the societal system is very far from being sustainable, because of the too fast exploitation of natural resources that it requires and does not consider the long-term consequences. By describing the exploitation here as being too fast, we simply mean that we are using resources at a faster rate than with which we replace them and for finite resources that we can foresee to use up storage before we have any replacing technologies or materials to substitute for them. It presumes that we can bypass the thermodynamic laws, because it presumes infinite resources, but it is of course impossible to violate nature's most basic laws. If we try, we will be defeated by nature. We cannot work against nature, only with nature, by accepting the most basic natural laws.

14.2 Use of Pigovian Tax to Promote the Three Rs

The depletion of natural resources can only be prevented by promoting the three Rs: Reduce, Reuse, and Recycle, but how do we do that when short-term profit seems to be the driving motivation in the economy? The answer is to convince politicians that fiscal incentives (i.e., carbon tax, resource depletion accounting, payment for ecosystem services, a green tax and what we often call Pigovian tax, etc.) are urgently needed to replace income tax. The idea is that when the fiscal incentives are introduced, they would encourage even the most liberal economists or capitalists to use the three Rs and even on a short-term basis, although they will of course fight against their introduction. The resulting economy is sometimes denoted a circular economy, because the resources circle and the economy follows the resources. Today with too low Pigovian taxes, the recycling rate is much too low. Twenty metals are recycled with a rate above 50% and 34 elements are recycled below only 1%.

Figure 14.1 illustrates the possibilities. Industries, farmers, and individual citizens will apply recycling and reuse to the extent that it is profitable, directly by reduction of waste treatment (see Jørgensen, 2006). The profit, P, of the production is the sales price minus the costs of raw

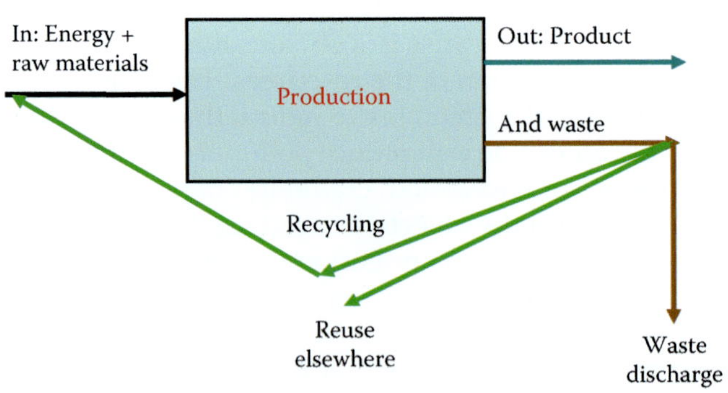

FIGURE 14.1
A production almost always has waste in addition to the product. The waste may be discharged to the environment, reused somewhere else, or recycled, which would save raw materials and perhaps energy.

materials including energy, minus the costs of production and minus also the costs of treating the waste according with the environmental laws:

$$P = S - RM - C - W, \tag{14.1}$$

where S is the sales price, RM the cost of raw materials + energy needed for the production, C the production costs, and W the costs of treating (discharging) the waste. The industries and farmers try to optimize profit, P.

Reduce is often profitable, particularly when the raw materials and energy costs inevitably will increase. To reduce often requires an investment, for instance, more insulation to reduce the loss of work energy as heat to the environment, which will save work energy. The question is, whether the reduced costs due to the reduced use of raw materials and energy can pay for the capital (interest) needed to realize the reduction. In many cases, it can even be very profitable to invest in reduction of the consumption of raw materials and energy (or rather work energy). It would even with the present prices of work energy pay to make what is called energy-saving investments in the United States. It is strange that investments in "reduction" have not been met with more response by the United States, because it is straightforward that it is profitable.

Reuse is profitable if the income by reuse sales plus the costs of treating the waste exceed the costs of making the reuse possible:

$$U - O + W \geq 0, \tag{14.2}$$

where U is the income by reuse sale and O is the cost of making the reuse possible.

Recycling is profitable if the value of the recycled raw material plus the waste treatment exceed the cost of recycling:

$$RC - CC + W \geq 0, \tag{14.3}$$

where RC is the value of the recycled raw material, CC is the recycling costs, and W is as above the costs of discharging the waste. In some cases it may be possible to add even the value of the obtained "green image."

Generally, the costs of discharging waste have increased considerably during the last decades, sometimes due to the introduction of "the polluter pays principle." Therefore, we have seen an increased interest by industry and agriculture to recycle or reuse. There is, however, still the tendency for states, regional administrations, and communities to try to attract industries by the reduction of waste treatment costs. It is not acceptable because the costs will then have to be covered directly or indirectly by the regional and local population.

Nonrenewable and partly renewable raw materials are, or are going to be, the limiting factor in our production during this century. In addition, we have seen that as the gross national product (GNP) is increasing, the environment

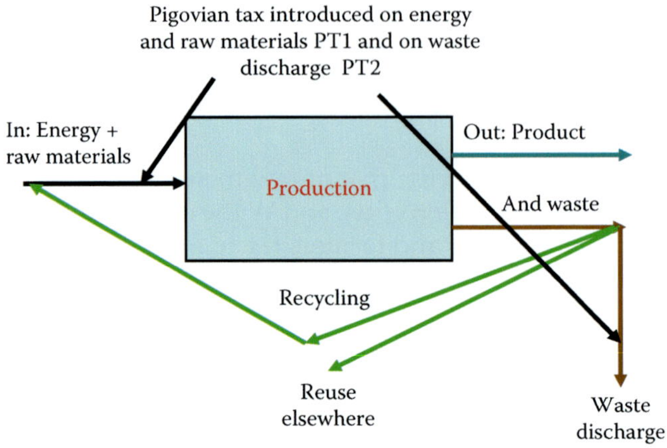

FIGURE 14.2
A production almost always has waste in addition to the product. The total global amount of raw materials, particularly if the raw materials are nonrenewable, are limited. Therefore, the introduction of a tax on raw materials and energy (PT1) and/or on waste discharge (PT2) will enhance the motivation to recycle and reuse.

is gradually deteriorating due to increased waste discharge. Recycling and reuse would improve the situation. These approaches require local labor keeping jobs and money within the community. We could enhance the application of recycling and reuse considerably by increasing the costs of raw materials and of waste discharge by introduction of an ecologically focused Pigovian tax on raw materials and waste discharge as shown in Figure 14.2. If we add a tax PT1 on raw materials and on nonrenewable energy sources and waste discharge PT2, the aforementioned equations would change.

Reduction of the consumption of raw materials and energy becomes of course more attractive by introducing Pigovian tax, because now the interest of the investment needed to realize the reduction has just to be lower than the costs of saved raw materials and energy with addition off PT1.

Reuse is now profitable if the income by reuse sale plus the costs of treating the waste plus the costs of Pigovian tax, PT2, exceed the costs of making the reuse possible:

$$U - O + W + PT2 \geq 0 \tag{14.4}$$

Recycling becomes profitable if: the value of the recycled raw material plus the waste treatment plus the Pigovian tax exceed the cost of recycling:

$$RC - F + W + PT1 + PT2 \geq 0 \tag{14.5}$$

Clearly, reduce, reuse, and recycling can easily be made more attractive by a suitable (high) tax PT1 and PT2. The Pigovian tax could be neutral revenue by reduction in other taxes such as income tax. That may even reduce the costs of reuse and recycling due to reduced labor costs and thereby it becomes even more attractive to recycle and reuse. The motivation for introduction of Pigovian taxes is therefore very strong.

It is possible to find numerous examples of the positive effect of such a tax policy or even of "the polluter pays principle." For instance, all industries in the Scandinavian countries have reduced their relative water consumption considerably during the past few decades due to the high cost for the discharge of wastewater. Particularly, the breweries have been successful in reducing their water consumption.

ILLUSTRATION 14.1

An industrial product costs $120 to produce. The raw material costs $50 and the production (labor, electricity, etc.) $40. The industry has not paid to discharge the waste, but a new environmental legislation implies that the waste discharge will cost the industry $25. Recycling makes it possible to reduce the cost of raw materials by 40% and eliminates 80% of the waste. How much is the industry able to pay for the recycling and still earn the same amount of money as before the new environmental legislation was introduced?

Solution

Forty percent reduced cost for the raw materials is $20 and 80% reduction of the waste discharge is $20. Still the industry has to pay $5 for waste discharge (20% of $25), which implies that the industry can pay 20 + 20 – 5 or $35 for the recycling per production unit.

Europe and most countries all over the world (except the Arabian oil countries and the United States) have introduced a significant tax on gasoline. The result is that the gasoline consumption per km is considerably lower in these countries than in the United States and the few other countries, where there is not a corresponding tax on gasoline. The Scandinavian countries have for a long time had a tax on oil and electricity and the result is that the energy consumption per capita is less than half the energy consumption in the United States although U.S. and Scandinavian countries have approximately the same gross national income per capita. So, clearly the introduction of a Pigovian tax implies a consumption reduction of natural resources.

It is, however, necessary to introduce a tax on *all* raw materials and on *all* discharges of waste to provoke a major increase of the sustainability. Lately, many European countries have discussed the possibilities to decrease the income tax, because the income tax may inevitably decrease employment. The effect on employment and sustainability would even be more pronounced by a transfer of tax from income tax to Pigovian tax. Most governments are, however, conservative and hesitate to make changes. The reduction of income tax and the increase of other taxes, however, can be made gradually over a period of 10–15 years to avoid these changes imposing significant net income changes for a part of the population, which may be the result of very major changes of the tax structure. When it becomes more attractive to reduce, recycle, or reuse, industrial cooperative networks will inevitably be formed to reduce the costs. Self-organization and self-regulation by industries and enterprises will lead to reduction, recycling, and reuse, and also to the formation of industrial networks with increased possibilities of reuse and recycling.

14.3 Cradle to Cradle

Reduce, reuse, and recycling is not easy to realize because modern society is producing more than 100 million different products based upon the order of more than 1 million different types of materials. Alone, the number of chemical compounds that we are using is about 100,000. How can we by end use of the more than 100 million different products extract the more than 1 million materials to be recovered, reused, and recycled? It is in this context, that the cradle-to-cradle principles point (Braungart and McDonough, 2008) toward a solution: all products should be followed from cradle to cradle—which means that all the used materials should be followed in detail from when they are considered in the design of the product, the products are produced, transported, used, and finally discharged, and further on the materials that are recovered and reused and recycled. Thereby, it will be possible to change the design of the products to facilitate the recovery, reuse, and recycling of the material. In other words, the possibilities to use the three Rs should be considered carefully in the design of the products, taking into account the entire lifetime of the products and the materials used to produce the products.

A number of important industries have understood the message and have changed the design of their product to facilitate reuse and recycling. Almost all parts of a scrapped car produced by Volkswagen can be recycled. It is easy to separate plastics, electronics, different kinds of metals, and so on. The Danish pump producer Grundfos has similarly made a design of their pumps that makes it easy to separate the parts for reuse or recycling, when the pumps are scrapped. In these cases, the changed design was introduced mainly to give the enterprises a green image. Introduction of a Pigovian tax will inevitably inspire many more producers to introduce design changes that facilitate the recycling.

The materials that are particularly difficult to recycle are the rare earth elements and the six platinum elements. They are used in small amounts for instance as catalysts and due to their magnetic properties in many important contexts. Because there is not yet a shortage of these elements, new and more effective recovery methods have not been developed, but it is strongly recommended to develop much better recovery methods and generally consider the three Rs in the use of these elements.

Cradle-to-cradle principles have not yet been applied very much up to now. There is a need for comprehensive research into the practical application of the principles. One basic reason for this might well be that the focus set by CtoC on durability and repair—as well as reuse of materials once they have been produced—is not viewed as compatible with an economy constructed around continuous growth.

In the end, the whole boils down to doing "with less." Recently, the focus has been on the important role of our high consumption of almost anything in our surroundings whether we actually need it or not. This relevance of this is only stressed by the fact that this excessive consumption does not lead to any more happiness.

EXERCISES AND PROBLEMS

1. Indicate the similarities and differences between LCA and the Cradle to Cradle approach.
2. The benefits by Pigovian tax to replace income tax is obvious, why is the use of a Pigovian tax only introduced modestly in the industrialized countries?

References

Braungart, W. and M. McDonough. 2008. *Cradle to Cradle*. New York: North Point Press, 230 pp.

Jørgensen, S.E. 2006. *Eco-Exergy as Sustainability*. Southampton: WIT, 220 pp.

15

Application of Environmental Legislation in Environmental Management

15.1 Introduction

To stop pollution that has a very unacceptable impact on nature and humans, health environmental legislation may be a very effective method. When we consider the value of ecosystems' services, it is obvious that the cost of pollution abatement easily can be profitable for the society and that such a cost is compensated by the value of the ecosystems' services that are maintained. In such cases, an environmental legislation that stops pollution-driven deterioration will with high probability be introduced, since it is the type of environmental legislation that is beneficial for the society to launch. Another question is who is paying for the cost of the pollution abatement. The general principle that the polluter should pay for the reduction or elimination of the pollution is often but not always applied, particularly if the society considers the activity that must pay for the pollution as very important for the society in general and eventually is decisive to create employment. Typical for this type of environmental legislation is the rules for treatment of wastewater discharged from industries or rules for the use of toxic substances, for instance, laws and regulation of pesticides. The legislation is based on environmental considerations and sometimes an ecological model is used to assess the applied rules.

Another type of environmental legislation is associated with the general principle of law. A polluter is responsible for the pollution he or she causes, because it would not be fair to let either society or another person take the responsibility. This type of environmental legislation is used in situations where there is a conflict in a pollution case: Who is responsible for the pollution? A typical example is legislation about waste disposal.

A third type of environmental legislation is associated with other legislation complexes. For instance, many countries have rules for energy commissioning of existing buildings and new buildings to reduce the consumption of energy for heating and air conditioning in buildings—it means to ensure a more environmental use of energy. There are many environmental considerations in a number of laws for instance legislation focusing on agriculture, road construction, deforestation, laws and regulation of food, and soil degradation.

15.2 An Example of Environmental Legislation

The environmental legislation today covers hundreds of thousand pages. Generally the legislation is different from country to country. Let us take as an example the marine environment, which is presently undergoing economic and social driven pressures for development and access to marine resources and activities (e.g., commercial fishing, aquaculture, tourism, recreation, and maritime transportion). In order to resolve these problems, policy makers worldwide seek to develop strategies to protect, conserve, and manage the marine environment. The United Nations Convention on Law of the Sea (UNCLOS, 1982) is the international basic legal framework that governs the use of the oceans and seas. UNCLOS establishes an international obligation to protect and use the resources of the marine environment sustainably as does the 1992 Convention on Biological Diversity (CBD, 2000), as emphasized by Parsons (2005). Borja et al. (2008) very well summarized different examples of initiatives developed at national or regional levels in Africa, Asia, Australia, Europe, and North America during the last decades:

i. In December 1998, Australia released an Oceans Policy (Commonwealth of Australia, 1999, 2006).
ii. The Canadian Parliament passed the Oceans Act, which came into force in January 1997, being Canada's Oceans Strategy released in 2002 (Parsons, 2005).
iii. In the United States, the Pew Oceans Commission, created in 2000, and the U.S. Commission on Ocean Policy, created by the Oceans Act of 2000, reported in 2004 (Granek et al., 2005).
iv. In Europe, the Water Framework Directive (WFD), which promotes the protection of continental, estuarine, and marine waters, was released in 2000 (Borja, 2004), and the European Marine Strategy (EMS) Directive, was presented in 2005 (Borja, 2006).
v. In South Africa, the National Water Act of 1998 and the developing Coastal Management Act were put in the form of the Integrated Coastal Management Bill.
vi. In the People's Republic of China (PRC), a substantial body of legislation exists to address environmental protection (Laws on Water (1988/01/21) and Environmental Protection (1989/12/26): Sea Water Quality GB 3097-1997, Environmental Quality for Surface Water GB 3838-2002, and Provisions for Monitoring of Marine Culture and Propagation Areas (2002/04/01).

The objectives of all these initiatives were to protect and/or restore the corresponding seas by ensuring that human activities are carried out in a sustainable manner, to provide safe, clean, healthy, and productive marine waters. In summary, they try to promote the sustainable use of the seas and conserve marine ecosystems. Hence, the main objective of these legislative measures and policies is to maintain a good environmental or ecological status for marine waters, habitats, and resources. The concept of environmental status takes into account the structure, function, and processes of marine ecosystems bringing together natural physical, chemical, physiographic, geographic, and climatic factors, and integrates these conditions with the anthropogenic impacts and activities in the areas concerned.

The aforementioned concept defines quality in an integrative way, by using several biological parameters together with physicochemical and pollution elements. This approach is intended to allow an assessment of the ecological status at the ecosystem level—"ecosystem-based approach" (EBA) or "holistic approach" methodologies (Apitz et al., 2006; Browman et al., 2004; Foster et al., 2005; Jennings, 2005; Nicholson and Jennings, 2004; Rudd, 2004), more effectively than can be done at a species (e.g., mussel biomonitoring or Mussel Watch) or chemical level (i.e., quality objectives). The EBA is defined as "a strategy for the integrated management of land, water and living resources that promotes conservation and sustainable use in an equitable way. The application of the EBA will help to reach a balance of the conservation, sustainable use, and the fair and equitable sharing of the benefits arising out of the utilization of genetic resources" (CBD, 2000). However, there are various interpretations of the EBA and its application almost always brings about confrontations and resistance among managers, proponents, and stakeholders (Morishita, 2008).

Following this approach, increasing attention has been paid to the development of tools for different physicochemical or biological (phytoplankton, zooplankton, benthos, algae, phanerogams, and fishes) elements of the ecosystems. However, very few methodologies integrate all the elements into a unique evaluation of status and performance of an aquatic system. The need for such integrative tools to assess the ecosystem quality is very important, both from a scientific and stakeholder point of view. The scientific challenge is to develop robust, simple, pragmatic, but scientifically sound methodologies, which can provide communities and decision makers with tools to define and monitor the evolution, current conditions, and biological performance of marine ecosystems and bioregions.

Such approaches must include (Borja et al., 2013, 2014):

i. Multidisciplinarity, inherent in the teams involved in their implementation.
ii. Integration of biotic and abiotic factors.
iii. Accurate and validated methods for determining ecological integrity.

iv. Accurate and validated methods for determining the extent and effect of human uses and impacts.
v. Adequate indicators to follow the evolution of the monitored ecosystems.
vi. Use of protected areas as a means of conserving and managing viable representative examples of marine environments especially coastal areas where the greatest anthropogenic inputs occur.
vii. Finally, there should be some early warning systems for abrupt changes in environmental conditions.

Some progresses have been done. For instance in the European Union (EU), environmental legislation is valid for all EU countries, but there is still also national legislation in the 28 EU countries, although the differences among the countries are minor. The principle is that there is a common EU legislation, but it has also allowed EU countries to have more strict legislation than the EU, provided that it does not interfere with the free competition in the EU. The United States has an almost similar practice, where there are states that have more severe environmental legislation than the general U.S. legislation. For instance, California has a few more strict environmental rules than the United States. The environmental legislation in industrialized countries is generally stricter than in developing countries and the emerging economies—Brazil, India, China, and Russia. In this context, it should be mentioned that there are a number of bilateral agreements about sharing water and other natural resources. These agreements, which are leading to legislation in the involved countries are important as instruments to avoid regional conflict.

15.3 Overview of Environmental Legislation

The coverage of environmental legislation can best be illustrated with the following overview of the various environmental laws:

I. *Air quality law* governs emissions of air pollutants into the atmosphere but also regulates the quality of the air inside buildings. Furthermore, it encompasses the protection of the ozone layer, acid rain, and climate change regulations. It includes the use of appropriate mitigation techniques. Air quality laws are often designed specifically to protect human health by limiting or eliminating airborne pollutant concentrations.
II. *Water quality laws* provide rules regarding the release of pollutants into water resources: surface water, groundwater, and stored drinking water. It includes drinking water regulations, which may be designed

solely with reference to human health. Furthermore, it encompasses restrictions on the alteration of the chemical, physical, radiological, and biological characteristics of water resources, and more broadly, aquatic ecosystems. Regulatory areas include sewage treatment and disposal, industrial and agricultural wastewater management, and control of surface runoff from construction sites and urban environments.

III. *Waste management laws* set rules for the transport, treatment, storage, and disposal of all manner of waste, including municipal solid waste, hazardous waste, nuclear waste, and many other types. Waste laws are generally designed to minimize or eliminate the uncontrolled dispersal of waste materials into the environment in a manner that may cause ecological or biological harm. They also include rules (laws) about how to reduce the generation of waste and promote waste recycling. Regulatory efforts include identifying and categorizing waste types and mandating transport, treatment, storage, and disposal practices.

IV. *Environmental cleanup laws* focus on rules for the removal of pollution or contaminants from environmental media such as soil, sediment, surface water, or groundwater. Unlike pollution control laws, cleanup laws are designed to respond to environmental contamination, and must often define not only the necessary response actions, but also the parties who may be responsible for undertaking (or paying for) such actions. Regulatory requirements may include rules for emergency response, liability allocation, site assessment, remedial investigation, feasibility studies, remedial action, postremedial monitoring, and site reuse.

V. *Chemical safety laws* provide rules about the use of chemicals, particularly man-made chemicals in modern industrial applications. As contrasted with media-oriented environmental laws (e.g., air or water quality laws), chemical control laws manage the (potential) pollutants themselves. Regulatory efforts include banning specific chemical constituents in consumer products (e.g., Bisphenol A in plastic bottles) and regulating pesticides (e.g., ban the use of DDT in industrialized countries).

VI. *Water resources laws* govern the ownership and use of water resources: surface water and groundwater. Regulatory areas include water conservation, use restrictions, and ownership regimes.

VII. *Mineral resources laws* cover the ownership of the mineral resources. Mining is also affected by various regulations regarding the health and safety of miners, as well as the environmental impact on nature of mining.

VIII. *Forestry laws* focus on activities in forestlands, with respect to forest management and timber harvesting. Ancillary laws may regulate forestland acquisition and prescribed burn practices. Forest management laws generally adopt management policies. Governmental agencies are generally responsible for planning and implementing forestry laws on public forestlands, and are commonly involved in forest inventory, planning, and conservation, and oversight of timber sales. Broader initiatives that may seek to slow or reverse deforestation are often included.

IX. *Wildlife laws* govern the potential impact of human activity on wild animals, whether directly on individuals or populations, or indirectly via habitat degradation. Similar laws protect plant species. Such laws may be enacted entirely to protect biodiversity, or as a means for protecting species deemed important for other reasons. Regulatory efforts may include the creation of special conservation status for instance for endangered species, prohibitions on killing, harming, or disturbing protected species, efforts to induce and support species recovery, including limitations on hunting, establishment of wildlife refuges to support conservation, and prohibitions on trafficking species or animal parts to combat poaching.

X. *Fish and game laws* regulate the right to pursue and take or kill certain kinds of fish, aquatic animals, and wild animals (game). Such laws may restrict the days to harvest fish or game, the number of animals caught per person, the species harvested, or the weapons or fishing gear used. Such laws may seek to balance preservation and harvest and to manage both the environment and populations of fish, aquatic animals, and game. Game laws provide a legal structure to collect license fees and other money which is used to fund conservation efforts as well as to obtain harvest information used in wildlife management practices.

All countries share the same Earth and it would therefore be natural to have a global environmental legislation but it has not yet been possible to introduce except for a few very crucial environmental problems. The best positive example is that all the industrialized countries have banned the use of Freon gases, which are responsible for the decomposition of the ozone layer that protects the humans against the health-damaging UV radiation. It was attempted in Kyoto in 1997 to come to a global agreement about the carbon dioxide emissions, but it has up to now failed and the result is that carbon dioxide concentration in the atmosphere is increasing more and more rapidly.

It is possible to classify legislation in laws that define rules that must be followed by enterprises, industries, agriculture, and the citizens and legislation that gives quantitative standards, for instance, maximum concentrations or amounts in discharged wastewater and in solid waste. Quantitative indications facilitate of course the interpretation of the legislation in concrete cases, while more qualitative indications often require an interpretation of the laws.

Environmental legislation is entangled in environmental policy and a number of international conventions resulting from a number of international meetings are often the basis for legislation in a number of countries (Hannam and Boer 2013). An illustrative example of conventions is the Rotterdam Convention (1999) about protecting humans and the environment from highly toxic chemicals. It has inspired in many countries, legislation on the control of highly toxic chemicals.

Even a very short overview of the environmental legislation would require several hundred pages and it would not be in accordance with the coverage of the other topics in this textbook.

References

Apitz, S.E., M. Elliott, M. Fountain, and T.S. Galloway. 2006. European Environmental Management: Moving to an ecosystem approach. *Integr. Environ. Assess. Manag.* 2, 80–85.

Borja, A., V. Valencia, J. Franco, I. Muxika, J. Bald, M.J. Belzunce, and O. Solaun. 2004. The water framework directive: Water alone, or in association with sediment and biota, in determining quality standards? *Mar. Pollut. Bull.* 49, 8–11.

Borja, A. et al. 2006. Red de seguimiento del estado ecológico de las aguas de transición y costeras de la Comunidad Autónoma del País Vasco. AZTI-Tecnaliareportto Dirección de Aguas, Departamento de Medio Ambiente y Ordenación del Territorio, Gobierno Vasco. 15 Vol, 838 pp. www.uragentzia. euskadi.net/u81-0003/es/contenidos/informe_estudio/red_osta/es_red_agua/2006.html.

Borja, A. et al. 2008. Overview of integrative tools and methods in assessing ecological integrity in estuarine and coastal systems worldwide. *Mar. Pollut. Bull.* 56, 1519–1537.

Borja, A. et al. 2013. Good environmental status of marine ecosystems: What is it and how do we know when we have attained it? *Mar. Polut. Bull.* 76, 16–27. http://dx.doi.org/10.1016/j.marpolbul.2013.08.042.

Borja, A. et al. 2014. Tales from a thousand and one ways to integrate marine ecosystem components when assessing the environmental status. *Front. Mar. Sci.* 1, 43–67.

Browman, H.I. et al. 2004. Perspectives on ecosystem-based approaches to the management of marine resources. *Mar. Ecol. Prog. Ser.* 274, 269–303.

CBD. 2000. *Ecosystem Approach. Fifth Conference of the Parties to the Convention on Biodiversity*. Nairobi, Kenya, May 2000. www.iisd.ca/biodiv/cop5.

Commonwealth of Australia, 1999. Australia's Oceans Policy, 2 vols, 104 pp. http://www.oceans.gov.au/the_oceans_policy_overview.jsp.

Commonwealth of Australia, 2006. A Guide to the Integrated Marine and Coastal Regionalisation of Australia Version 4.0. Department of the Environment and Heritage, Canberra, Australia. www.environment.gov.au/coasts/mbp/publications/imcra-4.html.

Foster, E., M. Haward, and S. Coffen-Smout. 2005. Implementing integrated oceans management: Australia's South East Regional Marine Plan (SERMP) and Canada's Eastern Scotian Shelf Integrated Management (ESSIM) Initiative. *Mar. Policy* 29, 391–405.

Granek, E.F., D.R. Brumbaugh, S.A. Heppell, S.S. Heppell, and D. Secord. 2005. A blueprint for the oceans: Implications of two national Commission Reports for conservation practitioners. *Conserv. Biol.* 19, 1008–1018.

Hannam, J. and B. Boer. 2013. Laws and Regulations: Soil. P. 1618–1620. In S.E. Jørgensen (Ed.), *Encyclopedia of Environmental Management*. Boca Raton, FL: CRC Press. 2890 pp.

Jennings, S. 2005. Indicators to support an ecosystem approach to fisheries. *Fish and Fisheries* 6, 212–232.

Morishita, J. 2008. What is the ecosystem approach for fisheries management? *Mar. Policy* 32, 19–26.

Nicholson, M.D. and S. Jennings. 2004. Testing candidate indicators to support ecosystem-based management: The power of monitoring surveys to detect temporal trends in fish community metrics. *ICES J. Mar. Sci.* 61, 35–42.

Parsons, S. 2005. *Ecosystem Considerations in Fisheries Management: Theory and Practice. Conference on the Governance of High Seas Fisheries and the UN Fish Agreement Moving from Words to Action*. St. John's, Newfoundland and Labrador, May 5, 2005. 44 pp.

Rotterdam Convention on the Prior Informed Consent Procedure for Certain Hazardous Chemicals and Pesticides in the International Trade. UNEP and FAO, January 1999.

Rudd, M.A. 2004. An institutional framework for designing and monitoring ecosystem-based fisheries management policy experiments. *Ecol. Econ.* 48, 109–124.

UNCLOS, 1982. United Nations Convention on the Law of the Sea signed at Montego Bay, Jamaica, on December 10, 1982, 202 pp. www.un.org/Depts/los/index.htm.

Part VI

Implementation of Ecological Management Tools and Follow the Recovery Process

16

Bringing into Action

16.1 Introduction

Environmental management must consider all seven steps presented in Chapter 1, simultaneously or expressed differently. All the tool boxes must be applied and integrated into a concerted effort to be able to go from a complex environmental problem to the solution. We need to integrate the entire spectrum of knowledge about the problem and its sources, effects and the knowledge gained by the use of the diagnostic tools which will lead us to a composite, combined application of the solution tool boxes. The goal is to solve the problem including the side problems and all derived problems, proximate as well as distanced. It is important in the process of solving one focal environmental problem not to create new and eventually overlooked environmental problems.

It is of course a difficult task to keep in mind all the seven steps and all the tool boxes in the development of—if not the very best solution then—at least a fully acceptable and appropriate solution. It is clearly necessary to maintain a holistic view and overview of the environmental and its reactions and processes originating from the problem. Integrated ecological and environmental management means that the environmental problems are viewed from a holistic angle considering the ecosystem as an entity and considering the entire spectrum of solutions including all possible combinations of proposed solutions. The integration in this context means that all the available tools and possibilities are taken into accounts and that all the problems are considered simultaneously.

The experience gained from environmental management from the last 40 years has clearly shown that it is important not to consider solutions of single problems but to consider *all or at least all major* problems associated with a considered ecosystem simultaneously and evaluate *all* the solution possibilities proposed by the relevant disciplines at the same time, or expressed differently: to observe the forest and not the single trees. The experience has clearly underlined that there is no alternative to an *integrated* management at least not on a long-term basis. Fortunately, we have seen in the previous chapters that new ecological subdisciplines have emerged and that they offer tool boxes to perform an integrated ecological and environmental management, provided that we are using a "macroscope," meaning applying a holistic integrated view. Following the seven steps (see Chapter 1) will facilitate the macroscopic view.

The use of ecological modeling has become much more important over the last 35 years, due to a need demand for quantification in ecology and environmental management. This is probably the most significant tool available for obtaining a quantitative, holistic overview of complex ecosystems such as lakes, rivers, wetlands, lagoons, or reservoirs, which is a prerequisite for selecting an optimum solution to the complex problems, facing them. Because solutions to the entire spectrum of environmental problems require quantitative estimation methods to assist in identifying a realistic trade-off between ecological and economic concerns, it is not surprising that ecological models have been increasingly used in environmental management efforts. Ecological indicators and assessment of ecosystem services are more recent tools than ecological modeling, but can supplement ecological modeling with slightly different views. They can work hand in hand with ecological models but can only partially replace ecological modeling if a completely quantitative understanding of the problem and the possible solutions are needed, but they can usually be applied with less effort and therefore less costs than ecological modeling.

We may assume that it is most often beneficial to apply all the three diagnostic tools to achieve the widest overview. Nevertheless, the final decisions about the implementation of the diagnostic tools are eventually an economic issue, and certainly a problem facing environmental agencies is how to efficiently distribute funds to enhance natural capital and improve ecosystem services (e.g., Hajkowicz et al., 2009; Pinto et al., 2014). In fact, according to Mysiak et al. (2002), "Environmental decisions require a spatial analysis of the impacts resulting from possible options in order to guarantee spatially balanced sustainable development." Hence, in the processes of sustainability assessments and/or environmental planning of ecosystems, it becomes fundamental to ensure the accurate interweaving of disciplinary approaches, and as already said, the three diagnostic tools must be considered for this task. The use of several approaches is always advised to achieve a better evaluation of ecosystem integrity combined with current human demands, perceived as key for the sustainable development of ecosystems.

In fact, sustainable development is a complex issue, involving strongly interconnected ecological, social, and economic aspects of an ecosystem, now and in the future (de Jonge et al., 2012). Eventually, according to Kiker et al. (2005), "Decision making in environmental projects can be complex and seemingly intractable, principally because of the inherent trade-offs between socio-political, environmental, ecological, and economic factors." Decisions affect different stakeholder groups in different ways, leading to inevitable trade-offs between them and between present and future generations. Decision-making involving such complex systems requires a logically well-structured process.

In this scope, multicriteria analyses (MCA) have been recommended as useful tools to ensure an integrated management of an ecosystem, allowing the incorporation of different sets of data (e.g., Villa et al., 2002; Pavlikakis and Tsihrintzis, 2003; de Jonge et al., 2012). MCA constitutes a stepwise process that allows comparing of decision alternatives with multiple and often complex impacts (Hermann et al., 2007). The information is often structured using a software tool, which aims to record proposed alternatives, while measuring and assessing their impacts (Hermann et al., 2007).

16.2 Recommendations on the Application of Integrated Environmental Management

From this discussion, several important recommendations may be deduced, as summarized in the following six points:

1. Expect that a proper environmental strategy will require a wide spectrum of approaches and techniques, which to a high extent must be based on the presented diagnostic tools.
2. Expect that proper environmental management will require the application of a combination of end-of-the-pipe technologies (environmental technology), ecotechnology, cleaner technology, and environmental legislation.
3. Correct timing in applying the various steps in environmental management efforts is extremely important; thus, it is recommended that a comprehensive environmental management plan be developed at a very early stage, in order to use the available resources in the most optimal manner.
4. It is usually very beneficial, particularly from an economic perspective, to consider prevention, rather than corrective adjustments, primarily because it is often very costly to restore heavily degraded ecosystems, if at all possible.
5. Because of the complexity of ecosystems and their problems, proper ecological knowledge about ecosystems is a prerequisite for ecologically sound environmental management programs; this is the only reasonable method for avoiding unexpected ecosystem responses. Therefore, all the ecological subdisciplines must be applied to all the environmental problems in the effort to find a proper solution.

6. Optimum solutions to environmental management problems are best obtained if all the ecosystems of the entire watershed and landscape are taken into consideration in developing and implementing management actions to avoid that the solution of the environmental problems of one ecosystem create other problems in other ecosystems. Remember in this context that all ecosystems are open systems.

Principles and *quantifications* are the keywords in our search for solutions to environmental problems. Chapter 1 has clearly presented how to go from (1) problems, (2) to sources (with an appropriate estimation of the quantities), (3) to impacts, where emissions are translated to emissions and concentrations, (4) determining the effect, (5) to diagnosis, (6) to selection among the possible solutions to the topic of this chapter, and (7) integration of these steps into an environmentally sound plan, based on the knowledge obtained in each step. The environmental plan will by following the seven steps inevitably be based on the answers to the following questions (see Jørgensen, 2000; Jørgensen and Nielsen, 2012):

A. What can be obtained by changing the methods that are applied today to produce the products, needed by society, to build and construct, to use products, matter, and energy? The answer builds very much in its widest sense on cleaner technology and production.
B. Could we use beneficially—both economically and environmentally—the three Rs, reduce, reuse, and recycle?
C. Would it be an advantage to decompose waste causing environmental problems to harmless or at least less harmful components, that in some cases even could be reused or recycled, but under all circumstances could be discharged without additional, harmful effects?
D. Is the solution to remove waste to another location where the deposition would be more harmless, or maybe even beneficial?
E. Can the deteriorated ecosystem be recovered by ecotechnological methods?
F. Can the problem be avoided by better ecological planning?

Figure 16.1 (Jørgensen, 2000) illustrates how a life-cycle analysis can be applied to integrate the various solution tool boxes, considering the entire life cycle for a product from its production to its final disposal. The figure illustrates well, therefore, the idea behind integrated environmental management. The arrows to the ecosystems presume that the ecological diagnostic tool boxes are applied and the sources indicated as the arrows are quantified and their impacts are known.

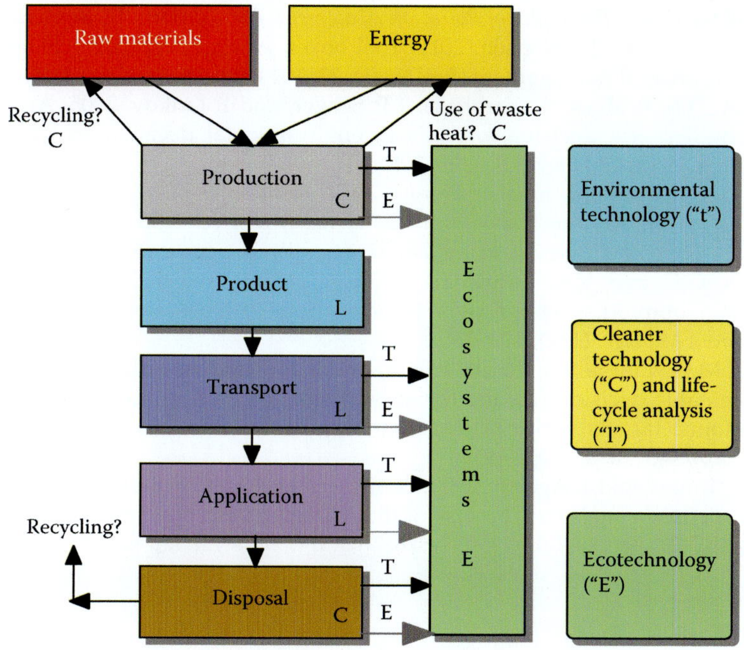

FIGURE 16.1
The arrows cover mass flows and the thin arrows symbolize the control possibilities. Point pollution is indicated as a thin black arrow to the ecosystems and nonpoint pollution is a thick gray arrow. Production includes both industrial and agricultural production. The latter is most responsible for nonpoint pollution. Environmental technology (indicated by T) is applied most to solve the point pollution problems and ecotechnology (indicated by E). Cleaner technology (indicated by C) is investigating the possibilities to produce the products by another method that causes less environmental problems and may facilitate reuse and recycling. The diagram is based on a life-cycle analysis—the product is followed from its production based on matter and energy to its final disposal as waste after its use. The diagram shows how the various solution tool boxes are needed in an integrated way to solve all the problems associated with a product from its production to its disposal.

References

deJonge, V.N, R. Pinto, and R.K. Turner. 2012. Integrating ecological, economic and social aspects to generate useful management information under the EU Directives "Ecosystem Approach." *Ocean Coast Manage*. 68, 169–188.

Hajkowicz, S., A. Higgins, C. Miller, and O. Marinoni. 2009. Is getting a conservation model used more important than getting it accurate? *Biol. Conserv*. 142, 699–700, 443.

Hermann, B.G., C. Kroeze, and W. Jawjit. 2007. Assessing environmental performance by combining life cycle assessment, multi-criteria analysis and environmental performance indicators. *J. Clean. Prod*. 15, 1787–1796.

Jørgensen, S.E. 2000. *Principles of Pollution Abatement*. Amsterdam: Eldsevier, 520 pp.

Jørgensen, S.E. and S.N. Nielsen. 2012. Tool boxes for an integrated ecological and environmental management. *Ecol. Indic.* 20, 104–109.

Kiker, G.A., T.S. Bridges, A. Varghese, T.P. Seager, and I. Linkov. 2005. Application of multicriteria decision analysis in environmental decision making. *Integr. Environ. Assess. Manage.* 1(2), 95–108.

Mysiak, J., C. Giupponi, and A. Fassio. 2002. *Decision Support for Water Resource Management: An Application Example of the MULINO DSS. Conference IEMSS 2002—Integrated Assessment and Decision Support*, Lugano: IEMSS, pp. 138–143.

Pavlikakis, G.E. and V.A. Tsihrintzis. 2003. A quantitative model for accounting human opinion, preferences and perceptions in ecosystem management. *J. Environ. Manage.* 68(2), 193–205.

Pinto, R., M.C. Cunha, C. Roseta-Palma, and J.C. Marques. 2014. Mainstreaming sustainable decision making for ecosystems: Integrating ecological and socio-economic targets within a decision support system. *Environ. Processes* (DOI 10.1007/s40710-014-0006-x).

Villa, F., L. Tunesi, and T. Agardy. 2002. Zoning marine protected areas through spatial multi-criteria analysis: The case of the Asinara Island National Marine Reserve of Italy. *Conserv. Biol.* 16(2), 515–526.

17

Are the Applied Management Tools Working?

17.1 The Recovery Process Is Followed

After completion of any project, it is of course necessary to investigate, whether the project has fulfilled its intentions. Similarly, for an integrated environmental management project, it is necessary to examine whether the planned results have been achieved. It implies that the following questions have to be answered:

1. Have the pollution and all types of impacts and their effects been reduced or eliminated to the extent foreseen?
2. Have the involved focal ecosystems (nature) recovered to the extent intended by the integrated environmental management project?
3. Are the ecosystems able again to offer the ecosystem services that the project intended to recover?
4. To which extent are the performed prognoses, for instance, by the use of ecological models, followed? If there are discrepancies, can they be explained?

The idea is as far as it is possible to determine, how successful the integrated environmental management project has been.

Obviously, to answer these crucial questions it is required to measure several quality parameters that assess the water quality, air quality, soil quality, and ecological functions of the focal ecosystems. If the environmental diagnosis has applied an assessment of ecological indicators or the ecosystems services, it would be obvious to determine the same indicators and ecosystem services after the project completion. If an ecological model has been developed, the same model now based on the conditions determined by accomplishment of the project should be applied and the model results can most probably answer the raised questions directly.

In this context, it should be considered to apply a continuous monitoring of the most important parameters (state variables in modeling context).

For instance, if a river with too low oxygen concentration is recovered by wastewater treatment, it would be beneficial to continuously monitor the oxygen concentration in the river at selected, important points, because oxygen concentration is possible to monitor by the use of relatively simple and cost-moderate equipment. It is in many cases, easy to follow other parameters, too, as for instance pH, conductivity, and temperature. A wide spectrum of continuous monitoring methods for air, water, and soil quality are fortunately available.

Furthermore, the applied and examined ecological indicators and ecological services should be determined directly, if possible, or indirectly. The lower level indicators, see Chapter 9, may often be easily determined directly, both before and after the integrated environmental management project. It is, however, often very important to assess the holistic indicators, as they often determine the key properties of ecosystems, such as the sustainability, the resistance, the buffer capacity, the resilience, and the biodiversity. These indicators should therefore be determined indirectly. It is often very crucial that these properties are regained at the same level after the project completion. In conclusion, it is strongly recommended to make the effort to determine these holistic properties to ensure that a full recovery has been achieved.

17.2 Project Adjustments

If the project has been successful and the purpose of the integrated environmental management been achieved, there is of course no need for project changes, but if the results of the project have not been satisfactory, adjustments of the project are needed. The choice of the needed changes will be based on the diagnostic tools, again, and on the five solution tool boxes presented in Chapters 11–15. It is probably necessary to go through all the steps of the project again and ensure that the required changes have been properly assessed. If the diagnosis is approximately the same as the previous one, the project management should be open to find alternative combinations of solutions based on the solution tool boxes. In principle, the needed changes could be in all the seven management steps, but most often it should be expected that the combinations of solution tools probably should be changed.

If the changes and adjustments of the project are major, an additional examination of the recovery process is required after the changes and adjustments have been implemented.

In addition to the more close examination of the recovery process and the success of the implemented integrated environmental management plan, a monitoring and measurement of environmental quality parameters take place in at least the industrialized countries to ensure that the environmental quality is maintained. For instance, the EU has an environmental quality program in all the member countries. Based upon these environmental quality determinations a survey of the environmental quality in all Europe is published by the European Environmental Agency in Copenhagen frequently to make it possible to conclude where it is necessary to make a particular effort to meet the EU environmental quality requirements.

Appendix

A.1 The Atomic Composition of the Four Spheres

	Atoms % in			
Element	Biosphere	Lithosphere	Hydrosphere	Atmosphere
H	49.8	2.92	66.4	v.l.
O	24.9	60.4	33	21
C	24.9	0.16	0.0014	0.04
N	0.27	v.l.	v.l.	78
Ca	0.073	1.88	0.006	v.l.
K	0.046	1.37	0.006	v.l.
Si	0.033	20.5	v.l.	v.l.
Mg	0.031	1.77	0.034	v.l.
P	0.030	0.08	v.l.	v.l.
S	0.017	0.04	0.017	v.l.
Al	0.016	6.2	v.l.	v.l.
Na	v.l.	2.49	0.28	v.l.
Fe	v.l.	1.90	v.l.	v.l.
Ti	v.l	0.27	v.l.	v.l.
Cl	v.l.	v.l.	0.33	v.l.
B	v.l.	v.l.	0.0002	v.l.
Ar	v.l.	v.l.	v.l.	0.93
Ne	v.l.	v.l.	v.l.	0.0018

v.l. = very low.

Index

F

G

H

I

For Product Safety Concerns and Information please contact our EU representative GPSR@taylorandfrancis.com Taylor & Francis Verlag GmbH, Kaufingerstraße 24, 80331 München, Germany

Batch number: 10722122

Printed by Printforce, the Netherlands